D1995365

ALGEBRAIC SEMANTICS IN LANGUAGE AND PHILOSOPHY

CSLI
Lecture Notes
No. 74

ALGEBRAIC SEMANTICS IN LANGUAGE AND PHILOSOPHY

GODEHARD LINK

CSLI PUBLICATIONS
CENTER FOR THE STUDY OF
LANGUAGE AND INFORMATION
STANFORD, CALIFORNIA

Center for the Study of Language and Information
Leland Stanford Junior University
Printed in the United States
02 01 00 99 98 5 4 3 2 1

Library of Congress Cataloging-in-Publication Data

Link, Godehard.
Algebraic semantics in language and philosophy / Godehard Link.
p. cm. — (CSLI lecture notes ; no. 74)
Collection of 14 essays, seven of which have been previously published.
Includes bibliographical references (p.) and index.
Contents: The logical analysis of plurals and mass terms — Plural — Hydras : on the logic of relative clause constructions with multiple heads — Generalized quantifiers and plurals — Je drei Äpfel = Three apples each : quantification and the German je — First-order axioms for the logic of plurality — Ten years of research on plurals : where do we stand? — Algebraic semantics for natural language — Quantity and number — The French Revolution—a philosophical event? — Algebraic semantics of event structures — The ontology of individuals and events — The philosophy of plurality — Mereology, second-order logic, and set theory — Appendix : a chapter in lattice theory.
ISBN 1-57586-091-0 (alk. paper). — ISBN 1-57586-090-2 (pbk. : alk. paper)
1. Grammar, Comparative and general—Number. 2. Grammar, Comparative and general—Quantifiers. 3. Semantics (Philosophy) 4. Language and logic. 5. Language and languages—Philosophy. I. Title. II. Series.
P240.8.L56 1997
415—dc21 97-46523
CIP

∞ The acid-free paper used in this book meets the minimum requirements of the American National Standard for Information Sciences—Permanence of Paper for Printed Library Materials, ANSI Z39.48-1984.

CSLI was founded early in 1983 by researchers from Stanford University, SRI International, and Xerox PARC to further research and development of integrated theories of language, information, and computation. CSLI headquarters and CSLI Publications are located on the campus of Stanford University.
CSLI Publications reports new developments in the study of language, information, and computation. In addition to lecture notes, our publications include monographs, working papers, revised dissertations, and conference proceedings. Our aim is to make new results, ideas, and approaches available as quickly as possible. Please visit our web site at
http://csli-www.stanford.edu/publications/
for comments on this and other titles, as well as for changes and corrections by the author and publisher.

Contents

Preface

This volume contains a collection of the author's papers on the algebraic approach to the characterisation of objects in language and philosophy.

The common *philosophical perspective* of the essays can be described as follows. Language is able to refer to a wide array of objects of different sorts, which differ from each other in their characteristic structural properties. The universe of linguistic or philosophical discourse is thus most naturally taken as a multi-sorted domain containing all those objects. However, the traditional set-up of Tarskian semantics, where the domain of individuals is just a non-empty set, does not *per se* represent the various relations that connect these kinds of object both internally and across different kinds. Algebraic semantics, then, deals with *structured domains* instead of flat ones: various algebraic relations model the structural properties of the entities in the domain.

The kinds of object that are investigated systematically in the essays are *concrete particulars* such as *plural entities*, *mass objects*, and *events*. However, the algebraic approach naturally applies also to other philosophical entities, like propositions, properties, relations, or situations. Thus, the volume complements algebraic work that has been done on the latter kinds of object.

The essays give a detailed motivation, both linguistic and philosophical, of the need to introduce concrete particulars of the above kind into the domain, and describe the structural properties that govern their behavior. A *first-order logical framework* is developed in which it is possible to quantify over these objects. Its usefulness in providing a compositional semantics for them is demonstrated. The framework exhibits most clearly the structural similarity between plurals and mass terms. Also, the interplay between plural objects and (plural) events is discussed.

The part of the theory which is developed most in the volume is the *Logic of Plurality, LP*. LP is a first-order mereology which can be viewed as a nominalistic version of monadic second-order logic. Its philosophical significance is explored in the last two essays.

Many chapters deal with *applications in linguistic semantics*. The main topics, which are treated extensively, are (i) plural determiners and plural quantification, and (ii) collective, distributive, and intermediate predication.

Three essays are concerned with the notion of *event*, both from a linguistic and a philosophical point of view. In particular, a separate chapter is devoted to the outline of a unified ontology of individuals and events.

The 14 essays have been written over a period of 15 years. Seven of them have already been published, but at widely different places with varying accessibility, and one of them in German. Inclusion in this volume should render them readily accessible now. Another chapter appeared before as preprint in a conference proceedings. The remaining chapters contain both unpublished material and recent new work that has been written for this book. Finally, a technical Appendix provides the necessary concepts from Lattice Theory that are used throughout the book.

The published essays has been left unchanged, except for minor corrections. Here is a list of the original sources. Chapter 1: 'The logical analysis of plurals and mass terms: A lattice-theoretical approach', in: *Meaning, Use, and Interpretation of Language*, R. Bäuerle *et al.* (eds.), de Gruyter, Berlin 1983, 302–323, reprinted with kind permission from Walter de Gruyter & Co.; Chapter 2: 'Plural', in: *Semantik. Ein internationales Handbuch der zeitgenössischen Forschung*, A. von Stechow and D. Wunderlich (eds.), de Gruyter, Berlin 1991, 418–440, English translation by the author of the German original, reprinted with kind permission from Walter de Gruyter & Co.; Chapter 3: 'Hydras: On the logic of relative constructions with multiple heads', in: *Varieties of Formal Semantics. Proceedings of the Fourth Amsterdam Colloquium*, F. Landman and F. Veltman (eds.), Foris, Dordrecht 1984, 245–257, reprinted with kind permission from Walter de Gruyter & Co.; Chapter 4: 'Generalized quantifiers and plurals', with 1 figure, in: *Generalized Quantifiers. Linguistic and Logical Approaches*, P. Gärdenfors (ed.), Reidel, Dordrecht 1987, 151–180, reprinted with kind permission from Kluwer Academic Publishers; Chapter 7: 'Ten years of research on plurals — Where do we stand?', in: *Plurality and Quantification*, F. Hamm and E. Hinrichs (eds.), Kluwer, Dordrecht 1997, 19–54, reprinted with kind permission from Kluwer Academic Publishers; Chapter 8: 'Algebraic semantics for natural language: some philosophy, some applications', in: *International Journal of Human-Computer Studies*, 43, 1995, 765–784, reprinted with kind permission from Academic Press, Ltd., London; Chapter 9: 'Quantity and number', in: *Semantic Universals and Universal Semantics*, D. Zaefferer (ed.), Foris, Dordrecht 1991, 133–149, reprinted with kind permission from Walter de Gruyter & Co.

Chapter 11, 'Algebraic semantics of event structures' appeared in the preprint volume of the *Proceedings of the Sixth Amsterdam Colloquium*, J. Groenendijk *et al.* (eds.), ITLI, Amsterdam 1987, 243–262. For compatibility of reference, it has also been left unchanged.

Chapter 2 was actually written in 1984, and had been around as an underground for some time. Chapter 5 dates back to 1986, and Chapter 10 to 1988. A first version of Chapter 6, together with the Appendix, was used as material for a joint course (with Jan Tore Lønning) on Algebraic Semantics at the Third European Summer School in Language, Logic and Information, Saarbrücken, Germany, in August 1991. The chapter has been considerably reworked since then. Chapter 12 is the most recent one and was written a few weeks ago. The first part of Chapter 13 is new, too, while the second part goes back to 1994. Finally, the content of Chapter 14 was presented at a symposium on second-order logic and plural quantification at the Annual Meeting of the American Philosophical Association (APA), Pacific Division, in April 1996.

Since the single chapters originated as independent papers there is a certain amount of inevitable overlap, which is true in particular of the exposition of the system LP of plural logic. I apologize for that; the reader is asked to skip repetitions as soon as he or she has picked up the sometimes varying logical notation that is used in a given chapter. However, the advantage is that the chapters can to a large degree be read separately.

The Bibliography has become rather long. There is a general and a specific reason for this. The general reason is that I had to include references from three fields: linguistics, logic, and philosophy; the specific reason is that the Bibliography contains all titles both from the survey article in the Handbook of Semantics (Chapter 2) and from a list of titles that was compiled for the European Summer School mentioned before. Also, I have at occasion included titles in the Bibliography that are actually not cited in the text; so the reader might want to look through it for additional relevant literature.

Since it is my conviction that research is basically a collective enterprise it is only natural for me to acknowledge my intellectual debt to the many colleagues, students, and friends who have helped me over the years to become clearer about the ideas developed in this book. Many of their names are listed in the acknowledgements to some of the individual chapters (Chapters 1, 4, 9, 11, 14 and the Appendix). In addition to and partial repetition of those individuals let me mention here Emmon Bach, John Barwise, Johan van Benthem, Ulrich Blau, George Boolos (†), John Burgess, Jaap van der Does, Marcel Erné, Claudia Gerstner-Link, Volker Halbach, Fritz Hamm, Stefan Iwan, Hans Kamp, Ed Keenan, Manfred Krifka, Fred

Landman, Hans Leiß, Jan Tore Lønning, Alice ter Meulen, Karl-Georg Niebergall, Barbara Partee, Michael Resnik, Mats Rooth, Remko Scha, Roger Schwarzschild, Hinrich Schütze, Peter Sells, Peter Simons, A. von Stechow, Wolfgang Stegmüller (†), Holger Sturm, Pat Suppes, Matthias Varga von Kibéd, Dietmar Zaefferer, and Ed Zalta.

I wish to thank the *Deutsche Forschungsgemeinschaft*, Bonn, Germany, for repeatedly sponsoring my research, most recently with a three months' grant for my stay at the Center for the Study of Language and Information (CSLI), Stanford University, from August through October, 1996. I had intended to finish the book by then, but it took me another year to be finally done with it. Due to the generous hospitality of the CSLI I was able to return this year, again for late summer and early fall, and to finish my work. I wish to express my sincere thanks to the staff of the CSLI, in particular, Michele King, and its director, John Perry, for granting me the privilege to be a frequent visitor there.

During my last two visits at the CSLI I had the pleasure to be a member of the Metaphysics Research Lab run by my friend Ed Zalta. I wish to thank him for being my continued discussion partner in philosophical and logical matters, for freely giving his time to help me with the intricacies of the NeXT workstation, and for having been such an enjoyable host.

Writing this book was an adventure for me in a particular respect: this is my first book that I wrote in English. At occasion I felt the pain of not being able to express myself in my mother tongue; but on the whole I hope that I managed to produce a text that doesn't sound outlandish at every turn. There are two chapters (10 and 12) which I felt I wanted to have checked by a native speaker. I wish to thank Peter Sells and Ed Zalta for carefully going through the material and suggesting stylistic improvements.

Many people have also provided technical advice and assistance in the course of the production of the book. I'd like to thank them all, in particular: Cleo Condoravdi for compiling a large part of the bibliography, CSLI's systems administrator Emma Pease for general computational support, and Tony Gee of CSLI Publications for the production of the figures in Chapter 2 as well as the splendid realization of the cover idea. Finally I wish to thank the editor of CSLI Publications, Dikran Karagueuzian, for his enthusiastic support and continued encouragement.

Last but not least, my special thanks go to my wife Claudia for all her love and support.

Stanford, California G.L.
October, 1997

ALGEBRAIC SEMANTICS IN LANGUAGE AND PHILOSOPHY

Introduction

If you walk up the steep path to the Acropolis in Athens and finally pass through the Propylaea you are not only struck by the majestic view of the Parthenon temple but also by the unrivaled beauty of the famous Caryatid porch of the Erechtheum. Those six Caryatids, also called *korai*, are draped female figures which are used as columns to support the roof of the porch. They seem to carry their load through the ages, in classical contrapposto and complete serenity.

The caryatids exemplify a collective property. A single one of them wouldn't be able to uphold the structure. There are two theoretical questions connected with this, one linguistic and one philosophical: First, how is collective predication to be treated in the semantic analysis of natural language, and second, is collective exemplification of properties a phenomenon that survives the logical-philosophical treatment of regimentation which natural language locutions are commonly submitted to? Are there "plural objects"?

To begin with the latter, imagine that, contrary to fact, the roof of the porch of the Erechtheum temple had originally been supported by a regular brick wall. Then the property of supporting the roof, call it P, would have been had by a single entity, the wall, which is as concrete an object as anything could be. Now subsequent architects found the structure rather unappealing and clumsy, so they took out more and more parts of the wall, first putting in windows, then creating more open space, and eventually ending up with replacing the remaining bricks with beautiful Ionian columns and finally with the caryatid figures. What exemplifies the property P now? According to some philosophers and linguists, it is now the *set* of those figures, which is an abstract entity. This is the set account of pluralities. The question is at which point in the history of the building the concrete thing turned into an abstract object.

It has been clear to me for a long time that the set account just couldn't be right. It is true that for a few decades now philosophers have used set theory as a modeling tool in their theorizing, but as a conceptual instrument in the history of ideas set theory is a real late-comer. Philosophers had typically been concerned with concepts rather than sets. While the notion of an object falling under a concept or property has always been in use, the proper abstractions necessary to arrive at a viable notion of extension (let alone one that could be used as a point of departure for the modern iterative notion of set) was hardly arrived at before Frege. It seems natural then to ask how pluralities had been conceptualized in the past, and how they should be treated today in a way that regains the pre-set-theoretic innocence. When we look at the rules of classical syllogistics we see that they constituted some kind of relational calculus. Much later it was Leibniz who formalized the relation of subsumption in such a way as to result in a regular algebraic framework.[1] Common to all those approaches is the notion of *part*. It is also basic in dealing with concrete particulars of our everyday world, an obvious fact that is reflected in the all-pervasive part locution in language speaking about things and their location in space and time. One of my claims in this book is that the part relation is also the proper tool for the analysis of pluralities. That involves the decision to treat pluralities together with particulars rather than with universals. The latter option would entail the assimilation of pluralities to properties and hence, in the language of set theory, to sets. By contrast, allowing "singular" objects being parts rather than elements or members of pluralities makes it possible to avoid the mysterious transition from the concrete to the abstract which was highlighted by our example above. Thus pluralities are viewed as objects of the same kind as individual things. The ontological claim here is what could be called "relative nominalism:" No matter what kind of entity you take to constitute your basic individuals (concrete things, properties, sets, events, qualia, tropes, or what have you) no extra commitment will be produced when admitting pluralities of those individuals in your ontology. Also, under such a view, an important parallelism between plural objects and mass-like objects can be perceived that will otherwise go unnoticed. A typical principle exhibiting this parallelism is the *cumulative reference property* that will be dealt with in the beginning chapters of the book.

The notion of part is of course highly ambiguous. I would like to distinguish between a rich and an austere notion. The rich notion is typically paraphrased by the term "constituent" hinting at the fact that parts in this sense are elements in a complex whole which is organized by a number of extra-logical relations. The austere version is the logical notion of part; its meaning doesn't go beyond what can be expressed in purely logical terms, so it doesn't involve extra-logical relations.

It is this latter notion of part that I will be mostly concerned with in the

[1] On Leibniz see, e.g., Lenzen (1990); Swoyer (1994); Zalta (1997a).

book. It lends itself to a mathematical treatment as an algebraic relation which obeys certain axioms. This is the source of the algebraic outlook underlying the present work.

Since algebra is the general theory of structures, the algebraic perspective also entails a kind of structuralist attitude towards reality as far as metaphysics is concerned. It doesn't deny that rich and highly sophisticated empirical knowledge is available about our external world. But that is just not part of metaphysics or ontology. Philosophers should confine themselves to a formal view of metaphysics. This theme is taken up in Chapter 12.

There are several places in the book where I try to explain the algebraic approach; see in particular Sections 1.1, 2.3, 11.1, and Chapter 8.

This book doesn't consist, however, of a number of purely philosophical reflections about topics of the kind just alluded to; quite the opposite is true. In fact, having arrived at a certain philosophical view concerning them I found it necessary to foster it by some serious research into the linguistic mechanisms of reference to individuals of various sorts, in particular, the mechanisms of plural reference. I also collected some cross-linguistic evidence that I thought were supportive of my view. To bear out the structural perspective I also made extensive use of all the formal tools from logic and mathematics that seemed to me relevant to my task. In the final chapters of the book I return to the philosophical issues involving problems of metaphysics, ontology, and the philosophy of mathematics.

I will now describe the content of the individual chapters, with some comments on their interrelations and, regarding the published essays, their significance in the light of current research.

Chapter 1 contains the my first account of the algebraic research program in linguistic semantics and philosophy. It develops a unified logic of plural and mass expressions, LPM, stressing the structural similarity between plural denotation and homogeneous mass denotation in terms of concepts from lattice theory.

In its philosophical part it argues for what I here called relative nominalism with respect to pluralities; it also contains a contribution to the ontology of concrete objects, by specifying a relation of constitution that admits of distinguishing a physical object from the portion of matter making it up. That involves a rejection of the ontological principle of No Coincidence, viz. that no two objects can be at the same place at the same time. The intuition that the stuff and the object it composes are really one thing is accounted for by introducing a homomorphic relationship between the algebraic structures of plural individuals and portions of matter which allows for "ignoring" information in so-called invariant contexts. There were some problems with this account that are discussed in Chapter 12, where the issue of constitution is taken up again in a purely ontological setting.

On the linguistic side, the chapter is written in the spirit of compositional semantics which was inherited from the Montagovian paradigm. In fact, it constitutes the attempt to save the principle of compositionality in the light of recalcitrant data involving plural constructions. Thus from a position of "pluralist minimalism" one could say that the theory is just a technical tool for securing compositionality. A central idea in this context was that of a compositional plural operator which later led to the introduction of other operators on verb phrases, in particular, the distributivity operator (see below).

The semantics that was given for the logic LPM contains a complete atomic Boolean algebra (with its zero element removed) as modeling the structure of pluralities. CAB algebras are known to be the models of Classical Mereology, so there is an obvious connection to the mereological theories in philosophy that were used in connection with the modern nominalist program, the outlook of which was quite different, though. The formation of pluralities involves taking joins of individuals only, so I found it more accessible to isolate this feature in later work in terms of the notion of semilattice. Such a semilattice, however, has to obey certain additional principles in order to conform with our intuitions about pluralities. It turns out that when these principles are specified we essentially come back to the CAB algebras. The logical issues involved here are described at length in Chapter 6.

Chapter 2 is a handbook article which tries to systematize what had been known about plural semantics at the time the article was written. Naturally, it heavily draws on the material presented in Chapter 1, leaving out, however, the mass term part of the system LPM; the remaining logic was henceforth called LP. A wide array of plural constructions is described in the chapter, which are eventually given explicit logical forms in the first-order framework of LP. The focus is on the various readings a plural sentence can get. In particular, there is the example of a construction involving two plural noun phrases, for which I gave seven different readings generated from scope ambiguity and the collective/distributive distinction. Distributive readings are marked in their logical form with the distributivity operator, D, which appeared here for the first time. The analysis of these examples led to a number of interesting developments later on, in particular, to the discovery of so-called partitional readings and their interaction with the monotonicity properties of various determiners; a discussion of these issues can be found in Chapter 7.

In *Chapter 3* a special problem for the compositional analysis is addressed which concerns certain relative clause constructions with more than one head noun and with a collective predicate in the relative clause. I call those constructions "hydras." It is shown in the chapter how these cases can be given a satisfactory treatment in the framework of plural logic.

Chapter 4 is one of the first papers in which plurality is treated in the framework of generalized quantifier theory. It is noticed there that in order to incorporate collective predication a new kind of filter construction is needed for the denotation of conjoined noun phrases like *John and Mary*. Instead of taking the usual intersection filter which actually represents the properties that are common to John and Mary individually the new filter is the (principal) filter containing all collective properties of the plurality $John \oplus Mary$.

Regarding the main theme of quantification, a distinction is made between spurious and genuine plural quantification. A construction is genuinely plural only when the quantifier ranges over pluralities and not just singular objects. An asymmetry is noticed between existential and universal quantification: while the former is quite common, the latter doesn't seem to be easily expressible in language.

The chapter also contains an adjectival analysis of numerals which is different in conception from the generalized quantifier view where numerals are treated as determiners. Finally, the phenomenon of "floated quantifiers" is addressed and analyzed in terms of the distributivity operator.

Chapter 5 is a case study of the German *je*-locution which acts like a distributivity operator. What is interesting here is that the quantificational element is not the determiner of the noun phrase specifying the quantifier scope but rather a particle attached to the "dependent" noun phrase. This device is called "anti-quantifier" by Choe (1987a) who discovered a similar phenomenon in Korean. The German data exhibit rather involved quantificational structures which can be expressed quite effortlessly in the language by using the *je* particle. Their analysis is given in terms of the D-operator.

Chapter 6 focuses on the purely logical properties of the theory of plurals, LP. A "genetic" approach is chosen in which a number of plausible axioms for pluralities are introduced and motivated in turn. First, the logical basis of LP is presented explicitly; it is a first-order free logic with abstraction and description terms. Then the axioms for characterizing pluralities are given; the resulting structures are called *plural lattices*. Apart from the axioms for the general notion of semilattice, there are four more axioms that seem to be specific to our conception of pluralities: (i) an axiom which excludes bottom elements (there is no "null entity"); (ii) an axiom called *atomic separation* which says that any two incomparable elements are "separated" by an atom, thereby making sure that every element has an atom below it; (iii) an axiom of *definable completeness* which guarantees a supremum for every non-empty set of individuals which can be defined in the theory; and finally (iv) an axiom which expresses the property that the semilattice is "maximally unfolded." A number of properties of the resulting system is discussed, and a comparison is made with a similar system of J.T. Lønning's. Furthermore, a connection is established with the logical and philosophical literature on mereology by discussing the relationship between the confusingly large number of axioms that can be found in various accounts. Finally, some important metalogical issues of the system are discussed. It is shown in particular that LP doesn't admit of a complete axiomatization with respect to full standard models. That means that LP, albeit formulated in a first-order language, is really second-order in nature. This will become important in the last two chapters of the book.

The proofs to a number of theorems are deferred to the Appendix.

Chapter 7 contains a survey of the semantic literature on plurals in the 1980s and early 1090s. From early on there was one problem with the adequacy of the plural structures of LP concerning the so-called intermediate level readings. A sentence like *The Leitches and the Latches met in the park* is ambiguous because it could mean either that the two families came together for a common activity or else that each family met separately in a place that happened to be the same park. Now the first, collective, reading can be expressed in LP whereas for the second there doesn't seem to be a device available because in a plurality there are no discernible subcollections "half-way down" towards the atoms. In order to cope with this problem Landman (1989) proposed a certain conception of "group formation" which in effect amounted to a usual set-theoretic solution. In this chapter I give reasons against adopting such a treatment.

The other main issue that is dealt with in the chapter is a discussion of readings for plural sentences, in particular the so-called *partitional readings*. The discussion focuses on interesting work by J. van der Does, in particular his (1992), where a number of new verb phrase operators are introduced. Since the analysis is carried out in the theory of generalized quantifiers, the interaction of plurality with determiners of varying monotonicity types is discussed. I give a representation of generalized quantifiers in the algebraic framework and defend an earlier analysis of mine that had been contested by van der Does.

Chapter 8 was written for a conference on the role of formal ontology in conceptual analysis and knowledge representation, the audience being mainly researchers working in artificial intelligence and knowledge representation. I took the opportunity to present the algebraic approach to semantics and ontology and to give a number of applications that have been dealt with elsewhere in the book in one form or another. Thus the chapter can serve as a kind of introduction to the ideas underlying the approach.

There is also a short section on the ontology of plurals that was prompted by the publication of David Lewis's book *Parts of Classes*, which proved highly relevant to my work. Lewis's theory and the nominalistic interpretation of second-order logic underlying it, which is due to Boolos, are discussed at greater length in Chapter 14.

Chapter 9 originated in a paper prepared for quite a different conference in 1988 which brought together logicians, semanticists and universal linguists to discuss issues of the universality of semantic concepts across languages. The present text focuses on ways in which the concept of quantity is realized in various languages, and on different modes of linguistic reference to concrete particulars.

The following three chapters deal with the notion of event and its characterization in the algebraic framework. *Chapter 10* gives a gentle introduction to the event conception whose technical semantic development is contained in Chapter 11. Its individual components are motivated in terms of examples drawn from a single and highly complex historical event, the French Revolution. The central feature is an algebraically structured domain of events in which individuals can play various roles, both severally and collectively; in this way the collective/distributive distinction can be reproduced in event theory by the contrast between a single event involving a collective agent and a sum of events involving atomic agents only. Apart from particular events, which can occur only once, also repeatable events are considered under the name *event type*. Event types are conceived here not as extensional classes of particular events, but as fine-grained intensional entities. The various components constitute what I call, for rather contingent reasons, the *Aether model*.

The technical details of this event conception are provided in *Chapter 11*. This chapter constitutes a variation on two themes that were prominent at the time the text was written, Situation Theory and Discourse Representation Theory (DRT). A notion of truth is defined for an event type in a given concrete but complex event which is called a chunk of the world. Then a translation procedure is sketched leading from natural language sentences to event types; this is done in a variable-free way by using "dependence levels" rather than DRT-style accessibility information. Instead of specifying a set of rules I illustrate the procedure with a number of examples where translations are given and evaluated semantically.

I realize that the paper reproduced in this chapter is unsatisfactory in various ways. I have included the text unaltered here since it keeps being referred to on occasion. I would like to point out, however, that I no longer defend the details of the interplay between events and their types as they were developed there although I still believe that the overall picture has some merit. The amendments and additions that I made are laid out, among other things, in Chapter 12.

Chapter 12 offers a unified philosophical account of the notion of concrete particular which draws on several ideas from previous chapters of the book. These entities come in two major varieties, as individuals and as events. Although in general not especially fond of reductionism I found it useful to view these two kinds of object as being special manifestations of just one underlying kind of entity, processes. Individuals are what I call *stationary processes*, whereas events are *classified processes*. Regarding the former, this view entails that individuals have temporal parts, and indeed I give reasons for embracing this much debated consequence. Also, I give a modified account of the relation of constitution that was first introduced in Chapter 1, in a way that criticisms of the earlier version are met. Concerning events, I try to argue that events are both coarse-grained and fine-grained depending on the context in which reference is made to them. Under the coarse-grained view, two events can be identified as having the same underlying process. The fine-grained view is similar to Kim's notion of event in that it involves an event type which corresponds to a property in Kim's theory. The transition from the fine-grained to the coarse-grained view is effected by "leaving out" information about the event which is irrelevant in the given context. The algebraic method is again applied throughout the chapter. A list with a number of commonly asked questions and puzzles about events, complete with detailed answers from the present perspective, concludes the chapter.

The chapter begins, however, with a reflection on philosophical methodology. In line with the general outlook of the book I argue for a completely formal conception of metaphysics which is concerned only with our most general conceptual tools that underlie the very structure of our thought. In the opening section I name a few of those tools and shortly comment on them in turn. Among them I count mereology which is the central theme in this book.

While the previous chapter was concerned with basic particulars and neglected the superstructure of pluralities, *Chapter 13* deals with some important philosophical problems of the concept of plural object. First the problem of ontological commitment is addressed. Penelope Maddy's view of concrete sets is discussed, and also the question of ontological innocence of forming mereological fusions. Regarding the *free-lattice fusions* that constitute pluralities I agree with David Armstrong that they are an ontological "free lunch," but doubts are raised against putting into the same category what I call *substantive fusions* (e.g., fusions of regions of space that describe our notion of physical space). The next section addresses the *counting fallacy* according to which what are two cannot be one. The counting fallacy has been a major obstacle for gaining a clear conception about the nature of pluralities. In another section the claim of an alleged inconsistency of the denotational view underlying plural theory, put forth by Barry Schein in a recent book, is refuted. The chapter concludes with a short discussion of cross-linguistic data from Chinese giving evidence for the fundamental nature of the part locution in language and hence, I submit, human thought.

The final *Chapter 14* deals with a topic in the philosophy of mathematics that has recently drawn wider attention, viz. the nominalistic interpretation of monadic second-order logic in terms of plural quantification. This interpretation was first advanced by George Boolos but was put to use by David Lewis (1991) in his reconstruction of classical set theory from purely mereological principles, which he calls *megethology*. The chapter first discusses Boolos's views, together with criticisms that have been advanced against them by Michael Resnik. It is shown that those criticisms can be answered by offering the concept of plural quantification available in plural logic. Drawing on the results from the discussion of the counting fallacy, and clearing up a point in Lewis about the supposed untenability of "singularism," an argument is given to the effect that the denotational view of pluralities is compatible with Boolos's position, and that it is the logic of plurality that is missing as a formal tool from Lewis's account. In support of this claim Lewis's concepts and principles of megethology are formalized in LP, showing that the usual standards of formal rigor can also be applied to the phenomenon of plural quantification, contrary to what seems to be suggested in Lewis's book.

The *Appendix* contains in a dense form elementary notions from lattice theory, such as relations, partially ordered sets, suprema and infima, semilattices and lattices, homomorphisms, and Boolean algebras. In a second part the structure of the plural lattice as it was defined in Chapter 6 is further explored. A proof is provided for the basic representation theorem of plural lattices according to which a plural lattice has as standard model the powerset algebra over the set of individual atoms minus the empty set (the result is well-known). Also, the property of being the most general lattice that can be spanned over a given set of atoms is characterized by a certain universal property in the sense of category theory, the *extension property*. Its equivalence with the axiom of "sup-prime atomicity," which is one of the plural lattice axioms, is proved in detail.

Chapter 1

The Logical Analysis of Plurals and Mass Terms: A Lattice-theoretical Approach

1.1 Introduction

The weekly *Magazine* of the German newspaper *Frankfurter Allgemeine Zeitung* regularly issues Marcel Proust's famous questionnaire which is answered each time by a different personality of West German public life. One of those recently questioned was Rudolf Augstein, editor of *Der Spiegel*; his reply to the question: "Which property do you appreciate most with your friends?" was

(1) "that they are few."

Clearly, this is not a property of any one of Augstein's friends; yet, even apart from the *esprit* it was designed to display the answer has a straightforward interpretation. The phrase (1) predicates something *collectively* of a *group* of objects, here: Augstein's friends.

As it is well known, collective predication is a rather pervasive phenomenon in natural language, as the following sample of sentences shows:

(2) The children built the raft.

(3) The Romans built the bridge.

(4) Tom, Dick, and Harry carried the piano upstairs.[1]

(5) The playing cards are scattered all over the floor.

(6) The members of the committee will come together today.

(7) Mary and Sue are room-mates.

(8) The girls hated each other.

There is a striking similarity between collective predication and predication involving mass nouns.

(9) a. The children gather around their teacher.

b. The water gathers in big pools.

Moreover, a characteristic feature of mass terms, their *cumulative reference property*[2] can be imitated by plurals.

(10) a. If a is water and b is water then the sum of a and b is water.

b. If the animals in this camp are horses, and the animals in that camp are horses, then the animals in both camps are horses.

All this has been observed and discussed in the literature although the noted parallelism has perhaps not been stressed too much.[3] As it can be seen from Pelletier's 1979 volume, however, there is much disagreement about the proper way of attacking the logical problems posed by plurals and mass terms. From a semantic point of view the basic question is: what do mass terms and plural expressions denote? Some have thought that in order to be able to give a satisfactory answer to this question it is necessary to give up or at last extend the underlying set theory and to define new kinds of objects, for instance *ensembles* (Bunt 1979) or *collections* (Blau (1981b)). I think, however, that we can retain the usual set-theoretic metalanguage and simply enrich the structure of our models as to account for properties like cumulative reference. On my view, such properties are also not secured by *defining* some plural or mass term denotations out of others through set-theoretic manipulations; they all should be recognized as

[1]Massey's example; see Massey 1976.

[2]See Quine 1960, p. 91; Bunt 1979.

[3]The main source for mass terms is Pelletier 1979; furthermore see Bennett 1979, Bunt 1979, Ter Meulen 1980, 1981. For the treatment of plurals and collective terms, see Massey 1976, Burge 1977, Blau 1981b, Hoepelman/Rohrer 1980, Scha 1981. The parallelism referred to is explicitly expressed in Bunt 1979.

simply being there. What we rather should try to discover, then, is the network of the various relations which they enter and through which they are tied together. In the case of group and mass objects this picture naturally leads to the notion of lattice structure,[4] an idea which is, again, not new: it is inherent in mereological predicate logic and the Calculus of Individuals as developed by Leonard/Goodman 1940 and Goodman/Quine 1947. However, its possible use in the present context has perhaps been obscured by reductionist ontological considerations which are, in my opinion, quite alien to the purpose of logically analyzing the inference structures of natural language.[5] Our guide in ontological matters has to be language itself, it seems to me. So if we have, for instance, two expressions a and b that refer to entities occupying the same place at the same time but have different sets of predicates applying to them, then the entities referred to are simply not the same. From this it follows that my ring and the gold making up my ring are different entities; they are, however, connected by what I shall call the *constitution relation*: There is exactly one portion of matter making up my ring at a time. A constitution relation C has been explicitly introduced into the discussion by Parsons 1979. Sharing his intuitions on this point I shall provide a similar 2-place relation, '$\triangleright$', for *constitutes* or *makes up*. Its semantic counterpart in the theory to be presented below, the "materialization" function h, lies at the heart of my reconstruction of the ontology of plurals and mass terms: individuals are created by linguistic expressions involving different structures even if the portion of matter making them up is the same. Consider the example from Blau (1981b) (imagine that there is a deck of playing cards on the table):

(11) a. the cards
b. the deck of cards

While the portions of matter denoted by (11a) and (11b) are the same, I consider the individuals as being distinct.[6] (11a) refers to the pure collection of objects, and in many contexts (11b), too, refers just to this collection. In general, however, the introduction of a collective term like (11b) is

---

[4]Recently, Keenan and Faltz 1985 and Keenan 1981 advanced a "Boolean approach" to the semantics of natural language. I feel very sympathetic with this enterprise which, unfortunately, I became aware of only a year ago (January 1981). It is reassuring to see similar techniques be successfully applied in other areas of semantics, too. I have to refer a concrete evaluation of these ideas to another occasion.

[5]On this point I agree with Ter Meulen 1981, I think. But I do not follow her in the conclusions she draws from this observation. The inherent lattice structure is independent of the philosophical motives that gave rise to the construction of mereological systems. For the role of nominal mass noun denotations see the remarks below, also footnote 11.

[6]I guess that in German, with *die Karten* vs. *das Kartenspiel*, the point might come out more clearly.

indicative of connotations being added enough for it to refer to a different individual; for instance, a committee is not just the collection of its members, etc. Note, by the way, that the transition to an intension function would be of no help here. There might be two different committees which necessarily consist of exactly the same members.[7]

It might be thought, then, that collective predication is just the context in which pairs of expressions like (11a,b) refer to collections and thus are coreferential. This is not so, however, as can be seen from the following example. Imagine that there are several decks of cards, a blue one, a green one, etc., lying on a table.[8] Then the two following sentences do not mean the same although *number consecutively* is a collective predicate (the German word is *durchnumerieren*).

(12) a. The cards on the table are numbered consecutively.

b. The decks of cards on the table are numbered consecutively.

By contrast, spatio-temporal collective predicates do refer to the pure collection, or, as I conceive it, the portion of matter making up the individual in question. Examples are *be-on-the-table*, *occupy*, etc. I call such predicates *invariant.* So the following a) sentences are indeed equivalent to their corresponding b) sentences.

(13) a. The cards are on the table.

b. The decks of cards are on the table.

(14) a. The stars that presently make up the Pleiades galactic cluster occupy an area that measures 700 cubic light years.[9]

b. The Pleiades galactic cluster occupies an area etc.

In the following I shall distinguish between pure plural individuals involved in (12) and collections in the portions-of-matter sense referred to in (13) and (14). The former I call (individual) *sums* or *plural objects*; they respect levels of "linguistic comprehension" as shown by (12). By contrast, collections do not, they typically merge those levels. Sums and collections are similar, however, in that they both are just individuals, as concrete as the individuals which serve to define them, and of the same logical type as these. The latter feature is important because there is no systematic type ambiguity inherent in predicates like *carry*, *build*, *demolish*, *defend*, etc. As to the question of concreteness of sums of concrete objects, I agree with

[7]The point was apparently made first by David Kaplan as Bennett 1979 reports.

[8]This is the situation originally analyzed by Blau in his 1981b paper.

[9]This is Burge's example, see Burge 1977.

the intuitions of those who say that an aggregate of objects like a heap of playing cards which can be shuffled, burned, etc., is simply not an abstract entity like a set (see Burge 1977, Blau 1981b). What is more important, however, is the fact that the set approach to plural objects[10] does not carry over to the case of mass terms, thus missing the structural analogy between the two cases. Inherent in the notions of a set is atomicity which is not present in the linguistic behavior of mass terms.

Before I go on to present my own approach I want to say something about what Ter Meulen calls *nominal mass terms*. Typical examples are stuff names like *gold* in sentences like *gold has the atomic number 79*. But there is also the time-honored sentence *water is widespread* in which the term *water* has apparently a somewhat different status. It seems to refer to the concrete "scattered individual" that you just find everywhere, hence Quine's analysis in terms of mereology. In this sense the sentence should be synonymous to *the water (on earth) is widespread*. Of the same type is the use of *gold* in *America's gold is stored in Fort Knox*. Here, again, a concrete object is referred to by *America's gold*, namely the material fusion of all quantities of the US gold reserve. So there can be no doubt that some notion of fusion is needed to account for definite descriptions involving predicative mass terms (the μ-operator defined below does just this). Genuine stuff names, however, are something else. Substances are abstract entities and cannot be defined in terms of their concrete manifestations. The question, then, is of what kind the connections are that are intuitively felt between substances and their quantities. Take water, for instance. A quantity is water if it displays the internal structure of water, that is H_20. But this relation is not a logical one. Or else we might look for substance properties which carry over to the quantities of the substance in question. Water is a liquid and yet, all concrete water might be frozen. So we have to go over to dispositional properties, getting more and more involved into our knowledge of the physical world ... What I am getting at is that *nominal mass terms do not seem to have a proper logic*. Be this as it may, this issue is completely independent of the lattice structure that governs the behavior of predicative mass terms and plural expressions; it is only this structure that I want address myself to in the present chapter.[11]

[10]This approach is the traditional one. In one form or another it can be found, for instance, in Bennett 1975, Hausser 1974, v. Stechow 1980, Hoepelman/Rohrer 1980, Scha 1981.

[11]Contributions to the problem of substance names can be found in Pelletier 1979 (in particular, Parsons 1979), Bunt 1979, and Ter Meulen 1980, 1981. Let me comment on the latter work, which is formulated in a Montague framework. The few remarks I made here will make it evident that I fully agree with ter Meulen in that nominal mass nouns cannot be *reduced to* predicative mass nouns. But for this very reason I fail to see any cogent argument for the kind of denotation ter Meulen wants to assign to these

Now what I am going to propose, then, is basically the following. First of all, let us take seriously the morphological change in pluralization, which is present in many natural languages, and introduce an operator, '*,' working on 1-place predicates P, which generates all the individual sums of members of the extensions of P. Such a starred predicate now has the same cumulative reference property as a mass predicate, it is closed under sum formation: any sum of parts which are $^*P$ is again $^*P$. This property gives rise to the introduction of a Boolean structure on the domain of discourse, E: technically, E becomes an atomic Boolean algebra which is taken to be complete so that every subset of E possesses a sum. Now let $\| \cdot \|$ be the denotation function in a model, and $\|P\|$ the extension of P. Then $\|^*P\|$, the extension of $^*P$, can be defined in terms of $\|P\|$ as the complete join-subsemilattice[12] in E generated by $\|P\|$. This construction is the mathematical expression of the closure property referred to above. The set A of atoms of E consists of the "singular objects," like this card, that deck of cards, etc. Among them are all the different portions of matter, like the water making up this ice cube.

Now, let a and b denote two atoms in A. Then there are two more individuals to be called below $a + b$ and $a \oplus b$. $a + b$ is still a singular object in A, the *material fusion of a and b*; $a \oplus b$ is the *individual sum* or *plural object* of a and b. The theory is such that $a + b$ constitutes, but is not identical with, $a \oplus b$. This looks like a wild platonistic caprice strongly calling for Occam's Razor. Language, however, seems to function that way. Take for a, b two rings recently made out of some old Egyptian gold. Then *the rings*, $a \oplus b$, are new, *the stuff*, $a + b$, is old.

terms at a given reference point (i.e., intension functions denoting in each world the set of concrete quantities of the substance in question). As it turns out, the arguments she puts forth in Ter Meulen 1981 really lend support only to the first, the critical, point (viz. that reduction is impossible). But what she then goes on to call a nominal mass noun's "extensional reference to an intensional object" (viz. the intension function referred to above) seems to me both syntactically and semantically misguided. For the inevitable doubling of syntactic rules is certainly unwelcome, to begin with. But what is more, those intension functions, even when lifted to still another intensional level as ter Meulen wants to have it, are simply not well motivated as substance name denotations. The statement, for instance, that two fictional substances can be differentiated (op. cit., p. 438) is not compatible with the principle of rigid designation introduced earlier (op. cit., p. 424). More generally, there are no rules that could justify intuitively valid inferences from contexts involving nominal mass nouns to contexts with their corresponding predicative terms — it is my view, anyway, that such inferences are not based on pure logic alone. I conclude from this that the problem of nominal mass nouns is best approached in a spirit of logical abstinence. Nominal mass nouns denote abstract entities, to be sure, and as such they are names of individuals just like *John*, *Munich*, and the rest. Beyond this minimal account things become notoriously vague.

[12]For this concept see, for instance, Grätzer 1978.

Sums are partially ordered through the intrinsic ordering relation '$\leq_i$' on E which is expressed in the object language by the 2-place predicate 'Π'. It is called the *individual part relation* (*i-part* relation, for short) and satisfies the biconditional

(15) $$a \, \Pi \, b \;\leftrightarrow\; a \oplus b = b.$$

The semantic counterpart of (15) is the well-known Boolean relation (where $\sqcup_i$ denotes the join operation on E):

(16) $$\|a\| \leq_i \|b\| \quad \text{iff} \quad \|a\| \sqcup_i \|b\| = \|b\|$$

Now if P is a 1-place predicate and $^*P$ the corresponding plural predicate let $^\star P$ (the *proper plural predicate* of P) be true of exactly the non-atomic sums in the extension of $^*P$. Then we can define the *sum* and the *proper sum* of the P's, $\sigma x Px$ and $\sigma^* x Px$, respectively, as the supremum of all objects that are $^*P$ and $^\star P$, respectively ('ι' is the description operator):

(17) a. $$\sigma x Px \;:=\; \iota x \, (^*Py \wedge \forall y \, (^*Py \rightarrow y \, \Pi \, x))$$

b. $$\sigma^* x Px \;:=\; \iota x \, (^\star Py \wedge \forall y \, (^\star Py \rightarrow y \, \Pi \, x))$$

$\sigma^* x Px$ carries the presupposition that there are at least two P's; in this case $\sigma x Px$ and $\sigma^* x Px$ coincide.

The *material fusion* of the P's, denoted by $\mu x Px$, becomes that object which constitutes the sum of the P's:

(18) $$\mu x Px \;:=\; \iota x \, (x \rhd \sigma x Px)$$

The μ-operator has the effect of merging levels of comprehension, as I called it above. Thus we will have $\mu x Px = \mu x Qx$, if P stands for *is a card of one of the card decks*, and Q for *is a card deck*. $\mu x Px$ and $\mu x Qx$ are not the same, however. If we have two decks of Bridge cards, for instance, $\sigma x Px$ contains 104 atoms whereas $\sigma x Qx$ contains only 2 atoms.

There is a second ordering relation, called the *material part (m-part) relation* and denoted by 'T'. It establishes a partial order on portions of matter, but only a preorder, called $\leq_m$, on the whole domain of individuals. Objects which are m-parts of one another are *materially equivalent* in that they have the same portion of matter constituting them. If a is an i-part of b then a is an m-part of b; symbolically:

(19) $$a \, \Pi \, b \;\rightarrow\; a \, \mathsf{T} \, b$$

In order to explain the meaning of 'T' more precisely let me supply the remaining concepts of the model structure to be defined below. In addition to the domain of individuals, E, there is a set D which is endowed with a join operation '$\sqcup$' making D into a complete, but not necessarily atomic, join-semilattice. D is partially ordered by its intrinsic ordering relation, $\leq$, defined by

$$(20) \qquad x \leq y \quad \text{iff} \quad x \sqcup y = y$$

D represents the set of all individual portions of matter in the model, and as such, it is considered as a subset of A, the set of atoms of E. The relation (20) stands for a material part-whole relation.[13] It can be used to order the individuals of E materially. For this I postulate for every model the existence of a semilattice homomorphism h from $E \setminus \{\mathbf{0}\}$ to D such that h is the identity function on D and commutes with the supremum operator. Then h is order-preserving, i.e., we have

$$(21) \qquad x \leq_i y \;\Rightarrow\; h(x) \leq h(y)$$

for all $x, y \in E$ with $x \neq \mathbf{0}$. I can now define the relation $\leq_m$ mentioned above by

$$(22) \qquad x \leq_m y \quad \text{iff} \quad h(x) \leq h(y) \qquad (x, y \in E \setminus \{\mathbf{0}\})$$

$\leq_m$ is the semantic counterpart of 'T'. Obviously, Π is coarser than $\leq_m$, a fact that is syntactically expressed by (19). In its intended interpretation, then, which is formally implemented into the model structure, $a \,\mathsf{T}\, b$ can be read as follows: "the portion of matter constituting a is m-part of the portion of matter constituting b."

One-place predicates are interpreted as subsets of $E \setminus \{\mathbf{0}\}$. (Here, the null element $\mathbf{0}$ is assigned the role of a dummy object to which no predicate applies and which can take care of denotation gaps arising in our description theory; for details see Link 1979, 5.2.) If P is a mass term the extension $\|P\|$ of P in a given model should be a set of portions of matter which is closed under joins. So $\|P\|$ becomes a subsemilattice of D. This completes the informal description of the Boolean model structure $\alpha B = \langle E, A, D, h\rangle$ and of the LPM-model $M = \langle \mathcal{B}, \|\cdot\|\rangle$. For precise definitions see below.

If P is a predicate we can introduce its *mass term correspondent*, called ${}^mP$; if $\|P\|$ is the extension of P in a model, define

---

[13] Let me point out that the notions of *material part* and *portion-of-matter* are only a paradigm case of fixing intuitions for *homogeneous reference*. The structural properties actually used (non-atomicity, cumulative reference) are also found to govern the behavior of abstract entities like events. Implications for modes of temporal reference are discussed in Hoepelman/Rohrer 1980.

(23) $$\|{}^m P\| \;:=\; \{\, x \in D \mid x \leq \sup h[\|P\|] \,\}$$

If P is already a predicative mass term we have, by this definition, $\|P\| \subseteq \|{}^m P\|$. Mass term correspondents occur in natural language.[14] Consider

(24) a. There is apple in the salad.

b. $\exists x\,(\,{}^m Px \,\wedge\, Qx\,)$

(24b) is a formalization of (24a), with P for *is an apple*, ${}^m P$ for *is apple*, and Q for *is in the salad*. From this we can infer, intuitively as well as formally:

(25) a. There are apple parts in the salad.

b. $\exists x \exists y\,(\,{}^* Px \,\wedge\, y \,\mathsf{T}\, x \,\wedge\, Qx\,)$

We have, in fact, the following theorem (cf. (T.14) below):

(26) $${}^m Pa \;\rightarrow\; \exists y(\,{}^* Py \,\wedge\, a \,\mathsf{T}\, y\,)$$

Returning now to the model structure consider once again the extension of a 1-place predicate P. It can contain atoms as well as proper i-sums. This general case of a *mixed extension* is exemplified by predicates like *carry the piano*: think of Tom, Dick, and Harry (together), Obelix (all by himself), etc. Common nouns and intransitive verbs like *die*, however, seem to admit only atoms to their extension. I call such predicates *distributive*:

(27) $$\mathbf{Distr}\,(P) \;\leftrightarrow\; \forall x\,(Px \rightarrow \mathbf{At}\,x)$$

(here the predicate **At** stands for the property of being an atom in the model). To illustrate take the intuitively valid inference from a) to b) in (28).

(28) a. John, Paul, George, and Ringo are pop stars.

b. Paul is a pop star.

This inference can be formally represented if we consider *pop star* as a distributive predicate P in the sense of (27). In this case the extension of ${}^*P$ is closed under non-zero i-parts (see theorem (T.10) below), so every atom of an i-sum which is ${}^*P$ is itself ${}^*P$, hence it is a P. In symbolic form the inference (28) looks like this.

(29) a. ${}^*P\,(\,a \oplus b \oplus c \oplus d\,)$

[14]See already Quine 1960, also Bunt 1979, Hoepelman/Rohrer 1980.

b. $\mathbf{Distr}\,(P)$

c. $b \, \Pi \, a \oplus b \oplus c \oplus d$

d. $^{*}Pb$

e. Pb

In a similar way we have the following valid inference with the distributive predicate Q for *die* (P stands for *animal*):

(30) a. The animals died. So every animal died.

b. $^{*}Q(\sigma^{*}xPx) \;\; \Rightarrow \;\; \forall x\,(Px \rightarrow Qx)$

For non-distributive predicates we have, of course, no such result, witness *carry*: $a \oplus b$ might be in the extension of *carry* while a or b alone is not. Notice that these predicates enter formalizations unstarred. To illustrate,

(31) a. The children gather around their teacher.

b. $Q\,(\sigma^{*}xPx)$ (with Q for *gather* etc.)

I think there is no harm in accepting this systematic difference: i.e., distributive predicates working on plural terms have to be starred, all the other predicates must not be. For distributivity seems to be a lexical feature. If we have a formal translation procedure the predicates have to be subcategorized accordingly.

There is one more operator, '^{T}', in the logic LPM to be presented below. $^{\mathsf{T}}P$, for a predicate P, is to be read as *partakes in* P. I introduce this operator in order to be able to distinguish between the plural terms *the children* and *all the children*. It seems to me that in *all the children built the raft* it is claimed that every child took part in the action whereas in *the children built the raft* it is only said that the children somehow managed to build the raft collectively without presupposing an active role in the action for every single child. This intuition enters the formalization of the two phrases given below. I want to stress, however, that the operator ^{T} can only be partially characterized in view of the essentially pragmatic nature of its intended interpretation. So let me formulate just the following meaning postulates for ^{T} which seem to be plausible:

(32) a. $\forall x\,(\,^{\mathsf{T}}Px \rightarrow \exists y\,(x \, \Pi \, y \wedge Py)\,)$

b. $\mathbf{Distr}\,(P) \rightarrow \forall x\,(\,^{\mathsf{T}}Px \leftrightarrow Px\,)$

I conclude the informal part of the chapter with some more formalizations of natural language sentences involving plurals and mass terms. Most of the principles governing their logic that have been mentioned in the literature come out valid in the system LPM below. Some of them are instantiated here, like Massey's plurality principles of symmetry, expansion, and contraction (see Massey 1976). What I did not treat in LPM, however, are any "downward" closure properties that are somehow felt to be present in the behavior of mass terms ("a part of water is still water"). Such principles can be added when a careful linguistic analysis has succeeded in giving them a form that takes care, in particular, of the *problem of minimal parts* (is every m-part of lemonade really lemonade again?). For some discussion of downward closure properties see, e. g., Bunt 1979 and Hoepelman/Rohrer 1980.

(33) a. A child built the raft.

b. $\exists x\,(Px \wedge Qx)$ — Px: x is a child; Qx: x built the raft

(34) a. Children built the raft.

b. $\exists x\,({}^{\star}Px \wedge Qx)$ — Px, Qx: dto.

(35) a. The child built the raft.

b. $\exists y\,(y = \iota x Px \wedge Qy)$ — Px, Qx: dto.

(36) a. The children built the raft.

b. $\exists y\,(y = \sigma^{*} x Px \wedge Qy)$ — Px, Qx: dto.

(37) a. Every child saw the raft.

b. $\forall x\,(Px \rightarrow Qx)$ — Px: dto., Qx: x saw the raft

(38) a. All the children built the raft.

b. $\forall y\,(y = \sigma^{*} x Px \rightarrow \tilde{Q}x)$ — Px, Qx as in (33); $\tilde{Q} := \lambda x(Qx \wedge \forall z(z \,\Pi\, x \rightarrow^{\mathsf{T}} Qz))$

(39) a. Tom and Dick carried the piano upstairs.

b. $P\,(a \oplus b)$ — a: Tom, b: Dick; Px: x carried the piano upstairs

(40) a. Tom and Dick carried the piano upstairs, so Dick and Tom carried it upstairs.

b. $P\,(a \oplus b) \;\Rightarrow\; P\,(b \oplus a)$ (*symmetry*)

(41) a. John and Paul are pop stars and George is a pop star, so John, Paul, and George are pop stars.

b. ${}^{\star}P\,(a \oplus b) \wedge Pc \;\Rightarrow\; {}^{\star}P\,(a \oplus b \oplus c)$ (*expansion*)

(42) a. John, Paul, George, and Ringo are pop stars, so Paul and Ringo are pop stars. (see (28))

b. ${}^{\star}P\,(a \oplus b \oplus c \oplus d) \;\Rightarrow\; {}^{\star}P\,(b \oplus d)$ (*contraction*)

(43) a. (All) water is wet.

b. $\forall x\,(Px \to Qx)$ Px: x is (a quantity of) water

(44) a. The water of the Rhine is dirty.

b. $Q\,(\,\mu x P x\,)$ Px: x is (a quantity of) Rhine water

(45) a. This ice cube is water.

b. $P\,\iota x(x \rhd a)$ a: this ice cube; Px: as in (43)

(46) a. The gold in Smith's ring is old, but Smith's ring is not old.

b. $Q\,\iota x\,(Px \wedge x \rhd a)\;\wedge\;\neg Qa$ Px: x is gold; Qx: x is old; a: Smith's ring

1.2 The Logic of Plurals and Mass Terms (LPM)

LPM is a first order predicate calculus with the usual logical constants '$\neg$', '$\vee$', '$\wedge$', '$\rightarrow$', '$\leftrightarrow$', '$\forall$', '$\exists$', the description operator denoted by 'ι', and the abstraction operator 'λ'[15] The syntactic variables are, for formulas of LPM, 'ϕ', 'ψ', 'χ'; for individual terms, 'a', 'b', 'c'; for variables, 'x', 'y', 'z'; for (1-place) predicates, 'P', 'Q'. These symbols can also appear with primes and indices. For metalinguistic (definitional) identity the symbol '$\doteq$' (':=') will be used.

The set of l-place predicate constants contains two specified subclasses: the set MT of *predicative mass terms* and the set DP of *distributive predicates.* MT and DP are taken to be disjoint sets.

As special primitive symbols of LPM we have a 1-place predicate symbol '$\mathbf{E}$', three 2-place predicate constants, 'Π', 'T', '$\rhd$', and two operators on 1-place predicates, '*' and '^{T}'. The intended interpretations are the following. $\mathbf{E}\,a$ stands for "*a exists*"; $a\,\Pi\,b$ for "*a is an individual part (i-part) of b;*" $a\,\mathsf{T}\,b$ for "*a is a material part (m-part) of b,*" $a \rhd b$ for "*a constitutes or makes up b;*" $^{*}P$ for "*the plural predicate of P;*" $^{\mathsf{T}}P$ for "*partakes in P.*" Now I introduce a number of defined expressions.

[15]For an outline of such systems ("PLIIKA") see Link 1979; notice that, in LPM, identity of individual terms need not be taken as primitive, but can be defined in terms of 'Π'

(D.1) $\quad P \subset Q \;\leftrightarrow\; \forall x\,(Px \rightarrow Qx)$

(D.2) $\quad P \equiv Q \;\leftrightarrow\; P \subset Q \,\wedge\, Q \subset P$

(D.3) $\quad P \oplus Q \;:=\; \lambda x\,(Px \vee Qx)$

(D.4) $\quad I_a \;:=\; \lambda x\,(x = a)$

Furthermore, the usual definitions for the following formulas involving R are assumed: *Refl* (R) ("R is reflexive"), *Trans* (R) ("R is transitive"), *Sym* (R) ("R is symmetric"), *Antisym* (R) ("R is antisymmetric"). We then have, with *PrO* (R) for "R is a preordering relation," *PO* (R) for "R is a partial ordering relation," and *Equ* (R) for "R is an equivalence relation:"

(D.5) $\quad \mathit{PrO}\,(R) \;\leftrightarrow\; \mathit{Refl}\,(R) \,\wedge\, \mathit{Trans}\,(R)$

(D.6) $\quad \mathit{PO}(R) \;\leftrightarrow\; \mathit{Refl}\,(R) \,\wedge\, \mathit{Trans}\,(R) \,\wedge\, \mathit{Antisym}\,(R)$

(D.7) $\quad \mathit{Equ}\,(R) \;\leftrightarrow\; \mathit{Refl}\,(R) \,\wedge\, \mathit{Trans}\,(R) \,\wedge\, \mathit{Sym}\,(R)$

2. *Defined predicate constants*. With two individual terms a, b, let $a = b$ stand for "a equals b," $a \sim b$ for "a is m-equivalent to b," $\mathbf{At}\,a$ for "a is an atom", and $\mathbf{Mp}\,a$ for "a is a portion of matter". Define

(D.8) $\quad a = b \;\leftrightarrow\; a\,\Pi\,b \,\wedge\, b\,\Pi\,a$

(D.9) $\quad a \sim b \;\leftrightarrow\; a \top b \,\wedge\, b \top a$

(D.10) $\quad \mathbf{At}\,a \;\leftrightarrow\; \forall x\,(x\,\Pi\,a \rightarrow x = a)$

(D.11) $\quad \mathbf{Mp}\,a \;\leftrightarrow\; a \rhd a$

3. *Defined predicate operators*. Let P be a 1-place predicate and a an individual term. The *proper plural predicate* $^{\star}P$ of P and the *mass term correspondent to* P, ^{m}P, are then defined as follows.

(D.12) $\quad {}^{\star}P\,a \;\leftrightarrow\; {}^{*}P\,a \,\wedge\, \neg\mathbf{At}\,a$

(D.13) $\quad {}^{m}P\,a \;\leftrightarrow\; \exists y\,(\,{}^{*}P\,y \,\wedge\, a \top \iota z(z \rhd y)\,)$

4. *Defined individual terms*. With 1-place predicates P and individual terms a, b, define the *i-sum of the* P's, the *proper i-sum of the* P's, the *i-sum of* a *and* b, the *material fusion of the* P's, and the *material fusion of* a *and* b, respectively:

(D.14) $\quad \sigma x P x \;:=\; \iota x\,(\,{}^{*}P\,x \,\wedge\, \forall x({}^{*}Py \rightarrow y\,\Pi\,x)\,)$

(D.15) $\sigma^{*}xPx \;:=\; \iota x\,({}^{*}P\,x \wedge \forall x({}^{*}Py \rightarrow y\,\Pi\,x))$

(D.16) $a \oplus b \;:=\; \sigma x\,(I_a \oplus I_b)x$

(D.17) $\mu xPx \;:=\; \iota x\,(x \rhd \sigma xPx)$

(D.18) $a + b \;:=\; \mu x\,(I_a \oplus I_b)x$

5. *Special abbreviative formulas.* Let P be a 1-place predicate.

(D.19) $\mathbf{Distr}\,(P) \;\leftrightarrow\; \forall x\,(Px \rightarrow \mathbf{At}\,x)$

(D.20) $\mathbf{M}\,(P) \;\leftrightarrow\; \forall x\,(Px \rightarrow \mathbf{Mp}\,x)$

(D.21) $\mathbf{Inv}\,(P) \;\leftrightarrow\; \forall x \forall y\,(x \sim y \rightarrow (Px \leftrightarrow Py))$

Distr (P) is to be read as "P is distributive" (i.e., P is true of atomic individuals only), **M** (P) as "P is a material predicate" (i.e., it is true of portions of matter only); **Inv** (P) means: "P is an invariant predicate" (i.e., it is closed under substitution of m-equivalent terms).

6. *Meaning postulates for* '^{T}'.

(MP.1) $\forall x\,({}^{\mathsf{T}}Px \rightarrow \exists y\,(x\,\Pi\,y \wedge Py))$

(MP.2) $\mathbf{Distr}\,(P) \rightarrow \forall x\,({}^{\mathsf{T}}Px \leftrightarrow Px)$

i.e., for all x, if x partakes in P, then x is an i-part of a y which is P (MP.1); if P is distributive, then for all x, x partakes in P just in case x is P (MP.2).

The semantics of LPM. I shall employ the usual set-theoretic terminology (see, for instance, Eberle 1970, Link 1979), with '$\subseteq$' for set inclusion and '$f[Y]$' for the f-image $\{\, f(x) \mid x \in Y \,\}$ of $Y \subseteq X$ under the function $f : X \rightarrow Z$. Moreover, some elementary notions of lattice theory are needed (see, e. g., Grätzer 1971, Sikorski 1969 for their definition): *partial ordering relation, partially ordered set* ("*poset*"), **1**-*element*, **0**-*element, atom, semilattice, join-semilattice, lattice, sublattice, ideal, principal ideal, complete lattice, Boolean lattice (algebra), atomic lattice, (semi-)lattice homomorphism.* The supremum of a subset Y of a poset X, relative to its ordering relation $\leq$, if it exists, will be denoted by $\sup_{\leq} Y$, the principal ideal generated by an element $x \in X$, where X is a lattice, by $(x]$. So we have $(x] = \{\, y \in X \mid y \leq x \,\}$. If X is a complete (semi)lattice, and $Y \subseteq X$, then $[Y]$ denotes the complete sub(semi)lattice generated by Y.

I now provide a model-theoretic interpretation of the formal system LPM.

(D.22) A *Boolean model structure with homogeneous kernel ("boosk")* is a quadruple $\mathcal{B} \doteq \langle E, A, D, h \rangle$ such that

1. E is a complete atomic (c.a.) Boolean algebra, with join operation $\sqcup_i$ and the intrinsic ordering relation $\leq_i$;
2. $A \subseteq E$ is the set of atoms in E.
3. $D \subseteq E$ is a complete join-semilattice with join $\sqcup$ and ordering relation $\leq$;
4. $h : E \setminus \{\mathbf{0}\} \rightarrow D$ is a semilattice homomorphism such that
   (i) $h\lceil D \doteq id_D$, i.e., $h(x) \doteq x$ for all $x \in D$;
   (ii) $h(\sup_{\leq_i}) \doteq \sup h[B]$ for all $B \subseteq E \setminus \{\mathbf{0}\}$.

It follows from this definition that we have, in particular, for all $x, y \in E \setminus \{\mathbf{0}\}$:

(47) $h(x \sqcup_i y) \doteq h(x) \sqcup h(y)$

(48) $x \leq_i y \Rightarrow h(x) \leq_i h(y)$

The homomorphism h induces a second ordering relation on $E \setminus \{\mathbf{0}\}$ which will be called the material part relation on $E \setminus \{\mathbf{0}\}$ and denoted by $\leq_m$:

(D.23) $x \leq_m y$ iff $h(x) \leq h(y)$ $(x, y \in E \setminus \{\mathbf{0}\})$

$\leq_m$ is only a preordering relation. It gives rise to an equivalence relation, defined by

(D.24) $x \sim_m y$ iff $x \leq_m y$ and $y \leq_m x$ $(x, y \in E \setminus \{\mathbf{0}\})$

and thus to a partition of $E \setminus \{\mathbf{0}\}$ into equivalence classes each containing all the individuals that are made up by the same portion of matter.

In the following definition of a model for LPM I am going to interpret only 1-place predicates, for simplicity (except for the special 2-place predicate constants). All the other many-place predicates receive their usual first order interpretation.

(D.25) A *model for* LPM is an ordered pair $M \doteq \langle \mathcal{B}, \|\cdot\| \rangle$ such that

1. $\mathcal{B} \doteq \langle E, A, D, h \rangle$ is a boosk (E is the *domain of individuals in* M, A the *set of atoms in* M, D the *set of portions of matter in* M, and h the *"materialization function" in* M);

2. $\|\cdot\|$ is a first-order assignment of denotations to the primitive expressions of LPM such that

(i) $\|a\| \in E$ if a is an individual constant;

(ii) $\|P\| \subseteq E \setminus \{\mathbf{0}\}$ if P is a 1-place predicate constant;

(iii) $\|P\| \subseteq A$ if $P \in$ DP

(iv) $\|P\|$ is a complete subsemilattice of D if $P \in$ MT

(v) $\|{}^{\mathsf{T}}P\| \subseteq \{x \in E \mid x \neq \mathbf{0} \,\&\, \exists y \in \|P\| \text{ such that } a \leq_i y\}$

(vi) $\|{}^{\mathsf{T}}P\| \doteq \|P\|$ if $P \in$ DP

The model is such that predicates are interpreted in the non-zero elements of E (condition (ii)). Conditions (iii), (iv) guarantee the special properties of the distributive and mass predicates, respectively. (v) and (vi) are designed to validate the meaning postulates (MP.1) and (MP.2) above, respectively.

Let $M \doteq \langle \mathcal{B}, \|\cdot\|\rangle$ be a model for LPM. The usual first order semantical recursion rules are assumed, where quantification is taken to run over the non-zero elements of the domain of individuals only. The truth and denotation conditions for the special primitive symbols of LPM are defined as follows ('1' stands for 'true'; bivalence is assumed):

(D.26) $\|Pa\| \doteq 1$ iff $\|a\| \in \|P\|$

(D.27) $\|\iota x Px\| \doteq d$ iff $d \neq \mathbf{0}$ and $\|p\| \doteq \{d\}$, and $\mathbf{0}$ otherwise;

(D.28) $\|\mathbf{E}\, a\| \doteq 1$ iff $\|a\| \neq \mathbf{0}$

(D.29) $\|a \,\Pi\, b\| \doteq 1$ iff $\|a\| \neq \mathbf{0}$ and $\|a\| \leq_i \|b\|$

(D.30) $\|a \,\mathsf{T}\, b\| \doteq 1$ iff $\|a\|, \|b\| \neq \mathbf{0}$ and $h(\|a\|) \leq h(\|b\|)$;

(D.31) $\|a \rhd b\| \doteq 1$ iff $\|b\| \neq \mathbf{0}$ and $\|a\| \doteq h(\|b\|)$;

(D.32) $\|{}^{*}P\| \doteq [\,\|P\|\,]_i$ (the complete $\sqcup_i$-subsemilattice generated by $\|P\|$).

A number of semantical facts can now be derived that give truth and denotation conditions for all the defined symbols of LPM.

(49) $\|P \subset Q\| \doteq 1$ iff $\|P\| \subseteq \|Q\|$

(50) $\|P \equiv Q\| \doteq 1$ iff $\|P\| \doteq \|Q\|$

(51) $\|P \oplus Q\| \doteq \|P\| \cup \|Q\|$

(52) $\quad \|I_a\| \doteq \{\|a\|\}$

(53) $\quad \|a = b\| \doteq 1 \quad \text{iff} \quad \|a\|, \|b\| \neq \mathbf{0} \text{ and } \|a\| \doteq \|b\|$

(54) $\quad \|a \sim b\| \doteq 1 \quad \text{iff} \quad \|a\|, \|b\| \neq \mathbf{0} \text{ and } h(\|a\|) \doteq h(\|b\|)$

(55) $\quad \|\mathbf{At}\, a\| \doteq 1 \quad \text{iff} \quad \|a\| \in A$

(56) $\quad \|\mathbf{Mp}\, a\| \doteq 1 \quad \text{iff} \quad \|a\| \in D$

(57) $\quad \|{}^*P\| \doteq \{\, x \in E \mid \exists X \subseteq \|P\| \,\&\, X \neq \emptyset \text{ such that } x \doteq \sup_i X \,\}$

(58) $\quad \|{}^\star P\| \doteq \|{}^*P\| \setminus A$

(59) $\quad \|{}^m P\| \doteq (\, \sup_\leq h[\|P\|]\,] \doteq \{\, x \in D \mid x \leq \sup_\leq h[\|P\|]\,\}$
$\quad \doteq \{\, x \in D \mid \exists y \in \|{}^*P\| \text{ such that } x \leq h(y)\,\}$

(60) $\quad \|\sigma x P x\| \doteq \sup_i \|P\|$, with $\sup_i \doteq \mathbf{0}$

(61) $\quad \|\sigma^* x P x\| \doteq \|\sigma x P x\| \quad$ if $\quad \|P\|$ has ≥ 2 elements, and $\mathbf{0}$ otherwise

(62) $\quad \|a \oplus b\| \doteq \|a\| \sqcup_i \|b\|$

(63) $\quad \|\mu x P x\| \doteq \sup_m h[\|P\|] \quad$ if $\quad \|P\| \neq \emptyset$, and $\mathbf{0}$ otherwise

(64) $\quad \|a + b\| \doteq h(\|a\|) \sqcup h(\|b\|) \quad$ if $\quad \|a\| \neq \mathbf{0} \neq \|b\|$

Let the notion of truth in a given model be defined in the usual way. A formula is *valid* if it is true in every model.

I now give, in a loose order, a list of theorems of LPM that come out valid under the above interpretation. I may mention, in particular, the *homogeneous reference properties* (T.11), (T.12), and the *existence and identity criteria* for the σ- and μ-terms, (T.16) – (T.21).

(T.1) $\quad a = b \leftrightarrow a \,\Pi\, b \wedge b \,\Pi\, a$

(T.2) $\quad a \,\Pi\, b \rightarrow a \,\mathsf{T}\, b$

(T.3) $\quad a = b \rightarrow a \,\mathsf{T}\, b \wedge b \,\mathsf{T}\, a$

(T.4) $\quad \mathit{Equ}\,(=) \wedge \mathit{Equ}\,(\sim)$

(T.5) $\quad \mathit{PO}\,(\Pi) \wedge \mathit{PrO}\,(\,\mathsf{T}\,)$

(T.6) $\quad \mathbf{Distr}\,(P) \rightarrow \forall x\, (Px \rightarrow \neg\, {}^\star P\, x)$

(T.7) $P \subset {}^*P$

(T.8) $\forall x\,(\mathbf{At}\,x \rightarrow (Px \leftrightarrow {}^*Px))$

(T.9) $\forall x\,({}^*Px \rightarrow x\,\Pi\,\sigma x Px \wedge \exists y\,(Py \wedge y\,\Pi\,x))$

(T.10) $\mathbf{Distr}\,(P) \rightarrow \forall x \forall y\,({}^*Py \wedge x\,\Pi\,y \rightarrow {}^*Px)$

(T.11) $\forall x \forall y\,({}^*Px \wedge {}^*Py \rightarrow {}^*P(x \oplus y))$

(T.12) $\forall x \forall y\,(Px \wedge Py \rightarrow P(x + y))$ for $P \in$ MT

(T.13) $\mathbf{M}\,(P) \rightarrow P \subset {}^mP$

(T.14) $\mathbf{Mp}\,a \rightarrow ({}^mPa \leftrightarrow \exists y\,({}^*Py \wedge a\,\mathsf{T}\,y))$

(T.15) $\exists x Px \wedge P \subset Q \rightarrow {}^*Q\,(\sigma x Px)$

(T.16) $\exists x Px \leftrightarrow \mathbf{E}\,\sigma x Px$

(T.17) $\exists x Px \leftrightarrow \mathbf{E}\,\mu x Px$

(T.18) $\mathbf{E}\,\sigma^* x Px \leftrightarrow \exists x \exists y\,(Px \wedge Py \wedge x \neq y)$

(T.19) $P \equiv Q \rightarrow \sigma x Px = \sigma x Qx$

(T.20) $\mathbf{Distr}\,(P) \wedge \mathbf{Distr}\,(Q) \rightarrow (\sigma x Px = \sigma x Qx \leftrightarrow P \equiv Q)$

(T.21) $P \equiv Q \rightarrow \mu x Px = \mu x Qx$

(T.22) $\mathbf{At}\,a \rightarrow a = \sigma x(x = a)$

(T.23) $\mathbf{Mp}\,a \rightarrow a = \mu x(x = a)$

(T.24) $\mu x Px = \iota x\,(Px \wedge \forall y\,(Py \rightarrow y\,\mathsf{T}\,x))$ for $P \in$ MT

(T.25) $a \sim b \rightarrow \mu x(x = a) = \mu x(x = b)$

(T.26) $\mu x Px = \iota x(x \rhd \mu x Px)$

(T.27) $\mathbf{Mp}\,(\mu x Px)$

(T.28) $a \rhd b \leftrightarrow \mathbf{Mp}\,a \wedge a \sim b$

(T.29) $\mu x Px \sim \sigma x Px$

(T.30) $\sigma x Px = \sigma x Qx \rightarrow \mu x Px = \mu x Qx$

(T.31) $\forall x \forall y\,(x\,\Pi\,y \leftrightarrow x \oplus y = y)$

(T.32) $\forall x \forall y\,(x\,\mathsf{T}\,y \leftrightarrow x + y = y)$

1.3 Applications to Montague Grammar

Let the basic *syntax* be as in PTQ.[16] The category CN of common noun phrases has to be subcategorized into MCN (mass noun phrases), SCN (singular count noun phrases), and PCN (plural count noun phrases). The quantifiers discussed are *a*, $\emptyset_{pl}$, *some, the, all the, every, all*; they are of category T/CN with suitable subcategorizations (term formation is done compositionally). We have the obvious plural rule

(65) $\quad \zeta \in P_{\text{SCN}} \;\Rightarrow\; \zeta_{\text{pl}} \in P_{\text{PCN}}$

where $\zeta \doteq$ *hand* becomes $\zeta_{\text{pl}} \doteq$ *hands*, $\zeta \doteq$ *child*, $\zeta_{\text{pl}} \doteq$ *children*, etc.

The logic of PTQ, TITL, is extended as follows. TITL′, or *extended TITL*, contains as new symbols all the special symbols of LPM. Additional *meaningful expressions* are: (i) $^*\zeta, ^\star\zeta, ^\mathsf{T}\zeta \in ME_\rho$ for $\tau = e$ ($\tau \in Type$, $\rho \doteq \tau t$ or $s\tau t$, $\zeta \in ME_\rho$); (ii) $\sigma u\phi, \sigma^* u\phi, \iota u\phi \in ME_\tau$, and $\mu u\phi \in ME_e$ ($\tau \in Type$, $u \in Var_\tau$, $\phi \in ME_t$); (iii) $\alpha \,\Pi\, \beta \in ME_t$ and $\alpha \oplus \beta \in ME_\tau$ ($\tau \in Type$, $\alpha, \beta \in ME_\tau$); (iv) $\alpha \,\mathsf{T}\, \beta, \alpha \sim \beta, \alpha \rhd \beta \in ME_t$, and $\alpha + \beta \in ME_e$ ($\alpha, \beta \in ME_e$).

The *semantics* for TITL′ has to specify the "boolification" of the sets of possible denotations since the i-sum operation applies to expressions of any type (m-sums are formed on level e only). Let E be a c.a. Boolean algebra, I, J sets, and $K = I \times J$. A *Boolean hierarchy of possible denotations over E with respect to K* is a family of sets $(D_{\tau,E,K})_{\tau \in Type}$ defined as in PTQ, but endowed with a c.a. Boolean structure for every D_τ as follows: (i) The sets $D_e = E$ and $D_t = 2$ are already c.a. Boolean algebras; (ii) $f, g \in D_{s\tau}$; then: $(f \sqcup_{s\tau} g)(k) := f(k) \sqcup_\tau g(k)$ ($k \in K$), similarly for the other operations; (iii) $f, g \in D_{\sigma\tau}$; then $(f \sqcup_{\sigma\tau} g)(x) := f(x) \sqcup_\tau g(x)$ ($k \in D_\sigma$), similarly for the other operations. The resulting structures are c.a. Boolean.

An interpretation for TITL′ is a 7-tuple $\langle E, A, D, h, I, J, \|\cdot\| \rangle$ such that (i) $\langle E, A, D, h \rangle$ is a boosk, (ii) $K = I \times J$ is as in PTQ, and (iii) $\|\cdot\|$ is the interpretation function that naturally emerges from a combination of TITL and LPM. As *rules of truth and denotation* provided by an appropriate recursive definition let me mention the following: (i) $\|^*\zeta\| \doteq$ the characteristic function of $[\{ x \in D_\tau \mid \|\zeta\|(x) \doteq 1 \}]_{\sqcup_\tau}$ ($\zeta \in ME_{\tau t}$); (ii) $\|^*\zeta\|(k) \doteq$ the characteristic function of $[\{ x \in D_\tau \mid \|\zeta\|(k)(x) \doteq 1 \}]_{\sqcup_\tau}$ ($\zeta \in ME_{s\tau t}$, $k \in K$); $\|^\star\zeta\| \doteq \|^*\zeta\| \sqcap_{\tau t} \chi'_{A_\tau}$, where $\zeta \in ME_{\tau t}$, A_τ the set of atoms in D_τ, χ_{A_τ} the characteristic function of A_τ, and "′" the Boolean complement ;

[16] Montague 1973a; the following notation is that of Link 1979 (which differs only slightly from the one used in PTQ) except that the interpretation function F is denoted by the "norm" symbol $\|\cdot\|$ here. I also ignore, for the present purpose, the individual concept language of PTQ.

(iv) $\|{}^{\star}\zeta\|(k) \doteq \|{}^{*}\zeta\|(k) \sqcap_{\tau t} \chi'_{A_\tau}$, ($\zeta \in ME_{s\tau t}$); (v) $\|\sigma u\phi\| \doteq \sup_i \|\lambda u\phi\|$, with $\sup \emptyset \doteq \mathbf{0}$, and $\|\sigma^* u\phi\| \doteq \|\sigma u\phi\|$ if $\lambda u\phi$ has more than one element, and $\mathbf{0}$ otherwise; (vi) $\|\alpha \,\Pi\, \beta\| \doteq 1$ iff $\|\alpha\| \leq_\tau \|\beta\|$ and $\|\alpha\| \neq \mathbf{0}$ ($\alpha, \beta \in ME_\tau$, and $\leq_\tau$ the intrinsic ordering relation of D_τ); (vii) $\|\alpha \oplus \beta\| \doteq \|\alpha\| \sqcup_\tau \|\beta\|$ ($\alpha, \beta \in ME_\tau$). The remaining basic symbols are characterized as in LPM.

The notion of distributivity introduced above can be extended to class-denoting expressions of any type. Call a $\zeta \in ME_{\tau t}$ *distributive* if it can be true of atoms of type τ only:

(66) $\quad \mathbf{Distr}\,(\zeta) \;\leftrightarrow\; \forall x\,[\zeta(x) \rightarrow \mathbf{At}\,(x)] \qquad (\zeta \in ME_{\tau t})$

Most of the basic count nouns like *child* are taken as distributive, similarly IV phrases like *die* or *see*. In accordance with what I did above the translation rule T4 involving distributive expressions has to be such that their translations always enter this rule under the star operator. While this seems to me the correct move from a technical point of view I am aware that it carries the empirical prediction that distributivity is a lexical property.

I now give *translations* for the quantifiers. Let $\mathbf{U}$ be the translation relation, and $\tilde{P} := \lambda y\,[P(y) \wedge \forall x[x \,\Pi\, y \rightarrow {}^{\mathsf{T}}P(x)]]$

1.	*a*, $\emptyset_{pl}$, *some*	$\mathbf{U}$	$\lambda Q \lambda P\, \exists x\,[Q(x) \wedge P(x)]$
2.	*the*	$\mathbf{U}$	$\lambda Q \lambda P\, \exists y\,[Q(y) \wedge \forall x[Q(x) \rightarrow x \,\Pi\, y] \wedge P(y)]$
3.	*all the*	$\mathbf{U}$	$\lambda Q \lambda P\, \forall y\,[y = \sigma^* x Q(x) \rightarrow \tilde{P}(y)]$
4.	*every*, *all*	$\mathbf{U}$	$\lambda Q \lambda P\, \forall x\,[Q(x) \rightarrow P(x)]$

a applies only to singular count noun phrases and $\emptyset_{pl}$ to PCN phrases. The quantifiers *some* and *the*, with one and the same translation, apply to both singular and plural phrases. As can be seen from 5. through 8. below it is only the incoming CN phrase which differentiates between the appropriate singular and plural readings. While this is obvious with *some*, a comment on the transformations under 7. is in order. The conjunct $\forall x\,[Q(x) \rightarrow x \,\Pi\, y]$ in the translation of *the* asserts maximality which is needed for the intended sum formation in the extension of $^*Q$. At the same time, however, it generalizes the usual uniqueness condition for definite descriptions. This comes out when *the* is applied to a singular count noun like *child*: The i-part x of y in 7. cannot be $\mathbf{0}$, and y is an atom because of $\mathbf{Distr}\,(child')$; so x equals y. I think this is a nice instance of strict compositionality in times in which this principle has come under heavy fire in view of all sorts of recalcitrant data (see below). Finally, 9. – 11. show that universal quantification is as in LPM where, again, the difference in meaning between *the children* and *all the children* is only partly characterized ($\tilde{P}$ being defined in terms of '${}^{\mathsf{T}}$').

5.	*a, some child*	**U**	$\lambda P\, \exists x\, [\text{child}'(x) \wedge P(x)]$
	some water	**U**	$\lambda P\, \exists x\, [\text{water}'(x) \wedge P(x)]$
6.	*(some) children*	**U**	$\lambda P\, \exists x\, [{}^\star\text{child}'(x) \wedge P(x)]$
7.	*the child*	**U**	$\lambda P\, \exists y\, [\text{ch}'(y) \wedge \forall x[\text{ch}'(x) \rightarrow x \,\Pi\, y] \wedge P(y)]$
			$= \lambda P\, \exists y\, [\text{ch}'(y) \wedge \forall x[\text{ch}'(x) \rightarrow x = y] \wedge P(y)]$
			$= \lambda P\, \exists\, [y = \iota x\, \text{child}'(x) \wedge P(y)]$
8.	*the children*	**U**	$\lambda P\, \exists y\, [{}^\star\text{ch}'(y) \wedge \forall x[{}^\star\text{ch}'(x) \rightarrow x \,\Pi\, y] \wedge P(y)]$
			$= \lambda P\, \exists\, [y = \sigma^* x\, \text{child}'(x) \wedge P(y)]$
9.	*all the children*	**U**	$\lambda P\, ally\, [y = \sigma^* x\, \text{child}'(x) \rightarrow \tilde{P}(y)]$
10.	*every child*	**U**	$\lambda P\, \forall x\, [\text{child}'(x) \rightarrow P(x)]$
11.	*all water*	**U**	$\lambda P\, \forall x\, [\text{water}'(x) \rightarrow P(x)]$

I may mention here that the incorporation of numerals like *one*, *two*, *three* into the fragment is straightforward. The extension of *three men*, for instance, is the set of all i-sums in $\|\text{men}'\| \doteq \|{}^\star\text{man}'\|$ which contain exactly three atoms. Sentences like *three men went to mow a meadow* receive a translation of the form $\exists z\, [(\text{three man})'(z) \wedge Q(z)]$.

The final topic I want to say something about is the vexing problem of relative clauses with more than one head noun.[17] Call those structures *hydrae*. Let me consider restrictive relative clauses only. First of all, there is a rather friendly type of hydra, like the following.

(67) the German or Austrian John met yesterday

(68) the cabinet member and mafioso who was deeply involved in the scandal

PTQ style relativization, i.e., CN modification, takes care of these pets yielding the obvious representations for (67) and (68), respectively:

(67′) the ((German or Austrian) such that John met him yesterday)
$\lambda P\, \exists y\, [\forall x[\text{German}' \oplus \text{Austrian}'(x) \wedge \phi[x] \leftrightarrow x = y] \wedge P(y)]$

(68′) the ((Cabinet member and mafioso) such that he was deeply involved in the scandal)
$\lambda P\, \exists y\, [\forall x[\text{Cabinet-member}' \odot \text{mafioso}'(x) \wedge \psi[x] \leftrightarrow x = y] \wedge P(y)]$

Here, ϕ and ψ are the translations of the relative clauses of (67) and (68), respectively; furthermore, $\zeta \oplus \eta \doteq \lambda x\, [\zeta(x) \vee \eta(x)]$ is the Boolean join and $\zeta \odot \eta \doteq \lambda x\, [\zeta(x) \wedge \eta(x)]$ the Boolean meet of ζ and η. This case

[17] Thanks to P. Staudacher and A. von Stechow for bringing the severeness of the problem to my attention. See v. Stechow 1980 and Janssen 1981 for thorough and critical discussions of the issues involved.

is simple because CN conjunction does not lead to plural structures, and there is only one determiner present.[18]

Next, let us conjoin two singular count nouns to form a plural phrase. Examples are *man and woman*, *boy and girl*, *husband and wife*, *landlord and tenant*. Such phrases can be true of a sum of two individuals one being of the first and the other being of the second kind. An appropriate translation, therefore, is of the form

(69) $(\zeta \, and \, \eta) \quad \mathbf{U} \quad \lambda z \, \exists x \exists y \, [\, \zeta'(x) \wedge \eta'(y) \wedge z = x \oplus y \,]$

With this sentences like *John and Mary are husband and wife* can be handled in a first approximation while the non-symmetric features of this sentence have to be taken care of by other clues. This is because the predicate (69), by itself, is necessarily symmetric reflecting the fact that pluralization has the force of group formation which typically gives rise to symmetric constructions.[19] Such is the following CN phrase:

(70) boy and girl who dated each other

With (69), the standard PTQ rule yields for this:

(71) $\lambda z \, \exists x \exists y \, [\, \text{boy}'(x) \wedge \text{girl}'(y) \wedge z = x \oplus y \wedge \text{dated.each.other}'(z) \,]$

Now term formation with one and the same determiner for both head nouns is again standard if we decide to have it distributed over the two nouns in a kind of copying process. So, for instance, *a boy and a girl who meet* becomes

(72) a ((boy and girl) such that they meet) $\mathbf{U}$
$\lambda P \, \exists z \, [\, \exists x \exists y \, [\, \text{boy}'(x) \wedge \text{girl}'(y) \wedge z = x \oplus y \wedge \text{meet}'(z) \,] \wedge P(z) \,]$

In a further step we admit conjunction of plural nouns. This move doesn't really add any new difficulty so I can write down an example immediately. Let θ be the translation of the phrase *students and professors who had met in secret*:

(73) $\theta \doteq \lambda z \, \exists x \exists y \, [\, {}^{*}\text{stud}'(x) \wedge {}^{*}\text{prof}'(y) \wedge z = x \oplus y \wedge \text{had.met.in.secret}'(z) \,]$

Adding a determiner as before, say the definite article, we get the term

[18] In German there seems to be a restriction to the effect that the two nouns have to agree in gender for this construction to be possible. The same article can then be related to both nouns.

[19] For the pair reading see the remarks below.

(74) the students and the professors who had met in secret
U $\lambda P \exists z [z = \sigma^* x \theta(x) \wedge P(z)]$

The present conception of conjunction of two CN phrases ζ and η is such that the extension of (ζ *and* η) contains i-sums of objects which are of kind ζ and η, respectively. Unlike the Boolean join this carries the presupposition that both sets of objects be non-empty. I think this is as it should be. Notice that in this case the overall sum is the same as the one which is formed in terms of the Boolean join, i.e. we have $\sigma x (\zeta' \oplus \eta')(x) = \sigma x (\zeta \; and \; \eta)'(x)$. What is not captured in either approach, however, is the proper pair reading which is dominant in sentences like *landlords and tenants who hate each other will always find something to argue about.* I find these constructions hard to treat at the moment though I am confident that they will finally lend themselves to an analysis which is compatible with the present framework. What is needed here is some notion or ordered pair in the object language, apart from the symmetrical sum operator.[20] It is only with such an additional instrument, it seems to me, that we can, in a close-to-language treatment, attack the most dangerous hydra lurking in the realm of pair reading and branching quantification.[21]

The last type of hydra I want to address myself to shows up in a sentence like the following

(75) All of the students and some of the professors who had met in secret were arrested after the coup d'état.

From a purely syntactic point of view it might look natural to try a T-S-analysis here: first conjoin the two unrelativized terms *all the students* and *some of the professors* and then modify the conjunct by the relative clause. However, I agree with Janssen 1981 in his sceptical attitude towards the semantical soundness of this approach in general, and the present case, I think, is apt to confirm these doubts. Even a generalized quantifier account along the lines of Barwise/Cooper 1981 is not easy to give because the set of properties denoted by the subject term of (75) is not the obvious intersection of two property sets. The point is that the collective predicate *meet.in.secret* does not distribute over the conjuncts.

Now there is no reason to leave the approach taken so far, it seems to me. We already have a clear understanding of the meaning of the plural CN phrase *students and professors who had met in secret*, which is expressed by the λ-term of (73); so in terms of this, what is the meaning of the above NP? Well, the set of properties that all the students of the group of students and

[20]The device of "multiple quantification" introduced in Thomason 1977 might be relevant here. Thanks to N. Belnap for bringing these unpublished notes to my attention.

[21]See Barwise 1979 for a lucid account of the latter phenomenon.

professors who had met in secret and some of its professors share. This set can be formally represented in TITL′. From the above group of individuals we have to pick out again the students and the professors, respectively, which is done by means of the Boolean meet $\odot$; we can then write (with $\alpha \doteq$ *student′ and $\beta \doteq$ *professor′):

(76) all the students and some of the professors who had met in secret
$$\mathbf{U} \quad \lambda P\,[\forall y\,[y = \sigma^* x(\alpha \odot \theta)(x) \to \tilde{P}(y)] \wedge \exists y\,[(\beta \odot \theta)(y) \wedge P(y)]\,]$$

In the syntax, then, we have to introduce the determiners *all* and *some* simultaneously. I hope that a rule to this effect does not look too outlandish to the eye of the syntactician.

Let me summarize what I consider to be the virtues of the present approach to plurals and mass terms. (1) The logic of plurals and the logic of mass terms share a common lattice structure the only difference being that the former leads to an atomic structure while the latter does not. (2) By means of the star operator pluralization and term formation involving plural constructions can be treated compositionally. (3) Plural terms (*the cards*) and collective terms (*the deck of cards*) are equivalent in that they are interchangeable in invariant contexts; this does not make them coreferential, however, in contrast to systems of the reductionist lot. (4) collective predication becomes possible in a unified way accounting for the fact that many predicates (e.g., *carry*) are not marked with respect to distributivity and can, therefore, have mixed extensions.

Acknowledgment. The basic idea leading to the present approach grew out of a seminar on the semantics of mass terms which I held in the summer of 1980. After that I had the opportunity to discuss various stages of the paper underlying this chapter with a number of people whom I wish to thank here for helpful comments, hints, and criticisms as well as for general support. Among these I should like to mention in particular Ulrich Blau, Paul Gochet, Barbara Partee, Christian Rohrer, Peter Staudacher, Arnim von Stechow, Alice ter Meulen, Matthias Varga von Kibéd, and Dietmar Zaefferer.

Chapter 2

Plural

2.1 Introduction

In the familiar Indo-European Languages like English, French, German, there is a predominant number of nouns that classify according to the distinction ±count. Examples of +count nouns are *horse*, *apple*, *task*, *set*; call them **individuating** or **count nouns** (**IN**). They stand for discrete objects that can be counted, and hence go with numerical determiners like *one*, *four*, *many*. −count nouns or **mass nouns** (**MN**) are natural kind words like *water*, *gold*, but also, differing from language to language, collective words like *furniture*, *garbage*, and abstracta like *solidarity*. The numerus distinction singular/plural is typically applicable to the former class IN only. Affixing the plural morpheme to INs produces nominals which express a **discrete multitude** ob objects of the same kind, viz. *horses*, *apples*, *tasks*, *sets*. With MNs, denoting non-discrete entities, the number distinction is neutralized: plurality as discrete multitude does not make sense here (cf. **three garbages*, **many solidarities*). So MNs could be called **transnumerals**. In Western languages they come along as singularia-tantum (*gold*) or pluralia-tantum (German *Möbel*, English *entrails*). The latter should be referred to as **syntactic plurals** ((Dougherty, 1970, p. 853); these also go along with semantic singulars (*scissors*) and can be triggered by mere agreement (e.g., INs pluralize automatically after decimal numbers, viz. *1.0 apples*). But we also have pluralization with transnumerals; it is accompanied by a rather item-specific modification in meaning expressing some kind of **abundance** (e.g., *"the Grand Coulee Dam where the waters run down"*) or **variety** (German *die übrigen drei Weine auf der Liste*, *wertvolle exotische Hölzer*). Finally, pluralization can separate itself from the semantic notion of counting; in Hungarian we have *hajókat láttam* (*ships I saw*) vs.

öt hajót láttam (*five ships I saw*). That means that the Hungarian plural suffix *-k-*, instead of being used with numerical modifiers, serves to form numerically unspecified aggregates only. In a sense, then, pluralization can have the semantic role of obliterating the lexically given number distinction.

In many languages there is also an inverse syntactic process: Starting from a transnumeral noun expressing some indefinite, non-discrete unity,an appropriate affix leads to **singulative** forms which stand for individuated samples of that unity (cf. Arabic *dabbān*: unspecified *fly*, vs. *dabbāne*: *a fly*). The German collective affix *-ge-* produces a related effect in that it turns a −count noun into +count (cf. *Holz* vs. *Gehölz*); but the case is semantically different. A nice illustration of the distinctions involved here is given by the German noun *Polizei*;[1] this noun itself is a transnumeral with a mass term meaning (cf. *Polizei umstellte das Gebäude*). It has a "singulative" IN *Polizist*, denoting individual policemen, with a regular object level plural *Polizisten*, but there is also a sortal plural *Polizeien* (e.g., *der Länder Bayern und Baden-Württemberg*) standing for different police organizations. For more examples drawn from a great variety of languages, see Biermann (1981).

Investigations across languages show that there is a universally rather stable hierarchy of classes of nouns which undergo pluralization with decreasing readiness. The classes are characterized by features that are ordered according to some fixed primacy relation on the semantic process of individuation in language. According to Smith-Stark (1974), reported in Biermann (1981), the hierarchy is SPEAKER > ADDRESSEE > KINSHIP NAME > RATIONAL BEING > HUMAN > ANIMATE. In languages without a plural morpheme the semantic role of forming discrete unities (that can then be counted by attaching numerals) is taken over by a more or less elaborate system of **classifiers** (e.g., in Chinese). Classifiers can also be found in English and German (cf. *30 head of cattle*, *30 Stück Vieh*; note that, typically, the classifier itself does not pluralize).

In summary, I conclude that most if not all languages possess grammatical devices for forming multitude-denoting terms. Pluralization is achieved either by a special plural morpheme, or by a suitable classifier system, or a combination of the two.

2.2 Types of Plural Constructions

In order to uncover the semantic properties of pluralization in more detail I confine myself to English and closely related languages like German. I

[1] Thanks to Manfred Krifka for drawing my attention to this.

am now going to discuss the main types of plural formation. "plural noun phrase" will be abbreviated as "plural NP" or simply "**PNP**".

2.2.1 Indefinite Plural NPs; Bare Plurals (BP)

Consider the following examples.

(1) a. Some friends dropped by.
b. A couple of days later John came back.
c. Children built the raft.
d. I met sm friends here.
e. At least some of the houses were completely destroyed.
f. There are only a few tickets left.

The indefinite PNPs under (1) intuitively refer to a numerically unspecified, yet concrete, collection of objects by way of existential quantification. The existential force is introduced by the determiners *some*, *a couple of*, *a few*, *several*, *various*. The same role is traditionally attributed to the **null determiner** $\emptyset$, "present" in (1c) (cf. Chomsky (1965)); it is considered as the plural variant of the indefinite singular article *a* (German *ein*). Stockwell *et al.* (1973) prefer the unstressed version of *some*, written *sm*, for the proper counterpart of *a*. A German analogue would be something like *mpa*(unstressed *ein paar*); see (1d). Example (1e) contains a partitive construction: There is a contextually specified collection of houses at least some of which are asserted to have been completely destroyed. For this construction, see Section 2.2.5. The existential reading admits *there*-insertion, just like the singular case; cf. (1f).

Indefinite PNPs are also used as predicate nominals; in this position they have to be BPs.

(2) The animals in this camp are horses.

(3) Some Jacobins were advocates.

(4) Kohl and Strauß have never been friends.

In (2) and (3) the plural in the predicate nominal seems to be a matter of mere agreement; the property it stands for applies to single individuals only. But (4) is different: the PNP *friends* is short for *friends of each other* and presupposes a plural subject. So this predicate nominal is seen to be irreducibly plural.

There is an important parallelism between sentences like (2) and the use of mass terms in predicative position. Compare

(5) a. If the animals in this camp are horses and the animals in that camp are horses then the animals in the two camps are horses.

b. If *a* is water and *b* is water then *a* and *b* taken together is water.

This shows that plural predicates and predicative MNs share the **cumulative reference property** (see the end of Section 2.3.2).

The rest of this section is devoted to the important topic of **Bare Plurals** (BPs). While in sentences like (1c) they are interpreted existentially BPs are most commonly used as subjects of generic sentences with quite a different meaning. Consider

(6) a. Horses are mammals.

b. Tigers are striped.

(7) a. In revolutionary France high positions were occupied by members of political clubs.

b. In revolutionary France there were high positions that were occupied by members of political clubs.

(8) a. Whales will be extinct before long.

b. Tranquilizers are widespread.

Sentence (6a) has a strict universal meaning: every horse is a mammal. (6b) can only be said to be "quasi-universal": most tigers (or all typical tigers) are striped. (7a) has again the quasi-universal reading: A high position during the French revolution was, as a rule, occupied by a member of some political club. Note that *there*-insertion, like in (1g), definitely gives the sentence an existential force, witness (7b). For a similar grammatical effect in Dutch see de Mey (1984). Now what about the sentences under (8)? Their BPs seem to have neither a (quasi-)universal nor a proper existential meaning. The sentences are collectively about the class of whales and tranquilizers, respectively. So the BPs function here as proper names of these classes, so to speak. Carlson (1977, 1978) prefers to speak of **kinds of things**, instead of their classes, to which BPs refer. In fact Carlson argues for a semantical analysis that treats all BPs as kind-denoting terms. I am not going into this general claim here; I shall rather draw on the wide range of data provided by Carlson's work to present and discuss some important aspects of BP constructions. Returning to generics, there are also **relational** BP sentences like those under (9) (see also Section 2.2.9).

(9) a. Unicorns have horns.

b. Beavers build dams.

Other generic constructions involving BPs are

(10) John repairs cars for a living.

(11) Deciduous trees become rare as you go higher up in the mountains.

Carlson (1977, 1978) contrasted the behavior of BPs with that of the singular indefinite article in combination with other grammatical phenomena like scope, opacity, and aspect. To begin with, BPs tend to take narrow scope in the presence of another scope-sensitive NP whereas a singular indefinite NP normally gives rise to a scope ambiguity. Compare the a)- and b)-sentences in (12) and (13); the possible readings are indicated in parentheses by giving the order of quantifiers, while $>$ stands for "is preferred to."

(12) a. Every girl read a book on horses. ($\forall\exists > \exists\forall$)

b. Every girl read books on horses. ($\forall\exists$ / $^{*}\exists\forall$)

(13) a. ?A dog was everywhere. ($\exists\forall > \forall\exists$)

b. Dogs were everywhere. ($\forall\exists$ / $^{*}\exists\forall$)

Carlson actually claims that the $\forall\exists$ reading is impossible for (13a) so that the pragmatically somewhat deviant $\exists\forall$ reading is forced. This would mean that (13a) and (13b) would not have a reading in common at all which is taken by Carlson as evidence that BPs cannot be counterparts of singular indefinite NPs headed by *a*. But if we vary (13a) a little bit both readings appear possible, as in (14),

(14) A strange voice was to be heard everywhere. ($\forall\exists$, $\exists\forall$)

Still the BP counterpart of (14) does not have the $\exists\forall$ reading. On closer inspection, however, even that seems to depend on the degree in which the concept referred to in the BP lends itself to individuation. Compare (15a) and (15b),

(15) a. CIA agents mined all Nicaraguan harbors. ($\forall\exists > \exists\forall$)

b. Fanatic ayatollahs gained control over all areas of public life. ($\exists\forall > \forall\exists$)

In (15a) the $\forall\exists$ reading is already possible (viz. a certain CIA mining detachment carried out the operation). In (15b) this reading even seems to be predominant. This is due to the rather contingent historical fact that the stereotype which most of us have connected with 'ayatollahs' for some time is something like a specific clique of high-ranked religious fanatics with political power that tends to act as a single group. Now replace 'ayatollahs' by 'mullahs' in (15b), and a scope switch is likely to ensue: now the sentence rather means that every single area of public life is under the surveillance of some watchful mullahs. So there is a highly pragmatic and culture-dependent "squish" here from narrow to wide scope going along with an increase in specificity of the subject NP. The almost continuous transitions are illustrated by the following list of sentences.

(16)	a.	Blood was spilt all over.	($\forall\exists$)
	b.	Sand was found in every part of the camera.	($\forall\exists$)
	c.	Ants were crawling around in every corner.	($\forall\exists$)
	d.	Migrant workers slept under every bridge.	($\forall\exists >> \exists\forall$)
	e.	Marines cut off all main roads.	($\forall\exists > \exists\forall$)
	f.	CIA agents operated efficiently everywhere.	($\forall\exists/\exists\forall$)
	g.	Revolting generals occupied all centers of power.	($\exists\forall$)
	h.	A military junta controlled all government positions.	($\exists\forall$)

I now turn to **opacity**. Consider (17).

(17) a. The reporter wants to talk to an eye-witness.
b. The reporter wants to talk to eye-witnesses.

Carlson claims, with respect to similar examples, that the a) sentence has both the opaque and the transparent reading whereas b) is only opaque. What he reports here is obviously the intuition about the dominant interpretation. It was noted by Kratzer (1980), however, that the context can also bring the transparent reading into focus; her example is

(18) a. Hans wollte Tollkirschen in den Obstsalat tun, da er sie mit richtigen Kirschen verwechselte.
b. [John was going to add belladonna berries to the fruit salad, mistaking them for regular cherries.]

Finally, I just mention two examples concerning **aspect**. Achievement verbs like *discover*, in connection with adverbial phrases like *for two hours*, admit indefinite NPs only as BPs in object position (since these refer cumulatively):

(19) John discovered {*a nice example, nice examples} for two hours.

When the adverbial changes, a specific (existential) reading becomes possible, viz.

(20) John discovered some especially interesting examples within two hours.

In summary, then, PNPs headed by an indefinite determiner have existential force. They stand for specific collections of objects and carry a clear numerical non-singularity presupposition. BPs are ambiguous: they may have existential force and even take wide scope over another operator, but this may depend on the contextually given degree of individuation of the concept they stand for, as we saw. Otherwise BPs denote a more or less homogeneous multitude which is not specified with respect to any "mensurative" dimension (numerical, spatiotemporal, classificational). In nominal position and in a generic context BPs behave like proper names (Carlson's *kind denoting terms*) or display a (quasi-)universal quantificational force, similar to the so-called *arbitrary objects* of Fine (1985). Thus, indefinite PNPs share a number of important features with indefinite mass nouns.

2.2.2 Definite Plural NPs

In first approximation definite PNPs, of the form *the PNP*, induce a universal force in sentences containing them, witness

(21) a. The soldiers died.

b. The real numbers can be expressed by non-terminating decimals.

This simply means that every soldier died or that every real number has a non-terminating decimal representation. But as in the indefinite case, exceptions to universality are possible without rendering the definite PNP sentence false.

(22) a. Jenny hates the WASPs (and yet she can't help being in love with Oliver Barrett IV).

b. The Germans suffer from a cleanliness fixation (and yet some pretty messy places can be seen in this country).

c. The tigers are striped (except for albinos).

So there is a vagueness involved here but the sentences can, in principle, be given a universal interpretation. Consider now

(23) The Romans built the bridge.

A universal paraphrase would sound bizarre in this case. The definite PNP rather stands for a group, in fact a whole people here, which is somehow taken to be collectively responsible for building the bridge. Scope considerations show even more clearly that *the PNP* should really be treated as a denoting phrase, just like its singular counterpart (see Scha (1981)). The following examples contrast definite PNPs with their *all* paraphrases.

(24) a. All Sansculottes hailed a Cordelier. $(\forall\exists > \exists\forall)$

b. The Sansculottes hailed a Cordelier. $(\exists)$

(25) a. Lucile doesn't believe that all Cordeliers are corrupt.

b. Lucile doesn't believe that the Cordeliers are corrupt.

(26) a. All Indulgents were banished or executed.

b. The Indulgents were banished or executed (I've forgotten which).

(27) a. Because they tried to stop the terror all Cordeliers were executed.

b. Because they tried to stop the terror the Cordeliers were executed.

The sentences in (24), describing Robespierre's nightmares, show the interaction with an indefinite NP. (24a) has the two familiar readings: some Sansculottes might have hailed Danton, others Desmoulins, etc.; or else there is the "Danton reading": One single member of the Club of Cordeliers, Danton, was hailed by everybody. (24b) has only the Danton reading; the definite PNP is insensitive to scope. (25) shows the same effect in connection with negation. The definite PNP version naturally means that the Cordeliers, according to Lucile's beliefs, are not corrupt. Negation is unable here to "split up" the group-denoting definite PNP. (26) concerns disjunction. Although intuitions are less stable here, the intended reading for (26b) seems quite natural (either the Indulgents were banned or they were executed). (26a) is different: the universal quantifier takes wide scope so that the Indulgents are asserted to have been submitted to a non-uniform treatment. Finally, from (27) it can be seen that definite PNPs lend themselves much more easily to backward pronominalization than universally quantified NPs do (if the latter is possible at all).

When definite PNPs are combined with collective predicates the universal paraphrase is no longer possible as (28) shows.

(28) The children gathered around their teacher.

Here the predicate *gather* is true of a certain contextually specified group of children. Now all this suggests that definite PNPs should formally be treated as definite descriptions, which would account for the linguistic fact that the definite article operates on singular and plural NPs alike. For this to work we have to introduce some kind of group object[2] into the semantics such that the usual existence and uniqueness conditions for definite descriptions can be met. Consider the PNP

(29) *the Ns*

where '*Ns*' stands for some pluralized noun N. Then the extension of N should contain at least two elements for (29) to denote. The identity conditions for (29) are more difficult to settle; this issue will be taken up in Section 2.3.

In its usual meaning, then, we take (29) as denoting some concrete collection of objects, however specified formally. But there is still the somewhat dubious status of the generic use of (29) that lurks in the background of most of the examples. The rest of this section is devoted to this use, by way of comparison with the indefinite case. Consider the sentence (30).

(30) a. A member of this club doesn't drink alcohol.
b. Members of this club don't drink alcohol.
c. The members of this club don't drink alcohol.

As Dahl (1975) observed, (30a) asserts a necessary connection with respect to some modal standard (here: club rules), between club membership and teatotallership. (30b) can be regarded as synonymous with (30a). (30c) makes the same universal statement, but it might just be a matter of accident that no club member drinks alcohol. This sentence is clearly about the actual members while (30a) and (30b) may also be about potential members of the club. While sentence (30c) doesn't make a statement about club membership under this reading, there is still something generic about it, but along a different, temporal dimension: It may express a habit which, contingently, all the club member have. This effect, however, is independent of the presence of the definite PNP.

The difference between the definite and the indefinite case is most clearly seen by considering so-called kind predicates like *be extinct*. Definite NPs,

[2]The word 'group' is used here and in the following in a non-technical sense to stand for whatever pluralities turn out to be. In particular, it is not intended to be a set, class, or "Landman group" [see Landman (1989)], and has nothing to do with the mathematical concept of group.

both singular and plural, admit **proper kind predication**, for which indefinite NPs cannot be used, at least not in the singular:

(31) a. The dinosaurs are extinct.
b. ?Dinosaurs are extinct.
c. *A dinosaur is extinct.

In (31a) the definite PNP stands for a certain kind (the species Dinosaur). the sentence makes a "historical" and hence non-generic statement about that species. It is, therefore, of a different quality as sentence (32a) below, where the predicate *is a ruminant* applies not to the species Cow itself but rather to its particular instances. Here we have a case of what can be called **derived object predication**. It is more natural with a BP or a singular indefinite NP (cf. (32b,c)). The definite article in (32a), in fact, is marginal and might be acceptable only due to a certain familiarity effect (this is true at least of its German counterpart *die Kühe sind Wiederkäuer*); in French, on the other hand, it has to be there since BPs are not admitted).

(32) a. The cows are ruminants.
b. Cows are ruminants.
c. A cow is a ruminant.

For more on the topic of generics, see Carlson (1982a), Pelletier and Schubert (1987a), Heyer (1987), Gerstner-Link and Krifka (1993), Gerstner-Link (1995), and Carlson and Pelletier (1995). The last reference contains a detailed research survey on generics.

2.2.3 Universally Quantified PNPs; Quantifier Floating

As we saw above definite PNPs may already carry universal quantificational force. What, then, is the difference between sentences like (21a) and (33)?

(33) a. All soldiers died.
b. All the soldiers died.

With distributive predicates like *die* sentences (21a) and (33a) seem to mean the same. As was observed above, however, the universal sentence and its definite PNP counterpart are not synonymous anymore when they involve a collective predicate like in (34).

(34) The soldiers withstood the enemy's attack.

This sentence does not imply that every soldier withstood the attack (there might have been casualties). Thus one difference is the degree to which the sentence admits a collective interpretation. In (33b) the presence of *all* stresses **totality**: it means that the members of some already familiar group of soldiers, which is provided by the context, died without exception. Some sort of scalar effect can also be involved, as illustrated by the following literary example:

(35) All the king's horses, and all the king's men couldn't put Humpty Dumpty together again.

This sentence can be interpreted both distributively and collectively since the verb *put together* is not marked for the collective/distributive distinction. So this is a good case for sorting out the meaning nuances involved. Consider (36), with the dominant reading judgments given in parentheses.

(36) a. All men lifted the rock. (distr > coll)

b. The men lifted the rock. (coll > distr)

c. The men all lifted the rock. (distr > coll)

d. The men each lifted the rock. (distr)

In (36c) and (36d) the quantifier is said to be *floated*. **Quantifier Floating** is a pervasive phenomenon in languages like English (cf. Postal (1974); Dowty and Brodie (1984)) and German(cf. Link (1974)) and serves mainly for accentuating scope relations in a sentence. Here it is seen to be also sensitive to the ±distributive distinction. I give two more examples. (37) shows that the floated quantifier *all*, unlike *each*, is still compatible with the collective reading when it is forced by a modifier like *common*. In such a case, emphasis is on totality. In (38) the quantifier is floated from a non-subject NP, with the same effect.

(37) a. The tenants all filed a suit against their landlord. (distr > coll)

b. The tenants {all/*each} filed a common suit against their landlord. (coll)

(38) a. Daddy gave his obnoxious kids a lecture. (coll)

b. Daddy gave his obnoxious kids all a lecture. (distr > coll)

2.2.4 Numerals and Other Plural Quantifiers

There is a great variety of further plural quantifiers which all serve as some kind of measuring device for PNPs. We have **exact quantifiers** like the numerals *(exactly) two, three, four, ..., at least n, at most n*; **mathematical quantifiers** like *finitely many, infinitely many*; and **vague quantifiers** like *about n, few, many, most, a lot of*.

The exact quantifiers have a precise meaning that can be expressed in a routine way (see, e.g., Blau (1983), Barwise and Cooper (1981)). While their semantics is clear, quantified PNPs do not seem to be completely determined in their syntactic structure. On the one hand, for instance, numerals are quantifying (cf. *three men met*); on the other hand they seem to behave like NP modifiers, which still admit a determiner in front of them (compare *all three apples* and *all red apples*). Now genuine quantifiers never pile up (viz. **all some N*); but usual adjective modifiers always go between the numeral and the noun (cf. *three nice apples* vs. **nice three apples*), which might suggest that the numerals are like quantifiers after all.

I propose the following approach here (see Chapter 4). Numerals play a double role within the prenominal domain. First there is a clear indication that they occur as adjectival modifiers with an intersective meaning. Consider the pair *three men lifted the piano* vs. *(any) three man can lift the piano*; if the numeral *three* is treated like a quantifier with an in-built existential force appropriate for the first sentence, then the universal (or generic) force of the second sentence cannot be accounted for. There is, however, a rather natural adjectival analysis for this sentence pair: the first says that there is a group of men with three individuals in it that lifted the piano; whereas the second means that any group of men with three individuals in it will be able to lift the piano. The syntactic problem that numerals and regular adjectives do not mix can be accounted for by introducing an extra node NUM for numerals. This leaves us with the problematic cases under (39).

(39) a. All three men left.

b. The three men left.

If *three* operated on men to create a set of groups each consisting of three men, then (39a) should be a universal quantification over all these groups of three. But the sentence says, of course, that there are exactly three men who all left. The same applies to (39b): there is a unique group of men that left, and the determiner appears numerically specified in a non-restrictive way (cf. also *the three of us*). This suggests that the numeral has two slots in the prenominal domain to go in: the adjectival NUM slot and one for a numerical specification of the quantifier. This is shown by sentences (40a,b), which are fine except for redundancy. Notice, however, that the specification of the quantifier is presuppositional in nature: sentences (40c,d) are out since the specifier three is at odds with that of the noun.

(40) a. all seven of the seven astronauts
b. all five of the quintuplets
c. *all three of the seven astronauts
d. *all three of the quintuplets

The semantical analysis of **vague quantifiers** is more involved. Consider the sentence

(41) Many students in this department study Hegel's *Logik*.

The vague determiner *many* can only be interpreted with respect to some relevant **standard** which can be (i) the size of the group (here: the students in this department; if the size is, say, a hundred, 80 students would be many); (ii) the character of the predicate (for the somewhat exotic subject in (41), 10 out of 100 students would be many); external information like the number of Hegel students last year, at other places, etc.

While the meaning of these quantifiers is vague and highly context-dependent there are some fixed meaning relations between them. Altham (1971) has the following equivalences:

(42) a. *many* = *not few* = *not nearly all not*
b. *not many* = *few* = *nearly all not*
c. *not many not* = *few not* = *nearly all*

But even these relations need qualifications. The first equality in each line presupposes a classical ("weak") concept of negation; and the second equality has to be interpreted with respect to some fixed standard of the kind referred to above.

The most important formal consequence of an adequate semantical treatment of vague quantifiers is the recognition that the quantifier and the noun it modifies have to be taken as a unit. This is why Barwise and Cooper (1981) call quantifier the whole quantified NP. Their seminal paper is a major source for the semantics of vague quantifiers; another is Blau (1983). See also Verkuyl (1981) and Hörmann (1983). Regarding the whole of Section 2.2.4, reference should also be made to the current work on **Generalized Quantifiers** which in linguistic semantics started with Barwise and Cooper (1981). See van Benthem (1983a), Gärdenfors (1987) with a bibliography therein, and Westerståhl (1988).

2.2.5 Partitive Constructions

In English, partitive constructions are NPs of the form determiner + *of* + NP, e.g., *some of the men*; in German we have either determiner + *von* + NP or determiner + $\text{NP}_{\text{genetive}}$ (*einer von den Anwesenden, einige der Anwesenden*). The semantic difference to purely quantifying NPs like *some men* can be described as follows: (i) a specific collection of objects, thought of as contextually given, is explicitly referred to by the NP following *of* (call it the **core NP** and the collection itself the **core**); (ii) the determiner then serves to pick out some appropriate part of the core. It follows that the core NP has to be definite (cf. **some of all men, *all of few books*); this is the **Partitive Constraint** (Ladusaw (1982)). It further follows that even if the core NP is plural (i.e., +count) particular non-discrete partitive determiners are admitted (*half of, a great majority of*; German *das meiste von diesen Büchern ist Ausschuß*; etc.). Thus a semantic representation of the partitive construction should provide a suitable core entity as well as a part-whole relation operating on the core. Notice that exactly these two pieces are also needed in floated quantifier constructions (QF constructions, for short). The similarity between the following two sentences is striking.

(43) a. All/many of the students joined the peace movement. (*Partitive*)

b. The students all/in great number joined the peace movement. (*QF*)

In German the partitive is still visible in some QF constructions. Compare

(44) a. *Die Mieter strengten alle/jeder einen Prozeß an.*

b. *Von den Mietern {strengten alle, strengte jeder} einen Prozeß an.*

c. [The tenants all/each filed a suit.]

Finally it was observed by Ladusaw (1982) that the present conception of the partitive construction accounts for a subtle difference, otherwise unexplainable, between the quantifiers *the two* and *both* in English. Note that we have *one of the two* but **one of both*. The reason is that the strictly distributive quantifier *both* cannot possibly stand for the core entity from which appropriate parts are taken. In short, the definite numerical quantifier *the two* individuates while *both* does not.

2.2.6 Coordinate Conjoined NP Structures

Plural structures also arise from NP conjunction. Consider the following sentences.

(45)
a. George and Martha are asleep.
b. George and Martha meet.
c. George and Martha got married.
d. George and Martha are husband and wife.
e. A boy and a girl played in the backyard.
f. The Leitches and the Latches come together.
g. Three men and four women joined.
h. Several men and a few women gathered.

(45a) is synonymous to *George is asleep and Martha is asleep*. The collective verb *meet* in b) does not admit such a paraphrase; nor does *got married* in c) when it is interpreted as *got married to each other*. (45d) is even more complex. Whereas in (45a – c) we can replace *George and Martha* by *Martha and George* without affecting the truth value it is also the order of the NPs conjoined that matters in (45d). Furthermore, the rest of the sentences shows that phrasal conjunction also applies to NPs other than proper names. But some restrictions have to be observed here. We do not have **John and no woman*, **few women and a few men (could lift this piano)*, **two violas and few violins (are playing in tune)*. (Barwise and Cooper, 1981, p. 194ff) give an explanation for this behavior. When NPs are interpreted as generalized quantifiers it can be seen that only those NPs can be conjoined that are "monotone" in the same direction, either both increasing or both decreasing.

2.2.7 Collective Nouns, Predicates, and Adverbs

There is a number of singular count nouns with plural meaning, called **collective nouns**. Like PNPs they stand foı recognizable group objects. Examples are *pair*, *duo*, *trio*, *couple*, *family*, *group*, *team*, *crew*, *audience*, *crowd*, *people*, *committee*, *assembly*, *faculty*, *congregation*, *cattle*. Here is also a list of **collective predicates** which only go with NPs that either express plurality (PNPs, collective nouns, coordinate conjoined NPs) or a homogeneous plenitude (mass nouns): *be numerous*, *plentiful*, *few*, *similar*, *alike*, *interspersed*, *perpendicular*, *parallel*, *uncountable*, *well shuffled*; *be friends*, *be classmates*; *meet*, *gather*, *come together*, *unite*, *separate*, *split* *up*, *spread out*, *disperse*, *swarm*, *scatter*, *surround*, *divide up*, *share*, *collide*, *outnumber*, *double in number*, *mix*, *combine*, and many more. But notice that the different categories of semantic plural are no perfect substitution classes for all the predicates listed here. Consider (46), for instance; here, the collective predicate can take definite PNPs, but no coordinate conjoined NP like *John, Bill, and Tom*, whereas *collide* is fine with such an NP but is hard to get with collective nouns.

(46) a. The men were numerous.
b. *John, Bill, and Tom were numerous.
c. John, Bill, and Tom collided.
d. ?The trio collided.

A verb or predicate is called **distributive** if it can only apply to regular individuals and never to groups. A distributive verbs can be made collective, however, by attaching to it **collective adverbs** like *together*, *simultaneously*, *unanimously*, *en masse*, *in chorus*, *at once*, *at the same time*. Other important cases of collective predication are reciprocal constructions (see Section 2.2.10).

2.2.8 Distributive vs Collective Predication

Consider again the sentences

(47) a. The soldiers are asleep.
b. The soldiers gathered.
c. Every soldier is asleep.

Unlike (47b), sentence (47a) involves an inessential use of the PNP in that it can be paraphrased by the singular universal sentence (47c). One way to deal with the situation is to acknowledge a systematic ambiguity of the NP *the soldiers*. But the difference between (47a) and (47b) is preferably traced back to the distributive/collective distinction in the predicate. On this view, *the soldiers* invariably stands for the group of soldiers given in a situation. An argument for placing the ambiguity in the VP rather than in the NP is the existence of sentences that have a conjoined VP with one collective and one distributive predicate, combining with a single PNP, as in *The Beatles split up and (each) started a solo career*. Thus, such a view lends itself to a smooth compositional interpretation.

In (47b), *gathered* is just a property of this group. Can a distributive predicate like *sleep* also be true of the whole group? In a derivative sense, yes, but for formal semantics to work we have to be more precise. Imagine there are three soldiers in the group, say a, b, and c; what then is to go into the extension of *be asleep* if (47a) is true? Should the group of the three be made an element, if only for "reasons of redundancy", over and above the single elements a, b, and c? Such a move would be no good, as the following example shows.

(48) The soldiers received 100 DM.

In our little model the paymaster has to issue just 300 DM to make (48) true, and not some fancy 700 DM (there are also three subgroups of two!).

Thus there is no way around making formally explicit the inessential use of plural NPs in combination with distributive predicates. Apart from the usual distributive predicate *sleep* we introduce a **derived predicate** **sleep* which can also take group denoting terms as arguments; that is, it applies (i) to all individuals who are asleep in a given situation, and (ii) to all groups of object that can be formed from the individuals sleeping in that situation. In the above model, the extension of **sleep* would consist of a, b, c, $\{a,b\}$, $\{a,c\}$, $\{b,c\}$, $\{a,b,c\}$.[3] Thus sentence (47a) should be interpreted as saying that the group of soldiers is in the extension not of *sleep* but of **sleep*, i. e., that it is made of individuals taken from the extension of *sleep*. But that just means that every soldier is asleep, which is what (47c) says. The critical case (48) can now be treated analogously: The group of soldiers is in the extension of **received.100.DM*, that is, it is made of individuals in the extension of *received.100.DM*. Thus every soldier gets 100 DM, and the paymaster pays an overall 300 DM and not more, just as it should be.

Now there is a large open class of predicates which freely admit both the distributive and the collective interpretation of their subject NPs. Let us call them **mixed predicates**. Consider

(49) a. Three men lifted the piano.

b. Three men lifted a piano.

The first sentence is two-ways ambiguous: it either refers to one single act of lifting the piano (collective reading) or there are three separate acts of lifting involved (distributive reading). How many readings does (49b) have? Here we have to be careful because of scope interference. This seems strange at the first blush since there should be no scope effect when

[3]For simplicity, the group objects are represented by sets here, but no systematic claim is attached to it; see Section 3 for a discussion of the problem of modeling those objects in an algebraic setting.

two existential phrases are combined (cf. *a man met a woman* which has just one reading); it is a remarkable fact that indefinite *plural* NPs do create scope differences. Indeed there are four readings (wide and narrow scope superimposed on the distributive/collective ambiguity) two of which happen to coincide: this leaves the following three different readings: (i) There is a group x of three men and a piano y such that x lifted y (one act of lifting, one piano lifted); (ii) there is a group x of three men each of whom lifted a piano (three acts of lifting, ≤ 3 pianos lifted); (iii) there is a piano x and a group y of three men such that each of them lifted x (three acts of lifting, one piano lifted). That shows, incidentally, that it is not enough to count the number of the pianos lifted for finding all the readings here. A fortiori, a statement about the total number of pianos lifted cannot serve as logical form for (49b), as it is sometimes assumed. This procedure is inadequate also on more systematic grounds: A numerical statement can at best be entailed by, but hardly be synonymous with, an assertion like (49b).

In a formal representation of (49) it has to be indicated whether the predicate is understood distributively or not. For this purpose, a **distributivity operator** 'D' on predicates is introduced which produces new predicates that may apply to groups and regular individuals alike. A group of objects, then, is in the extension of *lift* x if all individual objects in the group are in the extension of *lift* x, i. e. if every member of the group does the lifting. If the predicate *lift* x is applied to a group without modification then the intended reading is collective.

A further topic to be taken up in this section is a problem about "group levels". Recall (45f), here repeated as

(50) The Leitches and the Latches come together.

On one reading this means that the two families gather, which is not necessarily the same as the case where the members of the two families gather. Imagine also a red and a blue deck of playing cards lying on the table. When we say,

(51) The red cards and the blue cards are shuffled,

it does not mean that the two decks have been mixed, i. e. that the cards of the two decks together are shuffled. In either case, however, we do have the truth of both (52a) and (52b).

(52) a. The decks are on the table.

b. The cards of the two decks are on the table.

So collective predicates may be sensitive to the formation of an intermediate group level (e.g., *come together*, *be shuffled*), or they may not (e.g., *be on the table*). In Chapter 1 the latter predicates are called *invariant.* The way we are going to deal with the level problem is this: An intermediate level is introduced into the semantics as a separate formal object, and a meaning postulate for invariant predicates says that these "ignore" the group level. Notice that there does not seem to be an iteration of group levels realized in language, i.e., there are no predicates that distinguish groups of groups, groups of groups of groups, etc.

Now there is still another reading of (50), viz.: the Leitches gather at one place, the Latches at another. Here we have to "'shut off" the Leitch group and the Latch group and make the predicate *come-together* apply to the "sum" of the two groups. The removal of the superscript then leads to the assertion that the Leitches come together among themselves, and that the Latches do so, too.

In (50) the relevant subgroups are clearly marked syntactically. But Scha (1981) claims that also in sentences like

(53) Six boys gather.

there is a yet another collective reading, not discussed so far; it would, for instance, be satisfied by a model in which there are three groups of two boys gathering at different places, or else by a model in which four of the boys gather at one place, the rest of the boys at another. So Scha wants to say that the total group of six boys splits up into a *partition* each element of which gathers separately (because of the collective verb we have to make sure that the members of the partition are all non-singular). That the present reading in fact suggests itself in quite natural situations is seen from (54),

(54) Half a million people gathered throughout the country.

Note that the collective reading we had before is but a special case of Scha's: the one-element partition consisting of the total group itself. Let us call the new reading **partitional** and indicate it by the superscript 'P' attached to the predicate. In the most general case, then, a PNP gives rise to three different readings: the distributive, the simple collective, and the partitional reading.[4]

A final methodological remark. Upon reflection it seems preferable not to distinguish really between the collective and the partitional reading.

[4]Technical note: As will be seen from the formalization below partitions require higher-order quantification. This poses no problem for a Montague-style framework, of course, but first-order systems like LPM in Chapter 1 would have to be suitably extended.

Take *gather*; this predicate admits a continuous transition from one reading to the other. How far apart from each other have meetings to take place for them to count as different gatherings? Consider the "squish"

(55) Half a million people gathered in Germany/ in Bavaria/ in Munich/ on the October *Wies'n*/ in the Olympic stadium (?)/ on the *Marienplatz* (??)/ in the Bavarian *Staatskanzlei* (pragmatic star).

2.2.9 Relational Plural Sentences.

So far we have considered only 1-place collective predicates. Particularly interesting cases of plurality are 2-place predicates with PNPs at both argument places. Such constructions have been called *Elementary Plural (Relational) Sentences* (EPRS; Langendoen (1978); de Mey (1984) or *Dependent Plurals*. I propose here **Relational Plurals (RP)**. The following examples will be discussed:

(56) a. Four men lifted three tables.
b. 600 Dutch firms use 5000 American computers.
c. The women cooked for the prisoners.
d. The tenants hate their landlords.

(57) a. Unicorns have horns.
b. Beavers build dams.
c. Horses have four legs.
d. Antelopes gather near water-holes.

(58) a. The squares contain the circles. (see Figure 1)
b. The sides of rectangle 1 run parallel to the sides of rectangle 2. (see Figure 2)
c. The sides of rectangle 1 cross the sides of rectangle 2. (see Figure 2)

In the last section we saw that in principle, one indefinite PNP gives rise to three different readings, one distributive and two collective. Thus we should be able to count $3 \times 3 = 9$ readings when two such NPs are combined, as in (56a,b). But the discussion of (49b) has led us to recognize also a scope difference, which doubles the number of readings. Now I have already expressed doubts as to whether we should really distinguish the simple collective and the partitional reading. Be this as it may, I am going

Figure 1

(58a) The squares contain the circles.

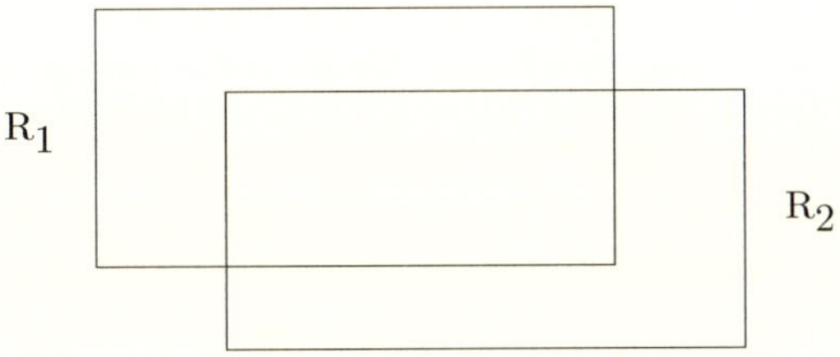

Figure 2

(58b) The sides of rectangle R_1 run parallel to the sides of rectangle R_2.
(58c) The sides of rectangle R_1 cross the sides of rectangle R_2.

to ignore the latter one here. This brings us back to eight different cases; but two of these coincide, as it turns out. An illustration is useful here; it is given for (56a) in Figure 3.

The logical forms given below yield exactly the results displayed here. For that it was necessary to apply the distributivity operator not only to elementary predicates like *lift* but also to whole VPs, that is, to arbitrary λ-expressions as their logical counterparts.

A peculiarity in connection with (56a) should be mentioned. When the numerals in the two PNPs happen to coincide a one-to-one reading comes to the fore which looks different from all the others. The sentence

(59) Four men lifted four tables.

most naturally means that there are four men and four tables, and each man lifted just one table. This brings us back to sentence (56b), which is Scha's example. He observed (Scha (1981)) that beyond all the readings we had so far there is still what he calls a **cumulative reading**: What counts here is just the total number of individuals involved on both sides of the relation. For (56b) this means, for instance, that the total number of Dutch firms with an American computer is 600, and the total number of American computers used by Dutch firms is 5000. The relation *use* can be as it may provided its domain has cardinality 600 and its range has cardinality 5000. A situation like the one described in (59) is a special cumulative reading where, in addition, the relation is a bijective function. Now again we might want to say that the cumulative reading is not a separate reading after all but rather describes a more detailed model of the doubly collective reading CC. So we touch on a methodological point of a quite general nature in linguistics here: Where exactly does the line of demarcation run between proper readings and mere models realizing a reading? But short of this generality we observe that there is an argument in favor of an independent cumulative reading. A genuine collective interpretation of (56b) would mean that the 5000 American Computers are common property of the 300 Dutch firms, but this, of course, is not the intended meaning. The cumulative reading of relational plural sentences is a separate phenomenon insofar as both the subject phrase and the object phrase are *thematic*: they function as *autonomously referring expressions* (Keenan), which can be seen from the fact that the numerals in both places give figures that have to be taken absolutely. By contrast, the object phrase *two tables* in the sentence *Three men lifted two tables* does not refer autonomously since the numeral *per se* is not sufficient indication of the total number of tables involved.

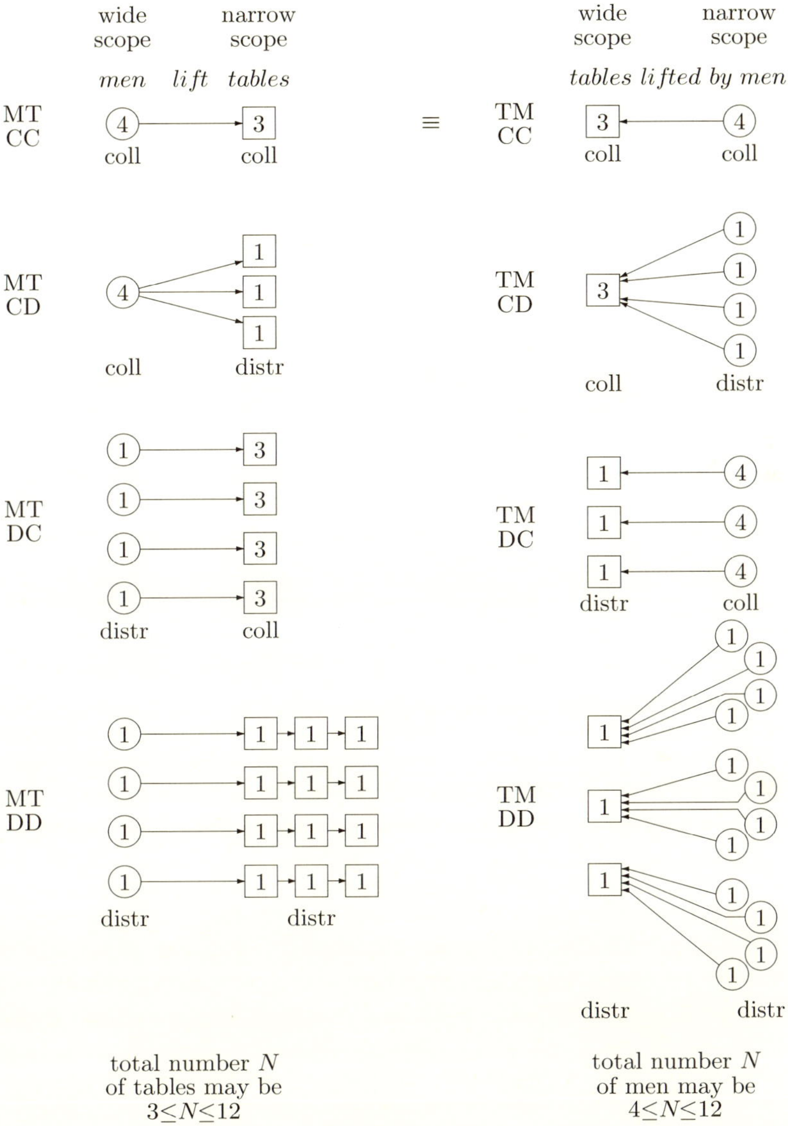

Figure 3

The 7 readings of (56a): Four men lifted three tables.

Next consider (56c). Here we have two definite PNPs which do not create a scope ambiguity. Thus there are just four readings (CC,CD,DC,DD). Sentence (56d) is different and more complex. The intended reading is that each tenant hates his or her own landlord. This can be easily formalized; what causes trouble is the fact that in carrying out the translation into logic we cannot proceed compositionally here in a superficial way. There are two groups of individuals involved but the second group is not independent of the first: rather it is the group of "function values" under the operation *is the landlord of*. I am not going to advance any particular proposal here how to get from the two functionally dependent groups down to the intended logical form; too many assumptions come into play that rather concern the general form of a logically based grammar.

Under (57) I have listed some important generic versions of RP sentences. Genericity is a separate subject with an increasing body of literature, already referred to in Section 2.2.1 above. So it might suffice here to make just one observation concerning those sentences. In generic contexts the (grammatical) number distinction appears semantically neutralized. Thus the plural form does not really refer to a plural phenomenon. That is why formalizations of the sentences in (57) will not be given here.

I turn now to the interesting examples under (58) which were discovered and analyzed by Scha (1981). They seem to ruin any approach according to which it is essentially the NPs that determine the quantificational structure of a sentence. Call the variable for the subject individuals x, for the object individuals y. Roughly, then, (58a) has the quantifier structure $\exists y \forall x$; (58b): $[\forall x \exists y \& \forall y \exists x]$; (58c): $\exists x \exists y$. But the verb is invariably flanked by two PNPs in all these cases. Thus they provide *prima facie* counterexamples to the claim that the quantifier structure is determined by the group/$\forall$ ambiguity of definite PNPs (Hausser (1974b); Bennett (1975)), which would predict a uniform $\forall x \forall y$ structure in the present case; but also Langendoen's proposal which favors the structure displayed by (58b) (Langendoen (1978)) cannot be upheld without qualifications. Scha's own solution consists of a recourse to meaning postulates for the three verbs involved; see (Scha, 1981, p. 497). Upon reflection this shows that the source of the seemingly "erratic" behavior of one of the most important determiners in language, the definite article, is really the predicate involved in each case. Notice to begin with that all three examples are taken from a rather mathematical context. The more technical the relational term the more precise the quantifier structure. *run parallel* is the clearest case: it denotes a two-place relation which is left and right total, whence the structure $[\forall x \exists y \& \forall y \exists x]$. In principle, *contain* is just the same case; but already here the context is loose enough as to allow vagueness in the first argument. The emphasis is only on: "no counterexample in the range of the relation"; this means

$\neg\exists y\forall x\neg$ or, equivalently, $\forall y\exists x$, which is just right-totality. If we vary the example not even this is necessary. Take

(60) The Montagues hate the Capulets.

This is true, according to the story told, even if we assume that Romeo and Juliet decided to stop hating each other's families. Similar considerations apply to Scha's third example: It looks like the meaning postulate $\exists x\exists y$ for *cross* just fits the mathematical example particularly well. In a less technical context the meaning of *cross* asks for more in order for a RP sentence involving it to become true. Thus we wouldn't want to say that the following sentence about America is true just because somewhere in the Midwest a highway crosses a railroad track:

(61) In America the highways cross the railroad tracks.

Now what moral can be drawn from all this? The discussion should have shown that the traditional semantical device of meaning postulates to too rigid to capture all non-semantic meaning nuances (pragmatic, contextual); they should not even be expected to do so. Semantics should stick to the invariant and stable meaning components. One of those is the distributive/collective distinction, which, in the case of two PNPs, leads to the four readings CC,CD,DC,DD. Some of them might be excluded by context or the special character of the relation at hand. Thus if we replace the verb *cook for* in (56c) by the achievement verb *release*, for instance, the DD reading is hard to come by:

(62) The women released the prisoners.

Now as far as Scha's examples are concerned, they are best subsumed under the doubly collective reading CC. It is not that double PNP constructions change their meaning all the time; rather, the CC reading leaves room for the various quantifier structures that might be triggered by the meaning of the predicate in specific contexts. As in other areas of semantic research it becomes apparent that the interpretation of a sentence is highly underdetermined by the context-invariant core semantics of a language. In this way the putative counterexamples from above can be put into a proper perspective.

2.2.10 Reciprocal Constructions and *respectively*

(63) a. George and Martha hate each other.

b. The professors hate each other.

c. The sea-elephants lay on top of each other.

d. The committee ?hates/hate each other.

e. The Leitches and the Latches hate each other.

f. Blücher and Wellington, and Napoleon fought against each other at Waterloo.

g. George and Martha expect each other to win.

h. Some relative of each villager and some relative of each townsman hate each other.

(63a) and (63b) mean that among the group denoted by the subject phrase there obtains a mutual relation of hate, where self-hatred is excluded **disjoint reference**). Thus, according to (63a), George hates Martha and Martha hates George; (63b) has the same interpretation: all pairs of professors are related by *hate*, except for the "diagonal" pairs $\langle d, d \rangle$ where d is a professor. That suggests to represent *each other* as a VP operator which generates collective predicates from the given VP and which can be defined in terms of the quantifier structure '$\forall x \forall y[x \neq y \rightarrow$'. But there is some vagueness involved that shows up already in (63b): the sentence can be true without each and every pair of two professors standing in the *hate* relation (we observed that also in connection with sentence (60)). In (63c) such a totality interpretation is even excluded by world knowledge. (Here we see again the limitation of meaning postulates.) Disjoint reference with collective nouns seems questionable, but with plural agreement in the verb they are more acceptable; cf. (63d). The next sentence (63e) is like (60) in that there is the reading "hate between the families." If we treat *hate-each-other* as a 1-place collective predicate as indicated we have to shut off the families as subgroups before the meaning postulate is evaluated; otherwise we get the undesired reading that even the internal family relations are dominated by hate. Intermediate group levels are also needed for the intended reading of (63f), viz. that Blücher and Wellington together fought against Napoleon; it seems doubtful, however, if the sentence can really express that reading. The last two sentences are meant to point in two different directions: (63g) shows that *each-other* constructions may involve complement sentences and, in fact, much more; for the complicated problems in English syntax created by these constructions see Dougherty (1974); Fiengo and Lasnik (1973); Higginbotham (1980) with references therein. Finally, sentence (63h) raises the question as to whether natural language contains proper instances of **branching quantification**; an excellent source for the logic of such constructions is Barwise (1979).

Langendoen (1978) reviews and analyzes many proposals about the semantics of reciprocity that have been put forth in the literature. Since in most cases *each-other* constructions can be reduced to RP sentences we may expect to encounter problems of the same kind as in the last section.

I now turn to one more topic that has to be mentioned in connection with plurality. Following the demand of the linear structure of language English has developed the *respectively* construction by which one is able to express simultaneously, so to speak, two (or more) instances of a two- (or higher-)place relation between major constituents. This is illustrated by the following sentences:

(64) a. George and Nick hate Martha and Honey, respectively.
b. George and Martha are drinking and dancing, respectively.
c. Charlie Brown, Linus, and Schroeder won the first, the second, and the third prize, respectively/in this order.

Here the semantic interpretation is clear, at expense of a clumsy syntax. The intended one-to-one connections between corresponding terms are easily established. In set-theoretic semantics they give rise to a representation involving Cartesian products. Notice that the terms within the conjoined phrases cannot independently be switched around. By contrast, normal group denoting coordinate conjoined NPs like in (63a) are symmetric. A non-symmetric behavior is also shown by sentences like (45d) above, here repeated as (65).

(65) George and Martha are husband and wife.

2.3 Ontology

2.3.1 The Power Set Model; Collections; Mereology

What kind of entities are plural objects? In the informal discussion of Section 2 I used the words 'group', 'collection', or 'aggregate' when I talked about these objects. Now suppose we take the notion of a group as given in the set-theoretic metalanguage, and address ourselves first to the way it enters our semantic relations. For instance, a group can be a member of the extension of a collective verb like *gather*; if extensions of 1-place predicates are subsets of the domain of individuals a group must be an individual. But obviously, such a group individual has an internal structure. Take the group consisting of Marat, Danton, and Robespierre, on the one hand, and the group consisting of Marat and Danton, on the other. The latter, call it a, is a subgroup of the former, say b. When we denote the relation "is a subgroup of" by $\leq$ we can write $a \leq b$. Now $\leq$ is a partial order as can be easily verified; since the standard model for it is just set inclusion, the first and most natural answer to the above question is: *plural objects denote sets of individuals* (cf. Hausser (1974b); Bennett (1975); Bartsch (1973), and others). This decision implies an adjustment in the structure of predicate extensions: *gather*, for instance, has now to denote sets of sets of individuals. But this introduces a rigid dividing line between collective and distributive predicates, the latter of which are still supposed to be true of simple individuals. As we saw above, however, there are many mixed predicates (like *lift a table*) that may apply to both simple individuals and groups at a time. So linguistic evidence forces us to take the next logical step: lift the status of simple individuals to the higher level by identifying them with their singleton sets (cf. Scha (1981)). Thus a model for a language with pluralities should consist of a **power set structure** 2^E, together with a denotation function f such that $f(t) \in \{\{d\} \mid d \in E\}$ for simple individual terms t, $f(g) \in 2^E$ for group terms g, and $f(P) \subseteq 2^E$ for 1-place predicates P. The subgroup relation is realized as the inclusion on 2^E, as envisaged above.

Now we could stop here and develop a logic of plurality on the basis of this simple idea. But do sets of sets have the right existence and identity conditions for them to serve as formal counterparts of our intuitive notion of a plural object? Blau (1981b) addresses himself to this question. The existence conditions for plural objects, here called **collections**, pose no problem; if P is a 1-place predicate the collection of the Ps exists if and only if P is true of at least one individual. Of course, the definite PNP the Ps seem to presuppose the existence of two or more objects, but Blau finds it convenient to include unit collections as a limiting case. As to identity, Blau claims that collections are **hyperextensional** in the following sense: identity of extension of two non-empty 1-place predicates P and Q is a sufficient, but not necessary, condition for their collections to be equal. For instance, with P for *(playing) card* and Q for *deck of cards*, we can say that the collection of the three Qs is on the table just in case the collection of the Ps, belonging to the Qs, is on the table. Now *be on the table* happens to be an invariant predicate as I called it above, and is not representative for the most general context in which a collection can appear. Are we entitled to sharpen the above assertion by concluding that identity of extension is also a necessary condition after all? On pain of contradiction we are not, as Blau shows (1981a:129, footnote). For illustration, roughly, it is inconsistent to assume that the predicate P which is true of just one entity, the collection of a and b, contains as many elements as the predicate Q which is true of two individuals, a and b (the collections of P and Q coincide!). Upon further reflection we are led back to the power set model. We have strong

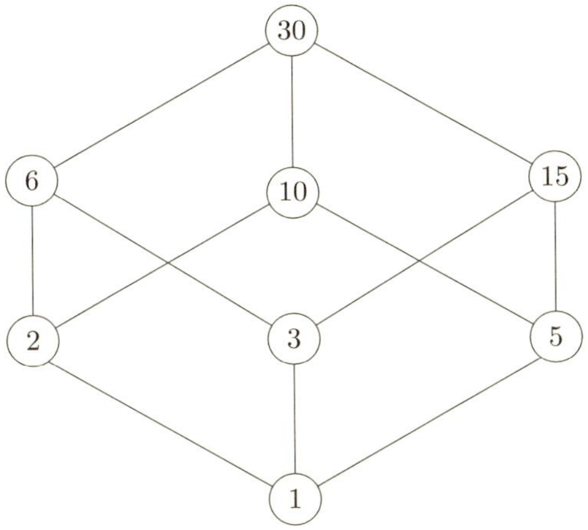

Figure 4a

The set of divisors of *30*.

intuitions about its being correct at least structurally. So, again, why not just identify plural objects with sets as this model says?

Well, I think there are both philosophical and linguistic arguments against such a move. Instead of repeating them here[5] I would rather like to illustrate the situation by a mathematical example. Consider the set of divisors of the number 30, $T_{30} = \{1, 2, 3, 5, 6, 10, 15, 30\}$. Would you say that this set is "'nothing else than" the power set of $\{2, 3, 5\}$? You wouldn't, if you are not initiated mathematically, even less so that 1 "is" the empty set, 2, 3, and 5 "are" singletons, and 30 "equals" the set $\{2, 3, 5\}$; in fact, the usual definition of natural numbers in set theory renders these assertions strictly false. And yet, they are all true under a rather precise interpretation: T_{30}, together with the 2-place relation "divides", which is called *lattice of divisors*, is order-isomorphic to the power set $2^{\{2,3,5\}}$, together with set inclusion. Figures 4a and 4b show this strict parallelism between T_{30} and the power set of $\{2, 3, 5\}$. T_{30} is what is called an atomic Boolean algebra over three atoms, of which the power set algebra over the set of these atoms happens to be the standard model (cf. Figure 4b). The power set "models" or represents the T_{30} but is still not identical with it.

Now consider three individuals, a = Marat, b = Danton, and c = Robes-

[5]See Chapters 1, 3, 13 and 14; also Blau (1981b).

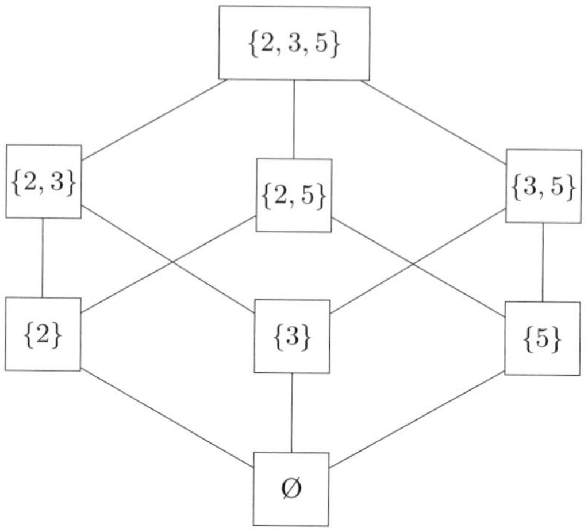

Figure 4b

The power set of {2, 3, 5}.

pierre, and the predicate P = *meet*. Suppose we say that if Marat and Danton meet P is true of the **individual sum** ("*i-sum*" for short) of a and b, denoted by $a \oplus b$; similarly with $a \oplus c$, $b \oplus c$. If Marat, Danton, and Robespierre all meet P is true of the sum of the three individuals, $a \oplus b \oplus c$. Arrange these compound individuals in a way shown in Figure 4c.

We see that the sums, including the atoms a, b, and c, are partially ordered by upward edges or combinations thereof, just as in Figures 4a and 4b. This is the subgroup relation I mentioned; call it Π, to be read as "*is an individual part (or i-part) of*." We symbolize this relation by $\leq_i$; its counterpart in the object language is the 2-place predicate Π. Now the set of these sums, together with $\leq_i$, is isomorphic to the power set algebra over the set $\{a, b, c\}$, minus $\emptyset$. It is my claim that the structure of Figure 4c incorporates all the intuitions there are about the behavior of plural objects in natural language, and that there can be no empirical evidence that is suited to show that, say, a is "really" a singleton set. For practical reasons (for instance, because people are "used to it") we could stick to the power set model as long as we don't forget that it is only a model. From both a philosophical and a methodological point of view, however, an axiomatic approach is to be preferred that takes the objects which language talks about as given and tries to characterize the structural relations between

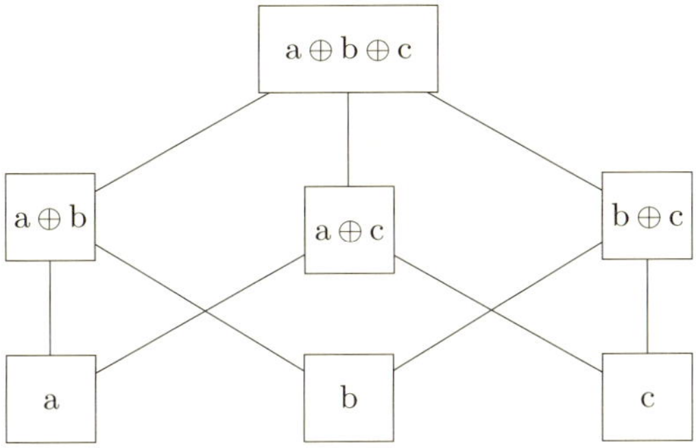

Figure 4c

The ordered set of i-sums over a, b, c.

the various individuals; this is the core of an **algebraic semantics** applied to language. Linguistically, the main argument in favor of this approach is the structural analogy between plurals and mass terms which tends to be blurred by the set-theoretic representation of pluralities.

Coming back to Figure 4c we note that it exemplifies an atomic Boolean algebra minus its zero element. Now it is known that these structures are the models for the part-whole relations of **mereology**; cf. Leśniewski (1927-31); Leonard and Goodman (1940); Eberle (1970); Simons (1987). But the logical and philosophical context in which mereological theories were investigated was quite different from the present one. Massey (1976) seems to have been the first to put mereological sums to use in connection with plurals. See also Bunt (1981) for his "theory of ensembles". A system developed in the spirit of mereology is also Lønning (1982). What these approaches have in common with the view about plurality advanced here, I think, is their emphasis on the *structural aspect* of the matter.

In Chapter 1 the domain of individuals in a model is no longer just a non-empty set but is endowed with an algebraic structure that governs the logic of plural sentences in language. In what follows I shall describe the essential features of this *lattice-theoretical* approach up to a point where a representative sample of the examples given above can be assigned a proper logical interpretation.

Finally, it should be mentioned that there have been others who use an algebraic approach to the semantics of language. There is the **Boolean semantics** in the sense of Keenan (1981); see Keenan and Faltz (1985);

Keenan and Stavi (1986). Furthermore, there is the Generalized Quantifier approach with its second order set structures that are inherently endowed with an ordering relation and hence carry an algebraic structure (for references see Section 2.2.4 above). However, neither of those theories were developed to deal with plural phenomena. For a discussion of plurals within the Generalized Quantifier framework see Chapter 4.

2.3.2 The Algebraically Structured Universe of Individual Sums and Groups

The formal language introduced here may be called the **Logic of Plurality, LP**. It is essentially the "plural part" of the system presented in Chapter 1. LP is a first order predicate calculus with the usual logical constants, a description operator 'ι' and simultaneous lambda abstraction (for the latter, see, e.g., (Link, 1979, p. 100)). The syntactic variables of LP are: for formulas, 'ϕ', 'ψ', 'χ'; for individual variables ranging over atoms and i-sums alike, 'u', 'v', 'w', 'x', 'y', 'z'; for (both elementary and complex) predicates, 'P', 'Q', 'R', 'S'. These symbols can also appear with primes and indices.

As special primitive symbols we have a 1-place predicate '**E**' for "exists" since the formal system is going to be a *free logic* in which in general not every individual term denotes an object in the domain. Other primitive symbols are '**PA**' for "is a pure atom;" the 2-place predicate constant 'Π' for "is an i-part of;" the 1-place function symbol '$\langle\rangle$' for "the group of;" and the plural operator on 1-place predicates, '*'.

Next we need the following *defined expressions*; a paraphrase is given after each definition.

(D.33) $a = b \;\leftrightarrow\; a\,\Pi\,b \wedge b\,\Pi\,a$

a is *identical with* b iff a is an i-part of b and b is an i-part of a.

(D.34) $\mathbf{At}\,a \;\leftrightarrow\; \forall x[x\,\Pi\,a \rightarrow x = a]$

a is an *atom* iff all i-parts of a are identical with a (notice that because of (TD.3) below, any i-part must be non-zero).

(D.35) $a\,{}^{\cdot}\Pi\,b \;\leftrightarrow\; a\,\Pi\,b \wedge \mathbf{At}\,a$

a is an *atomic i-part of* b iff a is an i-part of b and a is an atom.

(D.36) $\mathbf{PE}\,a \;\leftrightarrow\; \forall x[x\,{}^{\cdot}\Pi\,a \rightarrow \mathbf{PA}\,x]$

a is a *pure entity* iff all atomic i-parts of a are pure atoms.

(D.37) $\quad {}^{\star}P\,a \;\leftrightarrow\; {}^{*}P\,a \wedge \neg\mathbf{At}\,a$

a is a *genuinely pluralic* P iff a is a starred P, i. e. an i-sum consisting of Ps, but not an atom.

(D.38) $\quad \mathbf{Distr}\,(P) \;\leftrightarrow\; \forall x(Px \to \mathbf{At}\,x)$

A 1-place predicate P is *distributive* iff all Ps are atoms.

(D.39) $\quad {}^{D}P \;:=\; \lambda x \forall u[u\,{}^{\cdot}\Pi\,x \to Px]$

The *distributive version* of a 1-place predicate P denotes the set of all i-sums x such that all atomic i-parts of x have the property P.

(D.40) $\quad \sigma x Px \;:=\; \iota x[{}^{*}Px \wedge \forall y[{}^{*}Py \to y\,\Pi\,x]]$

The *individual sum* of the Ps is that i-sum x such that (i) x is a starred P (i. e. a sum of individuals all of which have P) and (ii) all starred Ps are i-parts of x. Thus $\sigma x Px$ denotes the *supremum* of all i-sums built up from elements in the extension of P.

(D.41) $\quad \sigma^{*} x Px \;:=\; \sigma x {}^{\star}Px$

The *starred individual sum* of the Ps is the individual sum of the genuinely pluralic Ps.

(D.42) $\quad a \oplus b \;:=\; \sigma x\, I_{a,b}\, x \qquad$ (with $I_{a,b} \;=\; \lambda x[x = a \vee x = b]$)

The *individual sum of* a *and* b is the supremum of the individuals a and b (which need not to be atoms).

The **semantics for LP** has to incorporate the idea of a algebraically structured universe. We have to endow the domain of individuals, E, with the partial order $\leq_i$ that expresses the i-part relation between individuals. Furthermore there is a 2-place join operation on E, denoted by $\sqcup_i$, which forms the i-sum of any two individuals in the domain. $\sqcup_i$ is a "semi-lattice" operation on E. It would be enough for the data treated here to stick to such a semi-lattice structure satisfying three additional intuitive requirements: it has to be *complete* (i.e., it has to contain an i-sum for every, not necessarily finite, non-empty collection of objects in E), *atomic* (everything can be "built up" by the sum operation from $\leq_i$-least parts), and in a certain sense *free* ("maximally unfolded"). It turns out, however, that these requirements already yield an entire complete atomic Boolean algebra minus zero (see Chapter 6). So for simplicity I assume that E is a complete atomic Boolean algebra (the zero element will not be in the domain, but will be assigned a special role, as we will see presently). To fix intuitions about this structure we just keep the power set model in mind. We have to provide some additional internal structure, however, to take care of the intermediate group level that was discussed above. These remarks lead to the following definition.

(D.43) A *Boolean model structure with groups* is a triple $\mathcal{B} = \langle E, A^\circ, \gamma \rangle$ such that

1. E is a complete atomic Boolean algebra, with A as the set of its atoms, $\sqcup_i$ as its join operation, and $\leq_i$ as its intrinsic ordering relation;
2. A°, the set of *pure atoms*, is a proper subset of A;
3. γ is an injective mapping from the set of all proper i-sums over A° into the set difference $A \setminus A^\circ$.

If $\langle E, A^\circ, \gamma \rangle$ is a Boolean model structure with groups, then call an element $d \in E$ a *pure object* if d is an i-sum that is built up from pure atoms only; otherwise d is called *impure*. So for instance, a pure object is denoted by *Blücher* $\oplus$ *Wellington*, and an impure one by $\langle$*Blücher* $\oplus$ *Wellington*$\rangle$ $\oplus$ *Napoleon*. γ is the group forming operation in the model structure that corresponds to the function symbol $\langle\rangle$.

(D.44) A *model* for LP is an ordered pair $M = \langle \mathcal{B}, \|\cdot\| \rangle$ such that

1. $\mathcal{B} = \langle E, A^\circ, \gamma \rangle$ is a Boolean model structure with groups; $E \setminus \{\mathbf{0}\}$, where $\mathbf{0}$ is the zero element of E, is called the *domain of E*.
2. $\|\cdot\|$ is the interpretation function on the constants of LP such that
   (a) $\|a\| \in E$ for an individual constant a;
   (b) $\|P^n\| \subseteq (E \setminus \{\mathbf{0}\})^n$ for an n-place predicate constant P^n.

Let $M = \langle \mathcal{B}, \|\cdot\| \rangle$ be a model for LP. I now give the truth and denotation conditions for the special primitive symbols of LP. (The complete recursive definition is standard; $\mathbf{0}$ serves as a "dummy object" for non-denoting terms (for this device in the semantics of free logic, see, e.g., the system PL1IKA in (Link, 1979, p. 102f)); quantification runs over non-zero elements only; note in particular that the formal counterpart of plural formation in language, the star operator, is given a precise set-theoretic characterization.)

(TD.1) $\|\mathbf{E}\, a\| = 1$ iff $\|a\| \neq \mathbf{0}$

(TD.2) $\|\mathbf{PA}\, a\| = 1$ iff $\|a\| \in A^\circ$

(TD.3) $\|a\, \Pi\, b\| = 1$ iff $\|a\| \neq \mathbf{0}$ and $\|a\| \leq_i \|b\|$

(TD.4) $\|\langle a \rangle\| = \gamma(\|a\|)$ if $\|a\| \in \mathrm{Dom}\, \gamma$, and $= \mathbf{0}$ otherwise

(TD.5) $\|{}^*P\|$ = the set of all i-sums of elements of $\|P\|$, i.e., the (complete) plural sublattice generated by $\|P\|$, if $\|P\| \neq \emptyset$, and $= \emptyset$ otherwise.

Using (TD.l) – (TD.5) the following semantic rules can then be derived.

(TD.6) $\|a = b\| = 1$ iff $\|a\| \neq \mathbf{0} \neq \|b\|$ and $\|a\| = \|b\|$

(TD.7) $\|\mathbf{At}\, a\| = 1$ iff $\|a\| \in A$

(TD.8) $\|a\, {}^{\cdot}\Pi\, b\| = 1$ iff $\|a\| \in A$ and $\|a\| \leq_i \|b\|$

(TD.9) $\|\mathbf{PE}\, a\| = 1$ iff $\|a\| \in [A^\circ]$, i.e., $\|a\|$ is an i-sum of pure atoms

(TD.10) $\|{}^{\star}P\, a\| = \|{}^*P\| \setminus A$

(TD.11) $\|\sigma x P x\| = \sup_{\leq_i} \|P\|$, with $\sup_{\leq_i} \emptyset = \mathbf{0}$

(TD.12) $\|\sigma^* x P x\| = \|\sigma x P x\|$, if $\|P\|$ has ≥ 2 elements, and $= \mathbf{0}$ otherwise

(TD.13) $\|a \oplus b\| = \|a\| \sqcup_i \|b\|$

(TD.14) $\|\iota x P x\| = \|\sigma x P x\|$, if $\|P\|$ has ≤ 1 elements

The LPM-theorems in Chapter 1 which contain only symbols of LP are also valid formulas of LP under the present interpretation. In particular I mention here the **cumulative reference property** for plurals; cf. example (5a) above. It says that if x and y are i-sums having the property ${}^*P$ then their join $x \oplus y$ also has the property ${}^*P$; thus the the join of any two i-sums of P-individuals is again an i-sum of P-individuals:

(66) $\forall x \forall y\, [\, {}^*P\, x \wedge {}^*P\, y \rightarrow {}^*P(x \oplus y)\,]$

2.4 Data Explained. Translation into Logical Form

I shall now give logical representations for the most important cases of plural constructions that I have discussed. 'LF' stands for "logical form;" the various readings receive numbers, like LF1, LF2, etc. In a Montague-style notation, I indicate the translation of a descriptive term of English into LP by attaching a prime to it.

2.4.1 Indefinite PNPs

(1a) Some friends dropped by.

LF1: $\exists x\,[\,{}^{\star}\text{friend}'(x) \wedge \text{dropped.by}'(x)\,]$ (collective)
LF2: $\exists x\,[\,{}^{\star}\text{friend}'(x) \wedge {}^{D}\text{dropped.by}'(x)\,]$ (distributive)
$\Leftrightarrow \exists x\,[\,{}^{\star}\text{friend}'(x) \wedge \forall u[u \,{}^{\cdot}\Pi\, x \rightarrow \text{dropped.by}'(u)]\,]$

(3) Some Jacobins were advocats.

LF: $\exists x\,[\,{}^{\star}\text{Jacobin}'(x) \wedge {}^{*}\text{advocat}'(x)\,]$
$\Leftrightarrow \exists x\,[\,{}^{\star}\text{Jacobin}'(x) \wedge \forall u[u \,{}^{\cdot}\Pi\, x \rightarrow \text{advocat}'(u)]\,]$

Because of the distributivity of *advocat* the second line for the logical form of (3) follows (cf. Chapter 1). Note that the plurality condition is retained through the use of '*'.

(49a) Three men lifted the piano.

LF1: $\exists x\,[\,(\mathbf{3}^{*}\text{man}')(x) \wedge \text{lift}'(x, \iota y\,\text{piano}'(y))\,]$ (collective)
LF2: $\exists x\,[\,(\mathbf{3}^{*}\text{man}')(x) \wedge {}^{D}\lambda u\,\text{lift}'(u, \iota y\,\text{piano}'(y))(x)\,]$ (distributive)
$\Leftrightarrow \exists x\,[\,(\mathbf{3}^{*}\text{man}')(x) \wedge \forall u[u \,{}^{\cdot}\Pi\, x \rightarrow \text{lift}'(u, \iota y\,\text{piano}'(y))]\,]$

The numerals '**1**', '**2**', '**3**', ... are adjective-like modifiers that pick out all those i-sums in an NP extension which have the indicated number of atoms. Thus, for instance, LF1 can be read as: "There is an i-sum of men that consists of three atoms and collectively lifts the piano." For a discussion see Section 2.2.4 above.

(49b) Three men lifted a piano.

LF1: $\exists x \exists y\,[\,(\mathbf{3}^{*}\text{man}')(x) \wedge \text{piano}'(y) \wedge \text{lift}'(x, y)\,]$ (collective)
LF2: $\exists x\,[\,(\mathbf{3}^{*}\text{man}')(x) \wedge {}^{D}\lambda u\,\exists y[\text{piano}'(y) \wedge \text{lift}'(u, y)](x)\,]$ (distributive)

$\Leftrightarrow \exists x\,[\,(\mathbf{3}^{*}\text{man}')(x) \wedge \forall u[u \,{}^{\cdot}\Pi\, x \rightarrow \exists y[\text{piano}'(y) \wedge \text{lift}'(u, y)]]\,]$

LF3: $\exists y\,[\text{piano}'(y) \wedge \exists x\,[(\mathbf{3}^{*}\text{man}')(x) \wedge {}^{D}\lambda u\,\text{lift}'(u,y)(x)]\,]$
$\Leftrightarrow \exists y\,[\text{piano}'(y) \wedge \exists x\,[(\mathbf{3}^{*}\text{man}')(x) \wedge \forall u[u\,{}^{\cdot}\Pi\,x \rightarrow \text{lift}'(u,y)]]\,]$

In LF3 *piano* has wide scope. The collective reading with *piano* having wide scope coincides with LF1.

2.4.2 Bare Plurals

(1c) Children built the raft.

LF: $\exists x\,[{}^{\star}\text{child}'(x) \wedge \text{built.the.raft}'(x)\,]$ (collective)

This is the unproblematic existential reading of bare plurals. For the representation of bare plurals in generic sentences see the references in Section 2.2.1.

2.4.3 Definite PNPs

(28) The children gathered (around their teacher).

LF: $\text{gather}'(\sigma^{*}x\,\text{child}'(x))$

(24b) The Sansculottes hailed a Cordelier.

LF1: $\exists y\,[\text{Cordelier}'(y) \wedge \text{hailed}'(\sigma^{*}x\,\text{Sansculotte}'(x), y)\,]$
LF2: $\exists y\,[\text{Cordelier}'(y) \wedge {}^{D}\lambda u\,\text{hailed}'(u,y)(\sigma^{*}x\,\text{Sansculotte}'(x))\,]$
$\Leftrightarrow \exists y\,[\text{Cordelier}'(y) \wedge \forall u\,[u\,{}^{\cdot}\Pi\,\sigma^{*}x\,\text{Sansculotte}'(x) \rightarrow$
$\text{hailed}'(u,y)]\,]$
$\Leftrightarrow \exists y\,[\text{Cordelier}'(y) \wedge \forall u\,[\text{Sansculotte}'(u) \rightarrow \text{hailed}'(u,y)]\,]$

(39b) The three men left.

LF1: $\text{left}'(\,\iota x(\mathbf{3}^{*}\text{man}')(x)\,)$
LF2: $\text{left}'(\,[\sigma/\mathbf{3}]x^{*}\text{man}'(x)\,)$

The logical form LF1 corresponds to the option of treating the numeral as an adjective. But as suggested above, we might prefer to treat it as a presupposing numerical specifier of the determiner. In this case the logical form of the sentence might be represented as under LF2. There $[\sigma/\mathbf{3}]xPx$ is short for $\iota xPx/\forall x[Px \rightarrow \text{three}'(x)]$, where the second slash is Blau's presupposing "prejunction" operator (see Blau (1985),(Link, 1987c, Appendix)). The term reads as "the unique x such that Px, where it is presupposed that all i-sums that are P contain exactly three atoms."[6]

[6]The notion of presupposition doesn't make sense, of course, in a two-valued system like LP. There are two ways to go here. We could either adopt some three-valued system, e.g., the one in Blau (1978), or else introduce logical forms in a context, where the context has then to take care of "truth value gaps." For a discussion of the connection between these two options, see Link (1987c).

2.4.4 *all*

(33a) All soldiers died.

LF: $^{*}\text{died}'(\,\sigma^{*}x\,\text{soldier}'(x)\,)$
$\Leftrightarrow \forall u\,[\,\text{soldier}'(u) \rightarrow \text{died}'(u)\,]$ (because of **Distr**(died′))

(24a) All Sansculottes hailed a Cordelier.

LF1: $^{D}\lambda x \exists y[\text{Cordelier}'(y) \wedge \text{hailed}'(x,y)]\,(\sigma^{*}x\,\text{Sansculotte}'(x))$

$\Leftrightarrow \forall u\,[u \,\dot{\Pi}\, \sigma^{*}x\,\text{Sansc}'(x) \rightarrow \lambda x \exists y[\text{Cord}'(y) \wedge \text{hailed}'(x,y)](u)\,]$
$\Leftrightarrow \forall u\,[\text{Sansc}'(u) \rightarrow \exists y[\text{Cord}'(y) \wedge \text{hailed}'(u,y)]\,]$

LF2: $\exists y\,[\,\text{Cordelier}'(y) \wedge {}^{D}\lambda u\,\text{hailed}'(u,y)\,(\,\sigma^{*}x\,\text{Sansculotte}'(x)\,)\,]$
$\Leftrightarrow \exists y\,[\,\text{Cordelier}'(y) \wedge \forall u\,[\text{Sansculotte}'(u) \rightarrow \text{hailed}'(u,y)]\,]$
(=24b)

In LF(33a) the plural operator had to be applied since died′ is a distributive predicate that cannot apply to proper sums. The mixed predicate hailed′ in LF1(24a), however, is not marked for distributivity; therefore the D-operator has to be attached to pick out the distributive reading.

2.4.5 Partitive

Some of the students

LF: $\exists x\,[\,x\,\Pi\,\sigma^{*}x\,\text{student}'(x) \wedge \ldots]$

43a all of the students

LF: $\forall u\,[\,u \,\dot{\Pi}\, \sigma^{*}x\,\text{student}'(x) \rightarrow \ldots]$
$\Leftrightarrow \forall u\,[\,\text{student}'(u) \rightarrow \ldots]$

2.4.6 Conjoined NPs

(45a) George and Martha are asleep.

LF: $^{D}\text{asleep}'(\,g \oplus m\,)$
$\Leftrightarrow \forall u\,[\,u \,\dot{\Pi}\, g \oplus m \rightarrow \text{asleep}'(u)\,]$
$\Leftrightarrow \text{asleep}'(g) \wedge \text{asleep}'(m)$

(45b) George and Martha met.

LF: $\text{met}'(g \oplus m)$

(45c) George and Martha got married.

LF: $\lambda x \exists u \exists v[x = u \oplus v \wedge \text{got.married.to}'(u,v)]\,(\,g \oplus m\,)$
$\Leftrightarrow \text{got.married.to}'(g,m)$

(45d) George and Martha are husband and wife.

LF: $\lambda xy\,[\,\text{husband.of}'(x,y) \wedge \text{wife.of}'(y,x)]\,(g,m)$
$\Leftrightarrow \text{husband.of}'(g,m) \wedge \text{wife.of}'(m,g)$

(45e) A boy and a girl played (in the backyard).

LF1: $\exists x\exists y\,[\,\text{boy}'(x) \wedge \text{boy}'(y) \wedge \text{played}'(x \oplus y)\,]$
LF2: $\exists x\exists y\,[\,\text{boy}'(x) \wedge \text{boy}'(y) \wedge {}^{D}\text{played}'(x \oplus y)\,]$
$\Leftrightarrow \exists x\exists y\,[\,\text{boy}'(x) \wedge \text{boy}'(y) \wedge \text{played}'(x) \wedge \text{played}'(y)\,]$

(45f) The Leitches and the Latches come together.

LF1: $\text{come.together}'(\,\sigma^{*}x\,\text{Leitch}'(x) \oplus \sigma^{*}x\,\text{Latch}'(x)\,)$
LF2: ${}^{D}\text{come.together}'(\,\langle\sigma^{*}x\,\text{Leitch}'(x)\rangle \oplus \langle\sigma^{*}x\,\text{Latch}'(x)\rangle\,)$

$\Leftrightarrow \text{come.together}'(\sigma^{*}x\,\text{Leitch}'(x)) \wedge \text{come.together}'(\sigma^{*}x\,\text{Latch}'(x))$

(45g) Three men and four women joined.

LF: $\exists x\exists y\,[\,(\mathbf{3}^{*}\text{man}')(x) \wedge (\mathbf{4}^{*}\text{woman}')(y) \wedge \text{joined}'(x \oplus y)\,]$

2.4.7 RP Sentences

For a 2-place predicate $\zeta(x,y)$, define ${}^{D1}\zeta(x,y) := {}^{D}\lambda u\zeta(u,y)(x)$ and ${}^{D2}\zeta(x,y) := {}^{D}\lambda v\zeta(x,v)(y)$. Let 'C' be short for "collective," 'D' for "distributive." 'MT' ('TM') stands for "*man* (*table*) takes scope over *table* (*man*)." Then 'MT-CD', for instance, means: "M takes scope over T, where the M position is to be read collectively, and the T position distributively;" and similarly for the other combinations.

(56a) Four men lifted three tables.

LF1: $\exists x\,[(\mathbf{4}^{*}\text{man}')(x) \wedge \lambda z\exists y[(\mathbf{3}^{*}\text{table}')(y) \wedge \text{lift}'(z,y)](x)]$ (MT-CC)
$\Leftrightarrow \exists x\exists y\,[(\mathbf{4}^{*}\text{man}')(x) \wedge (\mathbf{3}^{*}\text{table}')(y) \wedge \text{lift}'(x,y)]$

LF2: $\exists x\,[(\mathbf{4}^{*}\text{man}')(x) \wedge \lambda z\exists y[(\mathbf{3}^{*}\text{table}')(y) \wedge {}^{D2}\text{lift}'(z,y)](x)]$ (MT-CD)
$\Leftrightarrow \exists x\exists y\,[(\mathbf{4}^{*}\text{man}')(x) \wedge (\mathbf{3}^{*}\text{table}')(y) \wedge \forall v[v\,{}^{\cdot}\Pi\,y \rightarrow \text{lift}'(x,v)]]$

LF3: $\exists x\,[(\mathbf{4}^{*}\text{man}')(x) \wedge {}^{D}\lambda u\exists y[(\mathbf{3}^{*}\text{table}')(y) \wedge \text{lift}'(u,y)](x)]$ (MT-DC)
$\Leftrightarrow \exists x\,[(\mathbf{4}^{*}\text{man}')(x) \wedge \forall u[u\,{}^{\cdot}\Pi\,x \rightarrow \exists y\,[(\mathbf{3}^{*}\text{table}')(y) \wedge \text{lift}'(u,y)]]]$

LF4: $\exists x\,[(\mathbf{4}^{*}\text{man}')(x) \wedge {}^{D}\lambda u\exists y[(\mathbf{3}^{*}\text{table}')(y) \wedge {}^{D2}\text{lift}'(u,y)](x)]$ (MT-DD)
$\Leftrightarrow \exists x\exists y\,[(\mathbf{4}^{*}\text{man}')(x) \wedge \forall u[u\,{}^{\cdot}\Pi\,x \rightarrow \exists y\,[(\mathbf{3}^{*}\text{table}')(y) \wedge \forall v[v\,{}^{\cdot}\Pi\,y \rightarrow \text{lift}'(u,v)]]]]$

LF5: $\exists y\,[(\mathbf{3}^{*}\text{table}')(y) \wedge \lambda z\exists x[(\mathbf{4}^{*}\text{man}')(x) \wedge \text{lift}'(x,z)](y)]$ (TM-CC)
$\Leftrightarrow \exists x\exists y\,[(\mathbf{4}^{*}\text{man}')(x) \wedge (\mathbf{3}^{*}\text{table}')(y) \wedge \text{lift}'(x,y)]$ (=LF1)

LF6: $\exists y\,[(\mathbf{3}^*\text{table}')(y) \wedge \lambda z \exists x[(\mathbf{4}^*\text{man}')(x) \wedge {}^{D1}\text{lift}'(x,z)](y)]$ (TM-CD)
$\Leftrightarrow \exists y \exists x\,[(\mathbf{3}^*\text{table}')(y) \wedge (\mathbf{4}^*\text{man}')(x) \wedge \forall u[u \,\dot{\Pi}\, x \rightarrow \text{lift}'(u,y)]]$

LF7: $\exists y\,[(\mathbf{3}^*\text{table}')(y) \wedge {}^{D}\lambda v \exists x[(\mathbf{4}^*\text{man}')(x) \wedge \text{lift}'(x,v)](y)]$ (TM-DC)
$\Leftrightarrow \exists y\,[(\mathbf{3}^*\text{table}')(y) \wedge \forall v[v \,\dot{\Pi}\, y \rightarrow \exists x\,[(\mathbf{4}^*\text{man}')(x) \wedge \text{lift}'(x,v)]]]$

LF8: $\exists y\,[(\mathbf{3}^*\text{table}')(y) \wedge {}^{D}\lambda v \exists x[(\mathbf{4}^*\text{man}')(x) \wedge {}^{D1}\text{lift}'(x,v)](y)]$ (TM-DD)
$\Leftrightarrow \exists y\,[(\mathbf{3}^*\text{table}')(y) \wedge \forall v[v \,\dot{\Pi}\, y \rightarrow \exists x\,[(\mathbf{4}^*\text{man}')(x) \wedge \forall u[u \,\dot{\Pi}\, x \rightarrow \text{lift}'(u,v)]]]]$

(56d) The tenants hate their landlords.

LF: $^*\lambda u\,[\,\mathbf{At}\,u \wedge \text{hate}'(u, \iota y\,\text{landlord.of}'(y,u)]\,(\,\sigma^* x\,\text{tenant}'(x)\,)$
$\Leftrightarrow \forall u\,[\,u \,\dot{\Pi}\, \sigma^* x\,\text{tenant}'(x) \rightarrow \text{hate}'(u, \iota y\,\text{landlord.of}'(y,u)\,]$
$\Leftrightarrow \forall u\,[\,\text{tenant}'(u) \rightarrow \text{hate}'(u, \iota y\,\text{landlord.of}'(y,u)\,]$

2.4.8 Reciprocals

We set up the following *meaning postulate* for a 2-place relation R (for discussion, see Section 2.2.10 above).

(BP) $\text{EACH.OTHER}(R) \Leftrightarrow \lambda x \forall u \forall v\,[\,u \,\dot{\Pi}\, x \wedge v \,\dot{\Pi}\, x \wedge u \neq v \rightarrow R(u,v)\,]$

(63a) George and Martha hate each other.

LF: $[\text{EACH.OTHER}\,(\text{hate}')]\,(g \oplus m)$
$\Leftrightarrow \text{hate}'(g,m) \wedge \text{hate}'(m,g)$

(63h) George and Martha expect each other to win.

LF: $[\text{EACH.OTHER}\,(\lambda xy\,\text{expect}'(x, \text{win}'(y)))]\,(g \oplus m)$
$\Leftrightarrow \text{expect}'(g, \text{win}'(m)) \wedge \text{expect}'(m, \text{win}'(g))$

2.4.9 *respectively*

(64a) George and Nick hate Martha and Honey, respectively.

LF: $\lambda x_1 x_2\,[\,\lambda y_1 y_2\,[\,\text{hate}'(x_1,y_1) \wedge \text{hate}'(x_2,y_2)\,]\,(m,h)\,]\,(g,n)$
$\Leftrightarrow \text{hate}'(g,m) \wedge \text{hate}'(n,h)$

(64b) George and Martha are drinking and dancing, respectively.

LF: $\lambda xy\,[\text{drink}'(x) \wedge \text{dance}'(y)]\,(g,m)$
$\Leftrightarrow \text{drink}'(g) \wedge \text{dance}'(m)$

2.5 Outlook and Special Problems

Many problems involving plurals could not be treated here. I already mentioned generic constructions. For the closely related topic of mass terms see Chapter 1 of this book and Krifka (1989b). Plurals are also highly relevant for a theory of simple and multiple questions (see Zaefferer (1984); de Mey (1984); von Stechow and Zimmermann (1984)), as well as for *comparative constructions*, viz. sentences like *more plutonium barrels were shipped than freight documents were issued*. There is also a rather interesting problem for plural theory which arises in relative clause constructions with multiple heads, called *hydras* in Link (1984); see this reference and von Stechow (1980).

Furthermore, there is certainly much more to be said about dependent plurals. Here I only tried to treat them with the means available in a first order theory like LP. But there is evidence that a more appropriate approach would have to introduce functions to take care of the relational character of these constructions. There is also a close connection with the rather complex issue of *plural anaphora*, which had to be left out here; see Rooth (1987) and the standard textbook on Discourse Representation Theory, Kamp and Reyle (1993).

The topic of distributivity has been pursued to a greater extent in the recent literature. I mention Roberts (1987b); Gil (1988) and Choe (1987a). Choe investigates a particle in Korean, called *anti-quantifier*, that triggers distribution over a plural phrase. In Chapter 5 a similar particle is discussed, the German *je*; it is seen there as a linguistic counterpart of the distributivity operator.

Semantic theories which are based on the concept of event have become more and more the focus of attention. Plurals are also very important in an event-based semantics. An example is the influence of plural object phrases on the aspectual class of the VP. As already mentioned in Section 2.2.1, this phenomenon is discussed in Hinrichs (1985) and Krifka (1987). For an application of the algebraic approach to the theory of events, see Bach (1986); Krifka (1987) and Chapters 10–12.

Finally, plurals have been treated in knowledge representation systems dealing with natural language processing; see Schütze (1989); Eberle (1989); Allgayer and Reddig-Siekmann (1990); Link and Schütze (1991).

Chapter 3

Hydras. On the Logic of Relative Clause Constructions with Multiple Heads

3.1 Introduction

One of the leading methodological principles in the semantics of natural language that I subscribe to is this: *In giving formal representations of the sentences you are analyzing try to be faithful to their syntax as much as possible.* This principle is inherited from Montague's work and receives justification from the fact that valuable insights into the structure of processing and understanding language can be gained if close-to-language representations are established which go through a series of logical transformations and, eventually, yield a logically simple form of the semantic content (as a rule, first-order formulas). In its strongest form this maxim leads to the principle of strict semantic compositionality which, although probably too strong, is still a useful vantage point for analysis. I have been pursuing this line of research in the last couple of years, and my logic of plurality and mass terms, *LPM*, is developed in this spirit. Now every account of plurality is faced with the problem of treating term phrases like

(1) the boy and the girl who met yesterday

where you cannot distribute the relative clause over the conjuncts. Upon inspection I found that such constructions are interesting enough to deserve a special name. Since they have more than one head I like to call them *hydras*. (In Greek mythology, hydra is, as the reader will recall, that huge serpent-like sea monster with many heads). In what follows I am going to present some ideas of how these hydras can be analyzed within the LPM framework.

A key role in building up hydras is played by phrasal conjunction. Therefore, it is important to become clear about the logic of *and* in such constructions. It is my claim that *and* is multiply ambiguous when used to connect terms. Consider the following sentences:

(2)
a. George and_1 Martha are asleep.
b. George and_2 Martha meet.
c. George and_3 Martha are husband and wife.
d. George, Martha, Nick and_2 Honey are playing games with each other.
e. George and_4 Martha, and_2 Nick and_2 Honey hate and_2 humiliate each other.
f. George and_4 Martha, and_2 Nick and_4 Honey earn the same.

The $\textit{and}_1$ in (2a) can be called *Boolean* since term conjunction here is really shorthand for sentence conjunction. In (2b), $\textit{and}_2$ connects two proper names to form a term denoting what I call the *individual sum* (i-sum, for short) of the individuals in question. Let's dub it the *i-sum and*, to have a name for it. (2c) is again different; here the term *George* and_3 *Martha* denotes something like an ordered pair since it is the order of the conjoined terms that matters: (2c) is not the same as *Martha and George are husband and wife*. (2c) is really a kind of *respectively* construction: *George* is linked with *husband*, *Martha* with *wife*. (Note, however, that *husband and wife* is not simply an enumeration of two isolated predicates; connected by *and*, the constituent terms rather mutually "bind" their covert second argument place. Thus, strictly speaking, there is still another use of *and* involved here. For an analysis see below.) The subject term in (2d) denotes just the i-sum of the four individuals mentioned, whereas in (2e), a comma cannot replace the *and*. This sentence is rather about two groups, George and Martha on the one hand, and Nick and Honey on the other, and it says about the members of each group that they hate and humiliate each other. So we got three different *and*s here. $\textit{and}_4$ is pretty much the same as $\textit{and}_2$, though, and will, in the final analysis, be very similar to it. But some device of grouping is needed to distinguish this case from (2d). Finally, (2f) is like

(2e), but there is something collectively predicated here of the i-sum of the two groups.

With the arsenal of individuals i-sums, groups, and i-sums of groups at our disposal we can attack the following sample of hydras given in (3) – (10):

(3) The German or Austrian who lives next door wants to talk to me.

(4) The cabinet member and mafioso who was involved in the scandal resigned.

(5) A boy and a girl who met in the park have disappeared.

(6) The boy and the girl who dated each other are friends of mine.

(7) Students and workers who had gathered issued a resolution.

(8) All the students and some of the professors who had met in secret joined in underground activities after the coup d'etat.

(9) The landlords and the tenants who fight against each other will all come together tomorrow in an attempt to find a compromise.

(10) Some relative of each villager and some relative of each townsman who know each other well are asked to discuss ways of improving relations between their communities, which still suffer from the discovery of branching quantification.

3.2 The Frame of Analysis

The model theory I assume here recognizes as individuals not only objects like you and me but also *individual sums* of usual singular objects ($a \oplus b$ is the i-sum of a and b and $\sigma x P x$ the i-sum of all the Ps). Such i-sums are not sets of objects but rather made of the same stuff, so to speak, as their constituent parts. My reasons for not taking sets as the denotations of pluralities are at least fourfold and can be summarized as follows: They are (i) *practical*: If my kids turn the living room into a mess I find it hard to believe that a set has been at work, and my reaction to it is not likely to be that of a singleton set; (ii) *ontological*: due to a proper philosophical abstraction from experience as under (i) I came to believe that sums or collections of concrete objects remain concrete, i.e., more generally, sums are of the same kind as the objects they are made out of; (iii) *semantic*: Identification with sets leads to wrong predictions such as the truth of (11).

(11) Peter, Paul, and Mary have three elements;

(iv) *linguistic*: many predicates cannot be categorized according to the kind of argument they take (i.e. singular objects or pluralities). Thus the question

(12) Who turned the living room into a mess?

can be truthfully answered by "those two little brats" as well as by "my younger son all by his own". Therefore, the denotations of singular and plural objects should be treated on a par.[1]

Finally. having established individual sums in the ontology, I am going to introduce a further sort of object, the *group of individuals*, since it appears that we also refer to sums of groups as seen above. The group made up by a and b I shall represent by $\langle a \oplus b \rangle$ (for its exact nature as far as it is specified, see below). With this notation, and the Boolean meet of predicates, written '$\odot$', I can give a Montague-style analysis of the examples under (2). Let **U** be the translation relation. Then we have, with obvious abbreviations for the proper names:

(13) a. George and$_1$ Martha **U** $\lambda P\,[P(g) \wedge P(m)]$

[1]More explicitly, what I really mean here is this: my view of the semantics of predicate logic (or type logic, for that matter), when applied to the study of language, is such that we have an underlying set theory with urelements as our metalanguage. Now if individuals are urelements sums of individuals are just more urelements bearing a certain relation to their constituent parts. I am aware that this intuition is by no means necessary for foundational purposes (witness ZF set theory doing completely without urelements); but it suggests itself rather strongly when we think about the way we structure the external world, and it is mainly this world, and not the highly theoretical context of the foundations of mathematics that language is about. So (11) intends to hint at a category mistake: In English I am able to form the term *Peter, Paul, and Mary*, as well as the term *the set consisting of Peter, Paul, and Mary*. These terms are not the same: the latter, unlike the former, denotes an abstract entity which has a cardinality, is outside of space and time, etc. I am not saying that it is impossible to treat pluralities as sets as long as atomic individuals are also taken to be sets (so that the linguistic demands under (iv) are met); this is because the ordering structure of the underlying atomic Boolean algebra is fully represented by the power set algebra over the set of atoms. However, a more systematic argument drawn from the striking analogy between plurals and mass terms, put forth in Link (1983a) (the present Chapter 1), shows that the strategy of "ontological abstinence", as I like to call it, fits nicely into the algebraic approach to these phenomena, which, in the case of mass terms, cannot be avoided anyway without changing the underlying set theory (but see Bunt (1981) where just this latter alternative is chosen). Let me summarize: Ontologically, I favor urelements in the semantics of natural language, which can be further classified, among other things, into concrete and abstract objects; from a purely structural point of view, I only claim that whatever your favorite vehicles of denotation for the individual terms are, you'll hit on more objects of the same kind, algebraically related to the former, when you consider the denotations of pluralities.

b. George and$_2$ Martha **U** $\lambda P\, P(g \oplus m)$

c. George and$_3$ Martha **U** $\lambda R\, R(g, m)$

d. George, Martha, Nick, and$_2$ Honey **U**
$\lambda P\, P\,(g \oplus m \oplus n \oplus h)$

e. George and$_4$ Martha, and$_2$ Nick and$_4$ Honey hate and$_1$ humiliate each other **U**
$\lambda P\,[P(g \oplus m) \wedge P(n \oplus h)]\,(\hat{}\text{hate.e.o}' \odot \text{hum.e.o}')$

f. George and$_4$ Martha, and$_2$ Nick and$_4$ Honey **U**
$\lambda P\, P\,(\langle g \oplus m\rangle \oplus \langle n \oplus h\rangle)$

Case a) can be subsumed under case b) since I certainly want also to be able to treat *the children* in one way regardless of the kind of predicates that might be attached to the term. Thus, *the children* should generally denote the i-sum of the children; in the special case in which the predicate is distributive, as in (14a), the equivalence to (14b) is secured by a suitable meaning postulate.

(14) a. The children are asleep.

b. Every child is asleep.

In the same way, then, I get down from (2a) to

(15) George is asleep and Martha is asleep.

So translation (13a) can be dispensed with. I skip (13c) for a moment and go on to d), which is of the same kind as (13b); the difference is simply that four individuals are involved now that build up the i-sum $g \oplus m \oplus n \oplus h$ on the right-hand side. But in (13e) and (13f), we got something new, I think. (13f) is about a sum of two groups which is different from the pure i-sum under d) (think, for instance, of the annual income of two married couples in the sentence (2f) which is being analyzed here). To indicate the groups, I have put brackets around $g \oplus m$ and $n \oplus h$, respectively. The group $\langle g \oplus m\rangle$ is essentially like $g \oplus m$ except that it denotes again an atom in the domain of individuals. The concept of a model has to be slightly amended to incorporate this feature into the semantics. I shall say something about it presently. But let me first go back to (13e) (the sentence I made up just to get three different *and*s in) and point out that the Boolean predicate hate.each.other$'$ $\odot$ humiliate.each.other$'$ just happens to distribute down to the subsums $g \oplus m$ and $n \oplus h$. Hence the analysis analogous to (13a). But *each-other* constructions can refer to sums of groups also, as (16) shows. The liking is part of the external family relations; nothing is said about the internal affairs.

(16) a. The Leitches and the Latches like each other.

b. $\lambda P\, P\,(\,\langle \sigma x\, \text{Leitch}'(x)\rangle \oplus \langle \sigma x\, \text{Latch}'(x)\rangle\,)$

So the subject term in (16a) should be analyzed as is given in (16b). Now I propose to dispense with the translation (13e) just in the same way as in the case of (13a). The translation is not wrong but it leads to something that can also be got from structure (13f) just in case there is a predicate attached to it that distributes down to the constituent groups. But how do I recognize distributing predicates? Well, this is a question of context. Perhaps we might want to say, for instance, that in (16a), the predicate does in effect distribute: The Leitches live in perfect harmony, and so do the Latches. This is an ambiguity that I can account for, as it turns out, by either starring or not starring the predicate in question. There is even a third reading which is brought out more clearly if you put a floating quantifier in, like in

(17) The Leitches and the Latches *all* like each other.

Here we are referring to the sum $\sigma x\,[\text{Leitch}' \oplus \text{Latch}'](x)$ which is pure, that is, does not contain groups as parts.

Let me summarize: Conjoined term phrases are at least two-ways ambiguous: They either refer to pure individual sums or sums of groups.

But there is still case (13c) left open for analysis. I am not going to say much about it. Above I said that the term phrase here is not symmetric. So I tentatively propose to use higher-dimensional λ-abstraction to formalize the phrase *husband and wife*:

(18) $\lambda xy\,[\,\text{husband}'(x,y) \wedge \text{wife}'(y,x)\,]$

Accordingly. *Georges and*$_3$ *Martha* translates into the set of relations R such that $R(g,m)$, i.e., g bears R to m. If one puts the two things together he gets what is wanted intuitively.

I now explain roughly what the structure of the domain of individuals, E, in a model is going to look like. I want to represent groups in E, but I shall not define them. Rather I state the minimal conditions they must satisfy for them to play the intended role in our analysis. The condition is simply this: For every usual i-sum a, there is an atom $\gamma(a)$ that uniquely represents the group consisting of a. Since there is no need to iterate the process I shall distinguish between *pure* individuals (those that do not contain a group as i-part) and arbitrary individuals, viz. groups or sums with at least one group as i-part. Let A be the set of atoms of the set E which itself is taken to be a complete atomic Boolean algebra just as in

LPM. Then A is to consist of two disjoined sets, B, the set of pure atoms, and G, the set of groups. Further there exists a one-to-one function γ of the set of proper i-sums built up from elements of B, into C. This structure has simple finite models. Thus, if there are two pure atoms x,y we have three pure individuals, x,y,$x \sqcup_i y$, and at least one group $\gamma(x \sqcup_i y)$. Here $\sqcup_i$ is the semantic counterpart of $\oplus$. We have the semantic rule (with $\|\cdot\|$ standing for the denotation function of the model):

(19) $$\|\langle a \oplus b \rangle\| \;=\; \gamma(\,\|a\| \sqcup_i \|b\|\,)$$

Since reciprocal constructions with *each-other* are involved in half of the examples I should explicitly state a meaning postulate which eliminates the collective reciprocal predicate in favor of the underlined transitive verb.[2]

Meaning postulate for EACH-OTHER: Let α be a transitive verb, α' its translation in LPM. The collective predicate 'α *each other*' may translate into EO(α'). Let $x\,{}^{\cdot}\Pi\, y$ stand for 'x is an atomic i-part of y." Furthermore, for any individual a, let a' be b if a is a group of the form $a = \langle b \rangle$, and a otherwise. Then the following formula is valid in LPM and, *mutatis mutandis*, in a PTQ style intensional logic:

(20) $$[\mathrm{EO}(\alpha')](a) \;\leftrightarrow\; \forall x \forall y\,[\,x\,{}^{\cdot}\Pi\, a' \wedge y\,{}^{\cdot}\Pi\, a' \wedge x \neq y \rightarrow \alpha'(x,y)\,]$$

3.3 Analysis of the Hydras (4) – (10)

I assume the basic PTQ framework and notation,[3] together with the amendments necessary to incorporate plural expressions along the line of my LPM system.[4] The hierarchy of denotation sets is assumed to display a complete atomic Boolean structure for every type, with the set of i-sums of individuals being contained in the set of denotations of type e of entities. As above I use the symbol '**U**' for the translation relation; $\alpha\mathbf{U}\alpha'$ stands for "α *translates into* α'."

---

[2]I am aware that this rule does not cover all the *each-other* constructions possible. More complex examples are (i) *MIT linguists only quote each other's papers* or even (ii) *they read each other's books in each other's languages* (a sentence from Higginbotham (1980); for an extensive account on reciprocals within the generative grammar framework, see Fiengo and Lasnik (1973)). Here, the constructions with two or more argument places do not reduce to 1-place predicates. These cases should be analyzed in the context of *dependent plurals* like *unicorns have horns*. The latter phenomenon, although of utmost importance for a complete account of plurality, had to be ignored here.

[3]See Montague (1973a); Link (1979); Dowty *et al.* (1981).

[4]See Chapter 1.

Starting with sentence (3) we observe that the phrase *German or Austrian* is obviously the Boolean join of the CN phrases *German* and *Austrian*:

(21) German or Austrian **U** $\text{German}' \oplus \text{Austrian}'$
$\equiv \lambda x\,[\,\text{German}'(x) \vee \text{Austrian}'(x)\,]$

Since this does not produce a plural structure we may call (3) a pseudo-hydra. Adding a relative clause is standard here. Thus, the whole subject term is translated as follows:

(22) the ((German or Austrian)[1] such that he[1] lives next door) **U**
$\lambda P\,\exists y\,[\forall x\,[[\text{German}' \oplus \text{Austrian}'(x) \wedge \text{lives.next.door}'(x)] \leftrightarrow x = y] \wedge P(y)\,]$

The hydra of sentence (4) is translated in a similar manner, the Boolean meet $\odot$ replacing the join.

(23) the ((cabinet member and mafioso)[1] such that he[1] was involved in the scandal) **U**
$\lambda P\,\exists y\,[\forall x\,[[\text{cm}' \odot \text{maf}'(x) \wedge \text{involved}'(x)] \leftrightarrow x = y] \wedge P(y)\,]$

Genuine hydras, i.e., plural structures, start with sentence (5). I shall discuss the slightly more complicated sentence (6). Here it is only the restrictive relative clause that serves to pick out the unique individual sum of a boy and a girl that the sentence refers to. The structure is built up accordingly: In spite of the double occurrence of the determiner *the*, definite description is introduced only once:[5]

(24) boy and girl **U** $\lambda z\,\exists x\exists y\,[\,\text{boy}'(x) \wedge \text{girl}'(y) \wedge z = x \oplus y\,]$ (i)

(boy and girl)[1] such that they[1] dated each other =: ζ **U**
$\lambda z\,\exists x\exists y\,[\,\text{boy}'(x) \wedge \text{girl}'(y) \wedge z = x \oplus y \wedge \text{dated}'(z)\,]$ (ii)

The boy and the girl such that they dated each other **U**
$\lambda P\,\exists w\,[\forall u\,[\,\zeta'(u) \leftrightarrow u = w\,] \wedge P(w)\,]$ (iii)

According to (i), *boy and girl* denotes a set of sums each consisting of some boy and some girl (whence the existential quantifiers $\exists x$ and $\exists y$ in the translation). The logical type of this phrase is *et* and is thus subject to usual relative clause formation (ii) and term formation (iii). Here, the definite article is introduced; for comparison, *a boy and a girl* would translate into $\lambda P\,\exists w\,[\zeta'(w) \wedge P(w)] \equiv \lambda P\,\exists x\exists y\,[\,\text{boy}'(x) \wedge \text{girl}'(y) \wedge P(x \oplus y)\,]$, *every boy and girl* into $\lambda P\,\forall w\,[\zeta'(w) \rightarrow P(w)] \equiv \lambda P\,\forall x\forall y\,[\,\text{boy}'(x) \wedge \text{girl}'(y) \rightarrow P(x \oplus y)\,]$. Note that expression (24) (iii) is equivalent to

[5]I did not write up explicit syntactical and translation rules; their formulation shouldn't cause much difficulty in principle as soon as we got the semantics straight.

(25) $\lambda P \, \exists x \exists y \, [\forall v \forall w \, [\, [\text{boy}'(v) \wedge \text{girl}'(w) \wedge \text{dated}'(v \oplus w)] \leftrightarrow v = x \wedge w = y \,] \wedge P \, (x \oplus y) \,]$

so what I did here amounts to something like the simultaneous binding of two variables. I may note in passing that this is recognized by a number of people as also being the right tool to handle the Bach-Peters type sentences.[6]

In (7) we have pluralities being conjoined. The sentence refers to a sum consisting of some students and some workers which had gathered. Here is the analysis ('$\emptyset_{\text{pl}}$' stands for the null determiner in the case of indefinite plural phrases, and '*' is the plural operator in LPM):

(26) (students and workers)[1] such that they[1] had gathered =: ζ **U**
$\lambda z \, \exists x \exists y \, [\, {}^{*}\text{student}'(x) \wedge {}^{*}\text{worker}'(y) \wedge z = x \oplus y \wedge \text{had.gathered}'(z) \,]$ (i)

$\emptyset_{\text{pl}}$ **U** $\lambda P \, \exists z \, [\, \zeta'(z) \wedge P(z) \,]$
$\equiv$
$\lambda P \, \exists x \exists y \, [\, {}^{*}\text{st}'(x) \wedge {}^{*}\text{wo}'(y) \wedge \text{gather}'(x \oplus x) \wedge P \, (x \oplus y) \,]$ (ii)

The hydra of sentence (8) is more difficult to handle: The two heads have different determiners. Still the case is intuitively clear. There is a sum K of students and professors who had met in secret. Now some subsum consisting of all the students and some of the professors in K engaged in common activities. So the analysis should follow the same lines, with one novelty: The two determiners have to be introduced simultaneously. The translation of the present example, then, runs as follows.

(27) students and professors who had met in secret =: η **U**
$\lambda z \, \exists x \exists y \, [\, {}^{*}\text{student}'(x) \wedge {}^{*}\text{professor}'(y) \wedge z = x \oplus y$
$\wedge \text{had.met.in.secret}'(z) \,]$ (i)

Now the partitive construction hints at a preassigned unique student-professor sum with the property expressed in the relative clause:

the η **U** $\sigma w \, \eta'(w)$ (ii)

Next we pick out all the sums of students in $\sigma w \eta'(w)$; of such a sum the following predicate holds (with $a^{\downarrow} := \lambda x . x \, \Pi \, a$)

(28) $\alpha \doteq {}^{*}\text{student}' \odot (\sigma w \eta'(w))^{\downarrow}$

[6]See Thomason (1977) and Blau in unpublished work; for simultaneous λ-abstraction see(Link, 1979, p. 98–109).

This is because if v is an α, v is at the same time an i-sum of students and part of the sum K. Similarly with the professors among the η:

(29) $\beta \doteq {}^{*}\text{professor}' \odot (\sigma w \eta'(w))^{\downarrow}$

We are now ready to translate the complete subject term of our example (27).

all of the students and some of the professors who had met in secret $\mathbf{U}$ $\lambda P\, \exists x \exists y\, [x = \sigma u \alpha(u) \wedge \beta(y) \wedge P(x \oplus y)]$ (iii)

Remark. Upon reflection it seems to be no accident that it is very hard to find natural examples with two different determiners as in the sentence just treated. Since the two nouns are inseparably connected through the common tail of the hydra they cannot really undergo a completely independent quantificational binding any more. So I think that it is the partitive that makes the present construction possible at all.[7] There is, however, one type of phrases where there is no partitive involved.

(30) Three students and four professors who had met in secret

But these hydras lend themselves to a natural treatment within the present framework along the following lines. The number words *three* and *four* are not really quantifiers but adjective-kind modifiers which serve to pick out exact those subsums that contain three (four) atoms.[8] Only in a second step there is a quantification, which is existential and happens to be syntactically invisible as in example (26). Thus the analysis of (30) runs as follows.[9]

[7] A recent semantic account of the partitive construction in English can be found in Ladusaw (1982) which is formulated within the generalized quantifier approach as introduced into the linguistic literature byBarwise and Cooper (1981). I may note that Ladusaw also recognizes the ambiguity between what I called the Boolean and the i-sum *and*; but then his purely set-theoretical analysis immediately forces him to split up the type structure into an "entity level" and a "group level," which is an undesirable consequence because of the existence of mixed entity/group extensions of predicates (see Chapter 1). On my view, a sum denoting term like *John and*$_2$ *Mary* simply gives rise to a new principal filter with core set $\{j \oplus m\}$ which neither includes nor is included by either one of the two individual filters generated by the constituent parts of the sum (see also Link (1983b)). Once the inadequacy concerning the two levels is removed then Ladusaw's analysis, especially that of the distinction between *both* and *the two*, seems to be correct.

[8] Let ζ be an element of P_{CN}, $\zeta\, \mathbf{U}\, \zeta'$, ζ_{pl} its pluralization of category PCN ("plural CN"), μ a syntactical variable running over the number words *two*, *three*, *four*, etc., and N be any numeral > 1; then the expression '$\mu\, \zeta_{pl}$' is again of category PCN, and the translation μ' of μ, which is an appropriate N, denotes a transformation on the set of PCN denotations. The denotations of $(\mu\, \zeta_{pl})' \doteq \mu'(\zeta_{pl})' \doteq \mu'({}^{*}\zeta') \doteq N({}^{*}\zeta')$, then, is to be the set of all sums in the denotation of ${}^{*}\zeta'$ that are composed of exactly N atoms.

[9] For the first two steps see the preceding footnote; the third step is as in (27), (i).

(31) three students **U** $\mathbf{3}({}^*\text{student}')$

four professors **U** $\mathbf{4}({}^*\text{professor}')$

three students and four professors who had met in secret **U**
$\lambda z\, \exists x \exists y\, [\,[\mathbf{3}({}^*\text{stud}')](x) \wedge [\mathbf{4}({}^*\text{prof}')](y) \wedge z = x \oplus y \wedge \text{had.met.in.secret}'(z)\,]$

The next example contains a different problem. In our system i-sums are such that there are no in-between levels of comprehension that can be recognized. Now *landlords and tenants etc.* is a plural phrase that intuitively refers to a sum of groups each consisting of a landlord and a tenant; and it is only in those mixed groups of size 2 that there is a fighting going on (let us assume that at least the tenants practise solidarity among themselves). But if I translate the phrase into a simple sum, the mixed subsums cannot be recognized anymore. It is here now that the adjustment in the system envisaged above comes in. We have to close off the landlord-tenant sums as groups before the whole sum is formed. Here then, is the translation of the hydra in (9).

(32) (landlord and tenant)[1] such that they[1] fight each other =: ζ
U $\lambda z\, \exists x \exists y\, [\,\text{land}'(x) \wedge \text{ten}'(y) \wedge z = \langle x \oplus y \rangle \wedge \text{fight.e.o}'(z)\,]$ (ii)

landlords and tenants who fight against each other **U** ${}^*\zeta'$ (iii)

As can be seen from the last line pluralization comes in only at the very end.

At this point it may well be asked why the new group machinery is being introduced when the good old sets with just the right comprehension properties have been at our disposal all along. Well, I think, I have already given arguments that the plural objects should not be identified with sets. Moreover, they provide too much structure that is not needed here. There are, for instance, no predicates of sums of groups of groups, etc. I prefer to make just the minimal change in the ontology that is necessary to account for the facts.

Now how about the villagers and the townsmen? Well, since each obviously has wide scope the phrases *each villager* and *each townsman* are going to be introduced by a substitution rule for terms. It is like Montague's rule F_{10} except that the two terms are quantified in at the same time. The analysis which will be carried through presently shows that we come up with something that corresponds to the dominant first order reading of the example at hand. Cases of genuine branching quantification[10] I do not treat here. Finally, note that the analysis in terms of i-sums natu-

[10] Cf. Barwise (1979).

rally reflects the symmetry of the coordinate conjoined structure, a feature that has been explicitly pointed out by Hintikka and others.[11]

(33) (relative of he_0 and relative of $\mathrm{he}_1)^2$ such that they^2 know each other well $=: \zeta$ **U**
$\lambda z\, \exists x \exists y\, [\, \mathrm{rel}'(x, x_0) \wedge \mathrm{rel}'(y, x_1) \wedge z = x \oplus y \wedge \mathrm{know.e.o}'(z)\,]$ (i)

some relative of he_0 and some relative of he_1 such that they know each other well $=: \alpha$ **U** $\lambda P\, \exists z\, [\, \zeta'(z) \wedge P(z)\,]$ (ii)

$F_{0,1}(\,\text{each villager, each townsman}; \alpha\, [\mathrm{he}_0, \mathrm{he}_1]\,) \quad \doteq$
some relative of each villager and some relative of each townsman such that they know each other well **U**
$\lambda P\, \forall u \forall v\, [\, \mathrm{vil}'(u) \wedge \mathrm{tm}'(v) \rightarrow \lambda x_0 \lambda x_1 [\alpha'(P)](u)(v)\,] \quad \equiv$
$\lambda P\, \forall u \forall v \exists x \exists y\, [\, \mathrm{vil}'(u) \wedge \mathrm{tm}'(v) \rightarrow$
$[\mathrm{rel}'(x, u) \wedge \mathrm{rel}'(y, v) \wedge \mathrm{know}'(x \oplus y) \wedge P(x \oplus y)]\,]$ (iii)

3.4 Conclusion

The analysis has shown how some of the trickier cases of plurality can be accommodated in a rather natural way within the lattice-theoretical approach underlying LPM. As has become apparent I have stuck to a CN-S-analysis of restrictive relative clauses throughout; in fact, I think that the tension that is *prima facie* felt between the syntactically simple T-S-analysis and the semantically sound CN-S-analysis should be resolved in favor of the latter. I am confident that such a move does not engender too much syntactical adhockery because the only new type of rule, simultaneous quantification, is most likely needed in the grammar anyway. Finally, putting the present work into the perspective of the compositionality principle, it seems to me that another challenge to this principle, viz. the one posed by hydras, turned out to be no real threat to it after all. This is encouraging for those who remain undismayed by defeatist rhetoric like "do you really still believe in compositionality?," and continue to use some reasonable form of compositionality in their work.[12]

---

[11] See Hintikka (1973).

[12] For a representative overview over the problems and possible solutions as well as new relevant data connected with the principle see Partee (1984).

Chapter 4

Generalized Quantifiers and Plurals

4.1 The Logic of Plurals, LP: Review of the Basic Ideas

I would like to start with reviewing some of the basic intuitions behind the system LPM, the logic of plurals and mass terms, as it is described in Link (1983a).[1] Since I shall be concerned with plurals only in this chapter I'll skip the mass term part of the system. The remaining logic might be called LP.

LP is a first order logic introducing a sum operation for its individual terms. A sum term, say $a \oplus b$, is supposed to denote a new entity in the domain of individuals which is made up from the two individuals denoted by a and b. Call these $\|a\|$ and $\|b\|$, respectively. Then $a \oplus b$ does not denote the set consisting of $\|a\|$ and $\|b\|$ in LP but rather another individual of the same kind as $\|a\|$ and $\|b\|$.

So LP intends to assimilate plural objects to individuals rather than to sets of individuals. This approach is reminiscent of the nominalist project of reducing properties to particulars, and indeed, the purely formal devices used are quite similar (see, e.g., Eberle (1970) for the background on nominalistic systems). I would like to stress, however, that the rationale for LP is quite different. The philosophical task of the nominalists was to do away with a stratified ontology. The resulting picture was one in which, say, the term 'white' was to stand for just the sum of all white objects. For

[1] Chapter 1 of this book.

those who like me think that language is the primary source for forming intuitions about our ontology there is a clear reason why the nominalists' project failed: According to the way language works the world is simply not a huge heap of particles of just one sort. To put it more formally: there is more structure in the universe than just one part-whole relation of the kind the nominalists used.

Thus, classes or sets are of course recognized by LP as the denotations for predicate expressions. On the other hand, however, there is something most useful that's implicit in the nominalists' approach, and that is the *algebraic* point of view. It is this aspect of the philosophical tradition that is taken up in LP and put to use for the semantical analysis of plurality in language.

This picture naturally leads to a domain of individuals which is internally structured. All one really needs for LP is a set E together with a two-place operation on E forming individual sums (*i-sums* for short). Call this operation $\vee_i$. Then LP requires the ordered pair $\langle E, \vee_i \rangle$ to form a *semi-lattice*.[2] The syntactic counterpart of the i-sum operation is the circled plus $\oplus$. It is also convenient to demand that the lattice structure be complete, i.e., that E be closed under arbitrary i-sums. This allows us to introduce an abstraction operation for arbitrary 1-place predicates, called 'σ', to form individual terms of the form $\sigma x P x$. Such a term denotes the i-sum of all individuals that are Ps. 'σ' is actually a natural generalization of the usual ι-operator: If the extension of P is a singleton set then $\sigma x P x$ and $\iota x P x$ denote the same thing, viz., the unique element in this set.[3] Finally, the intrinsic ordering relation on E induced by $\vee_i$, called $\leq_i$, is a partial order and corresponds to the part-whole relation of the nominalists. Note, however, that most things that are intuitively called parts in our world, like my arm being a part of my body, are *not* i-parts in the sense of the $\leq_i$-structure. Not even group terms have i-parts, according to LP; the $\leq_i$-structure is only to take care of the pluralities. Thus, John is an i-part of the individual sum consisting of John and Mary, for instance, but a committee doesn't have i-parts in LP. This fact about group terms is certainly unsatisfactory from a semantical point of view, but formally, *committee* is singular, and therefore a committee is an atom in the lattice. A special semantics of group terms should add more structure here. Let me note that in what follows the term *group* is used as an informal synonym to the technical term *i-sum*.

Now what is the linguistic evidence in favor of this kind of set-up? The main trait of the system is that it has just one sort of variables for 'normal'

---

[2]Actually a semi-lattice of a special sort; see Chapter 6.

[3]One could take this as a definition of ι in terms σ. This is perhaps more natural than doing it the other way around, as in Chapter 1.

individuals and i-sums alike. This allows for genuine plural quantification where the quantifier runs over all the i-sums in the domain of individuals. With this device at hand the following linguistic phenomena involving plural constructions can be treated in a natural way:

1. There is a **recursive plural operator** in the semantics. It is denoted by '*'. As a prefix to a 1-place predicate P it forms all the possible i-sums from the members of the extension $\|P\|$ of P. In the case of the common noun *man*, for instance, we have the recursive rule $\|{}^*man'\| = \mathrm{cl}_{\oplus}(\|man'\|)$ (here, man' is the translation of the CN *man* in LP, and $\mathrm{cl}_{\oplus}$ stands for the described operation of closure under i-sums).[4]

2. There is just **one definite article** for both singular and plural NPs, not only in the lexicon, but also in the semantics. This is due to the fact, as was already hinted at above, that the ι-terms are only special cases of the σ-terms which denote uniquely specified i-sums. These are the suprema of the extensions of the predicates in question. If the predicate is *children*, then the supremum of all sums of children (which, according to 1., make up the extension of *children*) is the unique sum of all the children under consideration; and in the singular case, if the predicate is *child*, say, the supremum coincides with the atom in the semilattice that is the contextually specified child.

3. The existence of **mixed predicates** with respect to the distributive/collective distinction. Predicates like *lift a rock* can be true of i-sums and atoms alike. As a result, one expects there to be an ambiguity in a sentence like *three men lifted a rock*, and indeed, there is: there might be either three liftings involved or a single one depending on the distributive or collective interpretation chosen.

4. **Partitive Constructions** like *three of the students* are treated in terms of the i-part relation $\leq_i$. Its syntactic counterpart in LP is the special 2-place predicate constant 'Π', together with the variant '${}^{\bullet}\Pi$'. The former is read '... is an i-part of —', the latter '... is an *atomic* i-part of —'. So, for instance, the difference in LP between *three students* and *three of the students* would be this. Using the first phrase you introduce a sum of three students into the discourse, whereas with the second you pick out a sum of three from a contextually provided, already familiar group of students. How exactly the notion of familiarity is made precise is not for LP to spell out; Discourse Representation Theory and File Change Semantics naturally

[4]There should be made a distinction, however, between syntactic and semantic plurals. In many languages some plural forms are simply triggered by agreement or even more ideosyncratic features of the lexicon. Following a suggestion of Manfred Krifka, I rather like to think of the star as a kind of classifier expressing the semantic property of *countability*. Note, incidentally, that "singular entities" appear as atomic individuals in the extension of a starred predicate like ${}^*man'$.

come to mind here, and such a theory, together with LP, will be able to take care of it.

5. **Hydras**, i.e., relative clause constructions with multiple heads like *the man and the woman who met in the park*. I have treated these constructions elsewhere; see Link (1984). Basically, I introduce a complex nominal there of the form *man and woman* which denotes the set of all sums consisting of one man and one woman. These nominals can be modified by a relative clause containing a collective verb phrase like *meet* in a natural way; thus the phrase *man and woman who met in the park* denotes just the set of sums z of a man x and a woman y such that z met in the park. The final step is the assignment of a determiner like the definite article. I take it that although *the* appears twice in the above noun phrase it is only and precisely one sum z of the indicated kind that is picked out by the NP; but since sums are regular individuals, the normal ι-operator can just do that.

6. Simple **Plural Anaphora** behave much the same as in the singular case, and this parallelism is captured by the present account. Consider sentence (1), for instance.

(1) a. When John$_i$ got married he$_i$ was a student.

b. When (John and Barbara)$_i$ married they$_i$ were students.

The accounts in (1a) and (1b) run completely parallel; in (1b) the name phrase *John and Barbara* picks out a unique individual, the i-sum of John and Barbara, which can henceforth be referred back to by a plural pronoun. Note that the fact that the antecedent sentence contains the collective verb *married* and the consequent sentence the distributive verb *is a student* does not cause any problems in LP: First of all, we have to observe that the verb phrase that is really there is rather *were students*, and this is treated as the plural-starred version of *is a student* (ignoring tense). But then, the sentence (1b) says in its second part that the i-sum consisting of John and Barbara is a member of the set of i-sums each atomic part of which was a student. This is not only true given the truth of the claim made by (1b), but it also entails, as it should, that John was a student and Barbara was a student.

We also have, in the plural domain, both backward pronominalization and anaphoric linkage across sentence boundaries when the antecedent plural NP is a "singular NP" in the logical sense (see Barwise (1987)). Examples are:

(2) a. When they$_i$ arrived (John and Mary)$_i$ sat down and talked.

b. When they$_i$ arrived the friends$_i$ sat down and talked.

(3) a. (John and Mary)$_i$ finally met at John's place. They$_i$ sat down and talked.

b. The friends$_i$ finally met at John's place. They$_i$ sat down and talked.

(4) a. Their$_i$ teacher really likes (John and Mary)$_i$.

b. Their$_i$ teacher really likes the children$_i$.

For 'general' plural NPs like *all students*, *no students* or *most students* this is not possible, or at least intuitions seem to differ widely. For those who are able to get backward pronominalization with general plural NPs some kind of explanation might be offered by saying that *all students* and *no students* are really interpreted as *all of the students* and *none of the students*, respectively; but then, of course, *the students* is taken to denote a specific, predetermined (sum-)entity, and so there is something the pronoun can be hooked up on like in the case of a singular NP. Here are some examples.

(5) a. ?When they$_i$ arrived all students$_i$ sat down and started working.

b. When they$_i$ arrived all the students$_i$ sat down and started working.

(6) a. ?All students$_i$ were in by 9 o'clock. They$_i$ sat down to work.

b. All the students$_i$ were in by 9 o'clock. They$_i$ immediately started the discussion.

(7) a. ?No students$_i$ were to be seen in the lecture hall. They$_i$ had gathered outside.

b. Before they$_i$ register no students$_i$ are allowed to enter Maples Pavilion.

(example due to Ivan Sag)

(8) a. *Their$_i$ teacher likes all students$_i$.

b. *Their$_i$ teacher likes most students$_i$.

It is interesting, though, that also sentences of the following kind seem perfectly acceptable:

(9) (Every student)$_i$ was in by 9 o'clock. They$_i$ sat down to work.

Here the group serving as antecedent for the plural pronoun is not exactly denoted by any part in the first sentence; but it is conceptually obvious that along with processing the general singular NP *every student* the group consisting of all the contextually relevant students is formed in such a way as to make it possible for later discourse to refer back to it. We will observe a similar phenomenon below.

A real interesting problem with plural anaphora is this. Since plural phrases denote groups consisting of more than one atomic individual, there might be indefinite (and hence, in Barwise's sense, 'singular') plural NPs like *three researchers* which, when modified, display a scope behavior just like general NPs. Mats Rooth (p.c.) addressed this particular question, his paradigm example being (9),

(10) Three researchers with two microscopes (each) use them both.

It should be evident that an account of plural anaphora that is able to incorporate such data constitutes a challenging research goal on its own. Attempts to tackle these problems have started only recently; I trust, however, that a system like LP provides the proper framework for this task.

7. **Genuine plural quantification**. This topic will be addressed below. While there seems to be no problem in finding cases of existential quantification over sums universal sum quantification is harder to come by. The German particle *je* seems to provide examples for the latter. This is why I shall discuss some of the data involving *je* towards the end of the chapter. For a more detailed account on *je* see Chapter 5.

4.2 Lifting LP into the Generalized Quantifier Framework

Once we have gained a clear understanding of the way the semantics of pluralities works in language, the first order 'model' (in the sense Jon Barwise likes to use this term) of the logic of plurals has served its purpose; that is to say, there is no principled reason why we should 'overload' the role of first order logic in claiming any kind of priority for it regarding its role of providing logical forms in the grammar. But it *is* still highly useful to serve as a kind of *lingua franca* in which the basic ideas can be explained. This I have tried in the case of plurals, and I think possible confusions about the exact level on which pluralities should be located have not been that likely to occur.

Now I would like to take the next step and try to make my analysis more compatible with one of the prevailing frameworks of current semantic research, i.e., the theory of generalized quantifiers (GQ theory, for short; 'GQ' will also be used for the theory itself).

To begin with, in LP we have the 2-place function constant '$\oplus$' creating complex names of the form $a \oplus b$. Since in GQ names correspond to sets of sets, one of the first questions to ask is what these new names should be made to stand for; put less formally: what is the proper denotation for an i-sum term like *John*$\oplus$*Mary*? Consider first a simple example:

(11) a. John and Mary are students.

b. John is a student and Mary is a student.

Here it is obvious that (11a) is equivalent to (11b), so the filter denoted by the conjoined NP *John and Mary* should be just the intersection of the principal filters generated by $\|John\| =: j$ and $\|Mary\| =: m$, respectively. Thus we have the equation

(12) $\|John\ and\ Mary\| = \mathbf{F}_j \cap \mathbf{F}_m$

where

(13) $\mathbf{F}_x := \{X \mid x \in X\}$

Now being a student is a property that is shared by John and Mary, so its extension is indeed in the intersection of $\mathbf{F}_j$ and $\mathbf{F}_m$. But in a sentence with a collective predicate in it this is not true any more, witness (14).

(14) John and Mary met in Munich

Meeting in Munich cannot be a property of either John or Mary, so its extension is certainly not in the intersection filter. Rather I claim that the i-sum term *John*$\oplus$*Mary* gives rise to a third irreducible principal filter, i.e. the *sum filter*

(15) $\|John \oplus Mary\| = \{X \mid j \vee_i m \in X\} =: \mathbf{F}_{j \vee_i m}$

At this point it looks as if we had two ways of giving a denotation to such a simple conjoined NP like *John and Mary*. One could try to say that the conjunction *and* is ambiguous, and in a sense it is (see Link 1984 for the distinction between the Boolean and the i-sum *and*). But happily we are able to treat both (11) and (14) with the sum filter while playing the equivalence between (11a) and (11b) back to the fact that the predicate ... *is a student* is distributive, i.e., that it really applies to atomic individuals only. What this means in the present context is that the plural predicate involved in (11a), viz. the starred $^*student'$, inherits a kind of distributivity from its singular counterpart: it is closed under subsums, i.e. with any sum it contains also the atoms making it up. Thus its extension contains both j and m and in addition the sum $(j \vee_i m)$, and is therefore itself a member of the intersection of the filters $\mathbf{F}_j$ and $\mathbf{F}_m$ (see Link (1983a) for the details on distributive predicates). But this is the truth condition for the sentence (11b). What this all amounts to is that we start out with an analysis in terms of sums, and when we happen to meet a distributive predicate we are able to get down to the level of atomic predication like in (11b) (this is the same reasoning that underlies the analysis of a sentence like (1b) above).

It should be pointed out here, however, that the distributive or Boolean *and* within an NP is still present in examples like (16),

(16) John and every other student left.

This sentence should receive the normal GQ treatment since it doesn't really involve plural structures. That becomes obvious when we compare the examples under (17), (18), and (19).

(17) a. John and every other student is/?are from Stanford.

b. John and every other student invited his/*their parents to his/*their place.

c. *John and every other student dispersed.

(18) a. John and all other students are from Stanford.

b. John and all other students invited their parents to their place.

c. John and all other students dispersed.

(19) a. John and Mary are students.

b. John and Mary invited their$_{distr}$ parents to their$_{coll}$ place.

First we note that the NP *John and every other student* does not combine with collective verbs. It also requires a 'singular' VP like *is from Stanford.* The latter fact taken alone is perhaps not too convincing since one might argue here (and in the case of (11a) = (19a), for that matter) that the copula follows mere syntactic agreement rules. But pluralization is more pervasive as the VP *invited their parents to their place* shows. If *their* is meant distributively, which is the most natural interpretation for (18b), for instance, then it seems harder to account for the pluralization of the pronoun unless we want to say again that this is a purely syntactic phenomenon. Following a suggestion of Hans Kamp's, we might think of the above VP as having been 'pluralized' as a whole. This idea can neatly be expressed in LP by using the star operator in a way we did in connection with example (11). For that to work we have to introduce sum filters and not just intersective filters for the subject NPs. Incidentally, that also takes care of the mixed case (19b). This sentence means that John and Mary invite their respective parents to their common place. So we need the i-sum of John and Mary to serve as antecedent for *their*$_{coll}$; but at the same time the star operator is used to the effect that the 'singular' VP *invited his/her parents to their*$_{coll}$ *(!) place* gets distributed over the atomic individuals making up that sum. I'll come back to this example in section 5.

To finish up with this sketch of the lifted LP, there should be some formulation of the familiar *cumulative reference property* of plurals available in GQ. In LP this principle says that if x is a $^*P$ and y is a $^*P$ then $x \vee_i y$ is also a $^*P$. Now the principle comes out in GQ as a condition on the relation between the filters $\mathbf{F}_x$, $\mathbf{F}_y$, and $\mathbf{F}_{x \vee_i y}$. It could be formulated thus:

(20) **Cumulative Reference Property for Plurals in LP+GQ**: For all x and y in the domain, a starred predicate belonging to both $\mathbf{F}_x$ and $\mathbf{F}_y$ also belongs to the sum filter $\mathbf{F}_{x \vee_i y}$.

4.3 Plural Quantification

Let us now turn to quantifiers combining with pluralic nominals. I would like to distinguish between *spurious* and *genuine* plural quantification. Examples of the first kind are

(21) a. All men are mortal.

b. Most students ride a bike to school.

(21a) comes down to nothing but a normal distributive quantification, and it would be strange if a plural analysis would not render this obvious fact true (so Aristotle need not be rewritten). To be explicit, we could, of course, simply assume that the meaning of the GQ *all* directly equals that of *every*, that is, we could give it the usual GQ analysis and ignore the pluralic nature of the sentence by calling the plural form 'syntactic'. While this is a pragmatically reasonable move in the simple cases like (21a), it immediately prompts the question what the more involved cases are that do require a proper plural analysis. To answer the question I will later come back to a more detailed account of *all* in LP+GQ. It will include a discussion of the aspect of *totality* in the use of the quantifier *all* and give a reduction to the distributive case when the example allows for it.

A similar remark applies to sentence (21b); the vague quantifier *most* in there has a semantics which *measures* two amounts and takes their proportions: the first amount is the number of students riding a bike to school, the second is the overall number of students talked about. Their ratio is then compared with a certain number critical for making a statement involving *most* true. What is *not* counted here, to be sure, is the number of i-sums that can be formed from the underlying sets. So in neither case do we seem to have any kind of quantification *over i-sums*, i.e. genuine plural quantification.

But this is exactly the property that makes a semantic theory of plurals worth its name. Let's call it $\mathbf{Q}_\oplus$. We might attempt to produce the $\mathbf{Q}_\oplus$ effect by combining the quantifiers *all* and *most* with constructions that contain a collective predicate.

(22) a. (All) competing companies have common interests.

b. (All/most) children of same age stick together.

c. All (demonstrations/?demonstrators) disperse when tear gas is used.

(23) a. (Most/many) competing companies have common interests.

b. (All/most/many) parents of drug-troubled youngsters resent each other.

c. Most (demonstrations/?demonstrators) disperse when tear gas is used.

It looks like what one is talking about when using (22a) is any group of competing companies,[5] so that this would be an instance of universal $\mathbf{Q}_\oplus$ after all. There doesn't seem to be a stable pattern, though: replace *all* by *most* or *many* as in sentence (23a), and group quantification is fading. On the other hand, these sentences do sound impeccable. Sentence (23b) is much clearer in this respect; the quantification here is the intended $\mathbf{Q}_\oplus$, running over the sums of fathers and mothers of drug-troubled youngsters. In most of these examples a paraphrase like the following is helpful in getting at the right meaning:

(24) *In all/most/many cases, (a given group of) NOM VP*

[5] unless we are willing to read the sentence in an elliptic way, which could be paraphrased as *all companies that compete with some other company share certain interests with their competitors*. Here we do avoid quantifying over groups, but this paraphrase also comes out as a consequence of the more uniform $\mathbf{Q}_\oplus$ treatment.

(Here, *NOM* stands for the complex nominal in question, e.g., *competing companies.*) This paraphrase should also apply to (22c) and (23c) although with *NOM* = *demonstrators* the sentences are deviant. So I think we should at least allow for the possibility of $\mathbf{Q}_{\oplus}$ in connection with *all*, *most*, and *many* even if the LP-spurious reading is certainly predominant.[6]

Finally, what about sentence (22b)? Here the property of being of same age induces a partition on the set of children, and the quantifier runs over the children within each equivalence class rather than over the classes themselves. This is particularly obvious with the quantifier *most*. Then the sentence means that for all appropriate ages x most children of that age x stick together, and not the other way around, i.e. that for most ages x children of age x stick together. The above analysis can be upheld, however, by noting that (22b) is different in that the modifier *of same age* really takes scope over the initial quantifier just like *of a given age* does.

Going back to (22a) we note that the same $\mathbf{Q}_{\oplus}$ pattern holds for the corresponding generic statement with the bare plural *competing companies* in front, whether it is regarded as a strictly universal statement or not.[7] More examples of the generic kind are the following.

(25) a. *Tres faciunt collegium.*

b. Two's company, three is a crowd.[8]

These sentences introduce a new feature into the discussion, viz. the numerals, to which we will return presently. Barring these, the sentences do not contain a visible lexical item that induces the universal quantificational force. Clearly, they have this meaning though. The first says something like *any three* make up a *collegium*, and *any* can also be filled into the second sentence. So the determiner *any* would be another and perhaps an even better suited candidate for a universal sum quantifier than those that have been discussed so far. But we have to recall the fact that *any* as a universal quantifier has a rather limited range of applicability; it can go only with generic sentences or modal contexts. As we will see the German *je* has to be

[6]Especially with the vague quantifiers *most* and *many* the truth of a $\mathbf{Q}_{\oplus}$ reading might be difficult to assess, and one cannot even be sure whether it gives the right result at all. So imagine there are only seven firms that are in competition; then we would have to look already at $2^7 - 8 = 120$ proper i-sums to decide if the majority of them share interests internally.

[7]A generic statement is surely not synonymous with its strictly universal correspondent, but this is a separate issue; for the purpose of the present discussion I will speak of *universal force* in a broad sense to cover both the strictly universal quantification and the generic (or the universal modal) interpretation.

[8]Note that we don't have plural agreement here: a nice syntactical indication that plural objects form single entities; cf. also constructions like *there* **is** *two ways to view the problem.*

restricted in the same way for it to carry universal force. The quantifier *no* is not so limited, but it has to precede a numeral to bring out the intended $\mathbf{Q}_\oplus$ reading. In the presence of a numeral, therefore, not only *any*, but also *no* works fine, witness (26).

(26) a. Any two competing companies have common interests.

b. No two competing companies have/had common interests.

Plural NPs with numerals pose another, seemingly more serious, threat to the possible use of *all* as a genuine plural quantifier. Thus the sentence

(27) All three men were playing on the beat.

cannot possibly mean that every combination of three men out of a certain group of men performed together. Obviously some extra story has to be told here in order to save some minimum amount of compositionality in plural semantics; this will be done in the next section.

Now what about **existential plural quantifiers**? *several*, *a few* and *a couple of* have a meaning of something like *there is a sum of at least two* and so qualify for $\mathbf{Q}_\oplus$. *some* in front of a plural noun does so, too, but note that, when preceding a numeral it behaves rather like an adverbial modifier of the numeral than an inducer of the existential force. So

(28) Some fifty men gathered in the hall

means that approximately fifty came together. The numeral already carries the existential force, so adding *some* in the purely existential sense would be redundant. This fact about numerals is true in many cases and is accordingly reflected in their standard GQ treatment. However, the sentences under (25) are instances of a use of numerals where they carry universal force. In fact, we observe just the same flip-flop between existential and universal readings as in singular sentences like those under (29).

(29) a. A man came down the street.

b. A third-grader solves such a problem easily.

Thus, the standard GQ treatment of numerals seems to force us to acknowledge a systematic ambiguity here. This problem will be addressed presently.

In summary, the denotations in LP+GQ for some plural NPs without numerals are given below. Recall that E is the set of all individuals in the semi-lattice; let A be the set of atoms in E; for $X \subset E$, let $\sup_i X$ be the supremum of X (this is the denotation for the σ-term $\sigma x P x$, if $\|P\| = X$). The lower index '*sp*' indicates spurious plural quantification, and '$_\oplus$'

genuine $\mathbf{Q}_\oplus$. Note also that in the condition for $most_{sp}$ I used a measure function μ rather than cardinalities. This is to emphasize the analogy with *most*-phrases involving mass nouns: there the counting of quantities is out of question from the outset; rather one has to take proportions under an appropriate μ. In the discrete case, however, μ is just the cardinality of the sets involved (these are assumed to be finite).

(30) a. $\|the\ men\| = \{X \subset E \mid sup_i\|{}^*man\| \in X\}$

b. $\|some\ men\| = \{X \subset E \mid X \cap \|{}^*man\| \neq \emptyset\}$

c. $\|all_{sp}\ men\| = \{X \subset A \mid \|man\| \subset X\}$

d. $\|all_\oplus men\| = \{X \subset E \mid \|{}^*man\| \subset X\}$

e. $\|most_{sp}\ \text{men}\| = \{X \subset A \mid \mu(sup_i[\|man\| \cap X]) > \mu(sup_i[\|man\| \backslash X])\}$

4.4 The Treatment of Numerals

Consider the usual conception about numerals in the GQ framework, as it can be found, for instance, in van Benthem (1983a). We have several classes of infinitely many determiners; they typically consist of more than one lexical item which is characteristic of the syncategorematic approach of GQ accounts.

(31) a. *more than k A are B* $(k \geq 1)$

b. *(at least/no less than) k A are B* $(k \geq 1)$

c. *exactly k A are B* $(k \geq 1)$

d. *(at most/no more than) k A are B* $(k \geq 1)$

e. *less than k A are B* $(k \geq 1)$

f. *an even number of A are B*

Now what is interesting here is that those determiners all share the property of being *symmetric*,[9] i.e.:

[9]In correspondence with the author, Sebastian Löbner expresses doubts about the symmetry of natural language statements involving numerals, like *three girls chuckled*; he claims that it doesn't even make sense to turn such a sentence around, viz. *three chuckling persons were girls*, for instance. I take it that the same criticism applies to the simple indefinite sentence *a girl chuckled*, then. Now while turn-arounds are certainly awkward stylistically, the minimal semantical condition of symmetry is certainly preserved or else there is barely any semantics left at all. For another argument against symmetry, however, that involves the problem of vagueness, see Blau (1978).

(32) $D_E AB \Rightarrow D_E BA$

but this is the characteristic property of the (negated) existential quantifier. Thus the (negated) existential force is *built right into* the GQ-semantics of each and every numeral. I suggest that this is a clear case of missing a generalization; what is worse, the account is at variance with the existential/universal flip-flop we have observed above.

There is also another difficulty. Since the underlying domain E is now structured the determiners cannot be expected to satisfy the property of QUANTITY anymore, i.e. that they be invariant under every permutation of the domain. What comes to mind immediately as a remedy here is to require invariance under lattice-automorphisms, that is

(33) $D_E AB \Rightarrow D_E(\pi A)(\pi B)$ for all $\leq_i$-automorphisms π on E

While this move might be a natural step to take anyway as domains in current semantic theory tend to become more and more structured, the flip-flop problem is still there.

As an alternative it has long been suggested in the literature that numerals should be treated semantically as **adjectives**, that is, intersecting nominal modifiers. In LP this is achieved in a most natural way, by having *three children*, for instance, denote that subset of the set of all i-sums of children each member of which has exactly three atomic parts. If that view is accepted then what used to be a different determiner in GQ for each natural number, really consists of the unique (negated) existential quantifier plus a numerical specifier modifying the incoming nominal. This approach obviously takes care of the flip-flop problem. Consider a minimal pair exemplifying it.

(34) a. Three men lifted the piano.

b. Three men can lift the piano.

The first sentence expresses a particular 'historical' fact (in a certain context), the second can be used as a generic statement about the piano in question. It is really these different uses here that introduce the existential and the universal force, respectively. GQ theory cannot say that, as we saw, since it considers the (existential) quantificational force as inherent in the numeral. Under the present view, the phrase *three men* is not a GQ yet, it is simply a numerically specified nominal. But then, at which point does the quantification occur? Well, either one gives a zero determiner analysis, which is not very popular (but a convenient device to keep pragmatics out of the picture); or else one says that (i) combining *three men* with a VP in past tense results in an existential quantification or some corresponding discourse-theoretic process of the Kamp/Heim variety, and (ii) combining it with a present tense VP tends to invoke the universal $\mathbf{Q}_{\oplus}$ reading or some statement about the generic sum of three men, whatever one's account of generics might be.[10]

For the purpose of this discussion, let us suppose that there is indeed an existential zero determiner involved in sentence (34a) which is denoted by $\emptyset_{\exists}$. What, then, is the truth condition of (34a) in LP+GQ? Let us also ignore the collective/distributive distinction for the moment, which anyway creates an ambiguity only in the verb phrase, as I see it; then the sentence is true just in case the extension of the VP *lifted the piano*, i.e., the set of all i-sums that (collectively, say) lifted the piano, is an element of the GQ (over the lattice domain E)

(35) $\|\emptyset_{\exists} \textit{three men}\| = \{X \subset E \mid X \cap \|\textit{three men}\| \neq \emptyset\}$

But that condition means that there is at least one i-sum of three men that lifted the piano. Let us now separate cases. (i) The collective reading: the condition says that there has been at least one joint lifting with exactly three people involved; there might well have been another group lifting the same piano, but every act of lifting requires exactly three lifters. (ii) The distributive reading: the truth condition tells us here that there were three men each lifting the piano (how exactly this comes about is explained in more detail in section 5). Note that there is still a group of three men to start with. That is important for anaphoric purposes; thus sentence (34a), even under its distributive reading, could be followed by something like *...they then gathered to receive the promised reward*, with *they* referring back to (the i-sum of) the three men. In neither case, however, does the truth condition entail that the overall cardinality of all the men that lifted the piano is exactly three, as some GQ accounts would have it. This is, I think, as it should be.

Finally I would like to mention a syntactic argument in favor of the adjectival analysis: it accounts for the fact that another quantifier can be found in front of the numeral:

(36) a. Any two tires can be used.

b. No even number of members can vote on this issue to avoid ties.

c. Every three months John receives a letter from home.

However, an obvious counterargument from syntax might be that numerals cannot simply be plugged into the recursive adjective node in front

[10] Notice that I am deliberately vague about the theoretical options I mention in both cases since the choice between them is again a separate issue. All we have to make sure is that our account of numerals doesn't prejudge those choices in any way.

of a noun since we do not have strings like *pretty three girls*. But then I don't see any harm in postulating an extra node called *NUM*, say, preceding the adjective node.

A more serious problem with the adjective account is the by no means obvious status of the expressions that modify a numeral, like *at least*, *at most*, *exactly*, etc. What I would like to assume is that they are adverbial modifiers on the numeral, and as such they should still not carry a quantificational element. But while the sentence

(37) At most 50 people fit into this bunker

has a perfect universal $\mathbf{Q}_{\oplus}$ reading, it is hard if not impossible to combine *at most k NOM* with an existential quantifier. Thus, a sentence like (38a),

(38) a. At most three people are left in the house.

b. There exist no more than three people (who are) left in the house.

c. There are people left in the house, but no more than three.

does not commit a speaker uttering it to anybody being left in the house (think of a situation where the house is being evacuated and all but three people have been positively accounted for). It really says only that it is not the case that there are more than three people left behind. That seems to be true even of a formulation like (38b). To bring out the existential import one has to be explicit about it, as in (38c). My reaction to this problem is that it is too narrow to say that the non-generic use of a sentence with an adverbially modified numeral in the head of its subject NP induces existential force. It does so with *at least/exactly k NOM*, but in the case of *at most k NOM* it induces *negated* existential force, which does not carry existential import, after all.

By way of a short digression let me try to explain what I consider to be a reasonable, non-artificial way of approaching the kind of issue I have been discussing. It is of course the time-honored spirit of compositionality that gives rise to the adjectival analysis in the first place. Whatever its drawbacks might be, it certainly has the merit of taking syntax more seriously than the syncategorematic approach prevailing in GQ theory. But then, we have learned that we should not be too religious about compositionality, on pain of becoming rather artificial in our account. Applied to the situation at hand, that means that in trying to analyze the internal structure of the prenominal domain we should preserve the theoretical freedom to acknowledge the fact that sometimes different semantical functions can merge into a single construction like *at least three* or *at most three* (or *no*, for that matter). Thus the question of there 'really' being an empty determiner might seem, in this light, to become a rather academic point; furthermore, negated existential force is frequently lexicalized in language (witness *no*), so it is not surprising to find that *at most k*, which logically comprises *no*, tends to behave more like a numerically specified quantifier than a nominal modifier.

The idea of there being numerically specified quantifiers also helps to resolve another puzzle, already alluded to above, which has been with me for quite a while. We saw in connection with example (27) that the phrase *all three men* cannot be treated in a naive compositional way, when parsed in consonance with the adjectival analysis, i.e. *all (three men)*. Consider (39),

(39) a. all three men

b. all the three men

c. all of the three men

d. all three of the men

e. all three of the three men

In view of this, it is hard to decide which way the numeral goes. In (39e), which is fine except for redundancy, it even appears in two places. This suggests that there might be two slots in the prenominal domain for a numeral to go into: the NUM slot, which we had before, plus the position following certain determiners like *all*. But what is the role of the numeral here? Its specifying function is *presuppositional*, it seems to me. Consider the following examples.

(40) a. all (seven/*three) of the seven astronauts

b. all (five/*three) of the quintuplets

The respective numerical information given by the nominals *seven astronauts* and *quintuplets* can be redundantly repeated in the quantifier for reasons of emphasis, but it cannot differ in the two places. I take this to show that the role of the numeral after *all* is presuppositional rather than assertive in nature. The same double analysis should apply to the definite article, and indeed, both parsings, *(the three) men* and *the (three men)*, make semantical sense here. It now seems to me that the first parsing is even more natural than the second. Sentence (41a), for instance, is odd in a way in which (41b) is not. The reason is that the numerals, unlike the regular adjectives *white* and *black*, are preferably interpreted here in a non-restrictive way.

(41) a. The (three men) did it, not the (four men).

b. The white men did it, not the black men.

Much more natural is a situation where you have a group of three men over here and another group over there, consisting of four men. Then you can say,

(42) (The three) men over here did it, not (the four) men over there.

So we see that there was something after all to the numerical parametrization of the basic determiners that appears already in the GQ classic Barwise/Cooper (1981). But our discussion certainly sheds a different light on it. *Numerals are not quantifiers, although they help to specify the range of quantification in various ways.*

For reasons of illustration the following tree is given for the NP *all of the three surviving men.* In this example most of the prenominal nodes are filled. The lambda-term below (with x and y running over i-sums and P over sets of i-sums) expresses the semantics of the NP in LP+GQ. The slash indicates the numerical presupposition on the σ-operator.

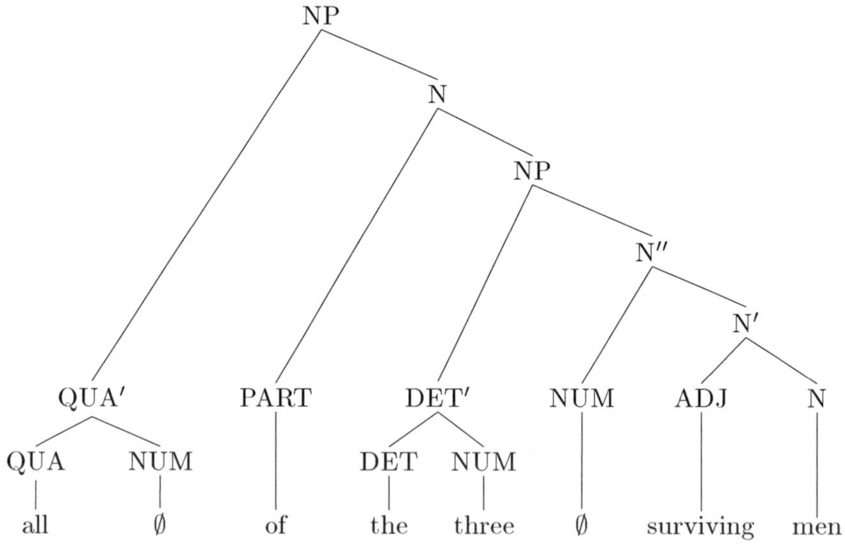

$$\lambda P\, \forall x \quad [[x^\bullet\Pi \quad (\sigma \;\; / \;\; 3)y \;\; {}^*[surv'(man')](y)] \rightarrow P(x)]$$

There is a problem with the semantic representation given by the λ-term. It doesn't seem to fit the case in which a collective predicate fills the

P-slot, as in (43a). This is because the universal quantifier in the λ-term runs over all atomic parts of the i-sum of surviving men in a distributive way, but a collective predicate cannot be true of atomic individuals.

(43) a. All of the three surviving men held a reunion.

b. ?All of the three surviving men founded a club.

c. All the men came together to discuss the problem.

d. ?All the men dispersed.

One might be inclined to argue that *all* really forces distributivity, whence the oddness of examples with 'hard' collective predicates like (43b,d). That is true in particular when *all* appears in floated position (see below). But suppose we accept the grammaticality of the other two sentences, (43a) and (43c). What is the role of *all* in those examples? It doesn't seem to add anything to the truth conditions, which are fixed already by the definite plural NP *the (three surviving) men*. Now while *all* looks redundant in the cases at hand, there are examples in which a difference in meaning is felt rather clearly. Compare

(44) a. The men were in serious trouble.

b. All the men were in serious trouble.

The second sentence focuses on the aspect of *totality*: every single man was in trouble. By contrast, (44a) might very well be true even if not every man finds himself troubled individually; it is more about the group than about its members. But the sentences (43a,c) show that if totality is the nuance in meaning that is added by *all* it is not restricted to contexts involving distributive predicates. Here the meaning cannot be, of course, that everybody came together, but rather that everybody *took part* in the gathering. It is exactly this property of *partaking* that cannot be analyzed further in a framework like LP or LP+GQ, for that matter. In Chapter 1 I tried to characterize it partly, at least, by means of a meaning postulate. It says something to the effect that if you have a distributive predicate P, partaking of P is already tantamount to falling under P. So if all the men are asleep the totality of the group of men in question, without any exception, is involved in the sleep-in: but since taking part in the activity of sleeping means to be asleep, everybody is in effect sleeping.

Let us go back to the problem of whether the above λ-term is compatible with incoming collective predicates. It is not as it stands; a complete account would have to say something like this. The partitive phrase *...of the three surviving men* is again a nominal that can be turned into a full

NP by certain determiners like *some* and *all*. Its denotation is the set of all i-parts of (the sum of) the three surviving men (so in general, the i-part relation, unlike the one expressed by the λ-term, should not be restricted to atomic parts). If the phrase is combined with *all* then the whole i-sum is formed again, but some meaning component is added which expresses totality.

I would like to settle, then, with the following account for the quantifier *all*: Unless it is LP-spurious, which is the most common case, its meaning is that of the definite article plus the aspect of totality, and only marginally the universal $\mathbf{Q}_{\oplus}$. Notice that two of these three senses of *all* reappear in the case of mass terms. There is clearly no analog there to the distributive or LP-spurious *all* since the theory of mass terms deals with non-atomic structures. But the other two senses are reflected. In Chapter 1 I likened *all* to the $\mathbf{Q}_{\oplus}$ sense in that I treated mass nouns as denoting quantities of stuff. So the sentence

(45) All water is wet

is taken to mean that all quantities of water are wet. By contrast, in Lønning (1985) *all water* is given the same meaning as *the water* (he too has to ignore the totality aspect), so this would correspond to the third sense of the pluralic *all*. I tend to think now that once totality is expressed in one way or the other, this latter sense is the more natural one. So the following example seems hardly to make sense when we think about it in terms of quantities.

(46) a. All water was gone.

b. The water was all gone.

Even the floated quantifier expresses totality here. This is interesting because the floated *all* in plural constructions seems to force distributivity when the incoming verb phrase allows for it. Thus, whereas the following a) sentence is ambiguous with respect to the collective/distributive distinction, the b) sentence can only be read distributively.[11]

(47) a. The children received 10 dollars.

b. The children all received 10 dollars.

The issue of floated quantifiers will be taken up in the next section.

[11] This observation has been made before; cf. Barbara Partee's sentence *The trees are (*all) denser in the middle of the forest* and Bill Ladasaw's *The students (all) voted to adopt the proposal*, quoted from Dowty&Brodie 1984.

4.5 Floated Quantifiers

As the metaphor still suggests a floated quantifier used to be thought of as having moved out of a noun phrase into various possible niches further down in the sentence. Dowty&Brodie (1984) have shown that this picture is really but a metaphor, and that there is a perfectly natural treatment of floated quantifiers as operators on verb phrases. Following a proposal by Sebastian Löbner we should rather view the phenomenon as an instance of what he calls *adverbial quantification.* This terminology becomes particularly plausible when we think of other quantificational VP operators like *partly, mostly, in their (overwhelming) majority, for their major part, in totality.* I shall adopt Dowty & Brodie's basic idea here, which can be expressed as follows. Take the quantifier *all* hooked up on a verb phrase, VP. Then it acts like a distributivity operator on VP, which might be written ^{D}VP. In Chapter 2 I gave the following semantic definition for it:

(48) $^D\mathrm{VP} := \lambda x \forall y[y^{\bullet}\Pi x \rightarrow \mathrm{VP}(y)]$

The D-operator was designed to take care of the distributive reading of *lifting* sentences like (34a). Note that it operates on possibly complex verb phrases; this expressive power is needed in order to get the scope relations right. As an illustration of this, consider (49),

(49) Three men lifted a piano.

Let **U** be the Montague-style translation relation between natural language and the second-order logical representation for LP+GQ, which I have been using in free variance with the set-theoretic notation. For the distributive reading of (49) we then get the following translation.

(50) $3\ men\, \mathbf{U}\, \lambda P \exists x[(3men)'(x) \wedge P(x)]$

$^D(lifted\ a\ piano)\, \mathbf{U}\, {}^D\lambda y \exists z[piano'(z) \wedge lifted'(y,z)]$

$= \lambda x \forall y[y^{\bullet}\Pi x \rightarrow \exists z[piano'(z) \wedge lifted'(y,z)]$

3 men lifted a piano

U

$\exists x[(3\ men)'(x) \wedge \forall y[y^{\bullet}\Pi x \rightarrow \exists z[piano'(z) \wedge lifted'(y,z)]]]$

By contrast, note how 'pluralized' verb phrases are treated. Our example was (19b), here repeated as (51) for convenience.

(51) a. John and Mary invited their$_{distr}$ parents to their$_{coll}$ place.

b. *John and Mary* **U** $\lambda PP(j \oplus m)$;

invited his/her parents to their$_{coll}$ place **U**

$$\lambda u[invited\text{-}to'(u, \sigma y\, parents\text{-}of'(y, u), \iota v\, place\text{-}of'(v, z))] \quad =: \lambda u Q[u, z]$$

The next step is to quantify $j \oplus m$ into the z-position. Only then the VP is pluralized, yielding the translation $^*\lambda u Q[u, j \oplus m]$. The translation of the whole sentence applies this starred λ-expression to $j \oplus m$. But that means that $j \oplus m$ is a sum whose atomic parts satisfy the unstarred predicate $\lambda u Q[u, j \oplus m]$, whence the final and intended representation for (51),

(52) $[invited\text{-}to'(j, \sigma y\, parents\text{-}of'(y, j), \iota v\, place\text{-}of'(v, j \oplus m))] \wedge$
$[invited\text{-}to'(m, \sigma y\, parents\text{-}of'(y, m), \iota v\, place\text{-}of'(v, j \oplus m))]$

The D-operator would not work here because the *parents-of* relation has to have an atomic individual in its second argument place for the definite description built from it to denote: the predicates which D operates on, however, start out with sum variables. A similar and perhaps clearer case of a dependent plural construction arising from VP-pluralization is the sentence *Before Christmas the Americans rob their savings account.*

After this digression let us return to the floated quantifier *all.* Its role is to transform a VP into a ^{D}VP, thereby removing the collective/distributive ambiguity. The subset of a given domain of individuals corresponding to (48) can then serve as an argument for a generalized quantifier like $\|$*the children*$\|$ which produces for (47b) a structure satisfying the expected reading (53).

(53) $\forall y[y^\bullet\Pi\sigma z^*child'(z) \rightarrow \mathrm{VP}(y)]$

This says that every atomic i-part of the sum of the children, i.e. every single child, received 10 dollars. The same analysis should be given for *each* in (54a); or more precisely: (48) is really the correct analysis for *each*, but for the floated *all* only when we neglect the complication discussed above concerning the aspect of partaking.

(54) a. The children each received 10 dollars.

b. The children received 10 dollars each.

c. Susan gave the children ten dollars each.

But how are we to handle (54b,c) now? Here the quantifier *each* is *shifted* as Dowty & Brodie call it, but it clearly belongs to the object NP *10 dollars* and doesn't seem to operate on the whole VP, at least syntactically. This is a feature, however, that we will have to handle anyway since it comes up in constructions involving *each other*, *same*, *different*, and the German *je.* So I shall assume that (54a) and (54b) are synonymous, i.e. that *each* continues to be a VP operator with the same distributivity effect.

As was just hinted at VP operators also handle cases of *disjoint reference* like the following *each other* construction:

(55) The students read each other's papers.

Call the operator here *DJR*; it takes a $\mathrm{VP}(x,y)$ with two argument places and converts it into a 1-place verb phrase $^{\mathrm{DJR}}\mathrm{VP}(x)$. we can define

(56) $^{\mathrm{DJR}}\mathrm{VP}(x) := \lambda x \forall y \forall z[y, z^{\bullet}\Pi x \wedge y \neq z \rightarrow \mathrm{VP}(y,z)]$

For our example (55) this comes down to something like this: The sum filter generated by *the students* contains a certain set of i-sums x such that for all pairs y,z of atomic i-parts of x with $y \neq z$ and all sums of papers w such that w belongs to z y reads w.[12]

So far it looks like all cases of *dependent plurals* could be treated this way. The harder cases, however, are those in which there is no plural to be seen on the surface. Consider the following sentence (57a) (cf. Dowty 1985).

(57) a. Every student saw a different movie.

b. The students all saw a different movie.[13]

In a strict reading of (57) no two students saw the same movie, so the correspondence between students and movies is one-to-one. Let us first try to analyze (57b). There are two VP operators coming into play here: the floated quantifier *all* has the effect described above; the word *different* gives rise to an *injectiveness clause* on top of the $\forall\exists$-structure of the verb phrase which is shown under (59) below. Everything 'lives on' a plurality that is provided by the plural NP *the students*. The interesting thing is that such a plurality has to be inputed also in (57a) as is clearly shown by the fact that the sentence can be continued with **they** *were all x-rated*. Now every $\forall\exists$-structure expresses a relation, and the plural pronoun *they* should obviously be linked to the i-sum made up from the *range* of this relation. Moreover the whole VP in (57a) is in need of a plural input. Intuitively, it

[12]There are a number of apparent counterexamples to the strictly logical explication of *each other* expressed by (56), e.g., *the girls combed each other's hair* or *the sea elephants were lying on top of each other*. But that simply means that depending on the semantics of the particular verb that *each other* combines with, the verb phrase can acquire a more or less loose collective meaning where the stable common element is the exclusion of the reflexive reading. (56) could of course be weakened appropriately. German, by the way, distinguishes between the strict and the loose sense of disjoint reference: *they fight each other* translates as *sie bekämpfen* **sich (gegenseitig)**, whereas the *sea elephant* sentence has to be rendered as *die See-Elefanten lagen* **aufeinander**.

[13]Mats Rooth informs me that in his English the plural *different movies* is required here.

is quite clear what this should be: The i-sum formed from the *core* of the filter denoted by *every student*, i.e. the set of students in question. In order to make this sum available as a possible antecedent to be referred to later on in the sentence I shall adopt a more dynamic picture of interpretation along the lines of Kamp (1981), Heim (1982), and Barwise (1985). By the time the VP gets processed the context has to have changed in such a way that the sum of students, call it w, is a parameter contained in it. Then the VP *saw a different movie* should denote the set of all x such that there is *a movie y seen by x* (call this predicate $SM(x,y)$) and *the movies seen are different with respect to the sum* $x \oplus w$ (call this predicate $SM\text{-}DIFF(x \oplus w)$). We have to include the x in the summing since according to the way *SM-DIFF* is to be defined this will also take care of external reference in cases like (58a):

(58) a. John saw a different movie.

b. John and Mary saw a different movie.

Here the parameter w will refer to persons other than John, and x will stand for John; analogously for (58b): x will stand for $John \oplus Mary$, and w for the group externally referred to when we read the sentence as saying that John and Mary saw a movie that is different from the ones seen by everybody else.[14] Of course, the more natural reading of (58b) is the internal one under which John saw a movie and Mary saw another one. But this is just a special case of the reading we have been after before (cf. (57b)). In these cases it doesn't hurt when x is an i-part of or even equal to w - the sum will be the same.

We are now ready to write down the semantics for the VP *saw a different movie.*

(59) $\lambda x[\exists y SM(x,y) \wedge SM\text{-}DIFF(x \oplus w)]$
where $SM(x,y) := [{}^*movie'(y) \wedge x\ saw\ y]$, and $SM\text{-}DIFF(x \oplus w) :=$
$\forall y, y' \forall z, z'[y, y'^{\bullet}\Pi x \oplus w \wedge SM(y,z) \wedge SM(y',z') \wedge z = z' \rightarrow y = y']$.

Summarizing we note that unlike *each other* which must have internal antecedents, *different* (and also *same* for that matter) can be bound both internally and externally. It shares this property with *each* and the German *je* to which we will now turn.

[14]To be precise the sum term in question has to be slightly amended to really yield the intended effect. We have to introduce a subgroup formed by John and Mary alone to set this sum off from the rest. Thus $x \oplus w$ has to be replaced in the predicate *SM-DIFF* by $\langle x \rangle \oplus w$; for the bracket notation see Link 1984.

4.6 The Case of the German je

Consider the following set of data. The b) sentences try to give English paraphrases of the German a) sentences.

(60) a. *Je drei Männer können das Klavier heben.*

b. Any three men will be able to lift the piano.

(61) a. *Der Verlag zahlt 10 Mark für je 1000 Anschläge.*

b. The publisher pays 10 marks for every 1000 strokes.

(62) a. *Die Kinder bekamen je drei Äpfel.*

b. The children got three apples each.

(63) a. *Er überraschte die Kinder mit je zwei Äpfeln.*

b. He surprised the children with two apples each.

(64) a. *Die Männer sprangen* { *in je* / *je in* / *alle in* / *jeder in* } *ein Auto.*

b. The men { jumped into [one car] each. / jumped [into one car] each. / all jumped into a car. / each jumped into a car. }

(65) a. *Je drei Äpfel waren faul.*

b. Three apples each were rotten.

(66) a. *Vier Männer hoben je drei Tische.*

b. Four men lifted three tables each.

(67) a. *Je vier Männer hoben je drei Tische.*

b. ??Four men each lifted three tables each.

(68) a. *Die Männer hoben je drei Tische.*

b. The men lifted three tables each.

(69) a. *Die Männer hoben jeweils drei Tische.*

b. The men lifted three tables at a time.

At first blush *je* looks like a regular quantifier with its normal place right in front of a prenominal string. But its meaning seems to vary widely, witness the three different paraphrases in (60) – (62). Sentence (60) is the only clear case of what I called above genuine plural quantification: *je* runs over all possible sums consisting of three men. But this interpretation is essentially linked to the modal or generic element in the sentence. As soon as we switch from *heben können* zu *hoben* (past tense) the meaning changes, and we get a case like (61) which I am going to address presently. This obviously leads to the conclusion that the universal plural quantification of (60) isn't really the effect of *je* in there after all; as we saw above already (viz. *tres faciunt colloquium*) the $\forall_{\oplus}$ effect has rather to do with the generic/existential flip-flop. But both *je* and the English *any* do force the generic interpretation and, in virtue of that, the universal reading. Universal plural quantification is also present in (61) but now there is a constraint on the sums that *je* is supposed to run over: they have to be disjoint in the sense that there is no overlap in the sets of atoms that the different sums are made of (otherwise you could be a rich man already by producing just 1001 strokes).

Beginning with (62) something else is going on. Rather than inducing $\forall_{\oplus}$ quantification, *je* now triggers what might be called *distribution over cases*. What exactly these cases are seems to be determined by the specific construction at hand. In (62) *je* behaves like a distributivity operator on the verb phrase with the same meaning as (48). This analysis is also evidenced by the fact that in the English translation we have the quantifier *each*, albeit shifted, achieving this effect. So the cases here correspond one-to-one to the children. This correlation seems to hold in all the sentences where a plural NP is around that can serve as antecedent; in (62), (64), (66) this NP is the subject phrase, but in (63) it is the direct object phrase. The latter fact shows already that (48) as it stands cannot be taken to cover all the *je* constructions here. Sentence (65) is even worse in that there is no reference group in there, and indeed, the sentence is quite strange when uttered out of the blue. But we can make sense of it in an suitable context: imagine several baskets filled with apples in a market-place; in each basket there are exactly three rotten apples. This would be a situation appropriate for uttering (65). The cases which *je* triggers the distribution over (here: the baskets) have to be recovered from the context. So just like in the case of *different* we have to allow for the possibility of *external reference*. The same applies to example (67) which is quite o.k. in German when uttered in a situation that exhibits suitable cases for the first occurrence of *je*; the second *je* then causes no extra problem since it can be internally related back to the subject phrase.

One comment on sentence (64). It is to show the somewhat capricious behavior of *je*. While there is clear evidence that it normally comes along as part of a NP it can also appear in something like a floated quantifier position (see the second case of (64a) where *je* is found outside of the PP *in ein Auto*). Note that in this case it may also be expanded to the adverb *jeweils* which roughly means *each time* or *at a time*. *jeweils* can replace *je* in the other cases, too, thereby bringing the external distribution over cases to the fore. Thus consider the minimal pair (68) – (69); while (68) is like (66) in forcing the distributive reading the distribution in (69) is external and compatible with a collective reading within the sentence.

In conclusion it seems to me that a fairly general formulation of the meaning of *je* constructions would be this. The interpretation of *je* triggers a *distribution over cases*; these cases can be retrieved either *internally* (through distribution over the atoms of a plurality given previously in the sentence) or else *externally* (by looking for an appropriate context providing the plurality). The common structure of a sentence S containing the phrase *je k N*, then, would roughly look like in (70). Here the expression $\Phi[v;x,y;(k\ N)'(z)]$ stands for the logical schema derived from S by 'quantifying out' the plural noun phrase PNP over which the distribution takes place; it might not be in there when it is supplied externally, but it contributes an argument place to Φ anyway, which is marked by the parameter v. As usual, primes provide a translation of natural language expressions into logical notation.

(70) $S' = PNP'(\lambda w \forall v[v^{\bullet}\Pi w \rightarrow \exists z \Phi[v; x, y; (kN)'(z)]])$

where the variable w is either

(i) $= x$ ('subject distribution'); or

(ii) $= y$ (distribution over a plurality in the direct object or some other NP other than the subject or the *je* phrase); or else

(iii) $\neq x, y$ (external distribution)

The English version of (68), for instance, would be rendered thus:

(71) $\text{PNP}' = (\textit{the men})' = \lambda P P(\sigma x\ men'(x))$

$\Phi = [man'(v) \wedge (3\ tables)'(z) \wedge lift'(v, z)]$

yielding the logical representation

$\forall v[v^{\bullet}\Pi \sigma x men'(x) \rightarrow \exists z[(3\ tables)'(z) \wedge lift'(v, z)]]$

The second case can be dealt with in a similar way. Finally, suppose sentence S = (65) is uttered in a context like the one described above. Then the PNP would be the given group of baskets. S is regarded as elliptical, and Φ is provided with an explicit extra relation, say $in'(z, v)$, where the parameter v stands for an individual basket. Schema (70) then yields the following representation for (65).

(72) $\forall v[v^{\bullet}\Pi\sigma x\, basket'(x) \rightarrow \exists z[(3\ apples)'(z) \wedge in'(z, v) \wedge rotten'(z)]]$

Acknowledgement. After having been presented to the Lund conference, this chapter has undergone quite a number of revisions during my stay at the Center for the Study of Language and Information, Stanford University, in the academic year 1985/86. I have to thank many people for 'talking plurals' with me and/or giving detailed comments on earlier versions of this chapter. I would like to mention in particular Johan van Benthem, Greg Carlson, Haim Gaifman, Claudia Gerstner, Erhard Hinrichs, Hans Kamp, Manfred Krifka, Sebastian Löbner, Jan Tore Lønning, Chris Menzel, Friederike Moltmann, Julius Moravcsik, Jeff Pelletier, Stanley Peters, Craige Roberts, Mats Rooth, Ivan Sag, Hans Uszkoreit, Dag Westerstahl, Dietmar Zaefferer, and Ed Zalta. I am most indebted to Claudia, Dietmar, Mats, and Stanley, though, for being my continued discussion partners on these matters at the CSLI.

Chapter 5

Je drei Äpfel—three apples each: **Quantification and the German** je

There are at least two different means in language for expressing the fundamental quantifier structure '$\forall\exists$'. To begin with, there have to be two noun phrases involved; call them NP_1 and NP_2. The first construction is the familiar one: we have a general NP_1, headed by an explicit universal quantifier like *every*, which is followed by an indefinite NP_2 at some point later in the sentence. There is, however, a second construction that may be used to express a $\forall\exists$ structure. It typically involves a definite *plural* NP_1 and an indefinite NP_2 which contains a particle that triggers a distribution of the content of NP_2 over that of NP_1. An example is the sentence (1a), which is obviously synonymous to the usual $\forall\exists$ sentence (1b).

(1) a. The winners received a hundred dollars each.
    b. Every winner received a hundred dollars.

That is, we have a group of winners each of whom receives \$100. Choe (1987) calls NP_2 the *distributive share (DstrShr)* and NP_1 the *sorting key (SrtKy)*. While I'd like to adopt his basic analysis of the "anti-quantifier" structure displayed in (1a) I prefer to reserve the term *(distributional) key* for the quantificational element alone, at the same time renaming Choe's SrtKy (distributional) domain. The key can either be part of the domain NP_1, as in (1b), or else part of the DstrShr NP_2, as in (1a). In the latter position the key functions as *anti-quantifier*, so-called because the key is found in the DstrShr undergoing the distribution, rather than in the domain which determines the range of quantification. It is the German anti-quantifier *je* that I'd like to focus on in this chapter.

Historically, *je* derives from Germanic *aiw* (time), which, incidentally, also serves as root for *each*. It occurs in a number of compounds, viz., *jeder*, *jedweder*, *jeglicher* (Engl. *every*), *jemand* (Engl. *someone*, *any one*), *jemals* (Engl. *ever*), *jedenfalls* (Engl. *in any case*), *jeweils* (Engl. *each time*, *at a time*, *respectively*). Closest to *je* in meaning comes *jeweils*, which marks distribution over times, occasions, or cases; *jeweils* can often be regarded as a variant of *je*, but see below for a discussion of differences between the two items.

5.1 Different Uses of *je*

1. The most important use of *je* is as **anti-quantifier**, i.e. as key on the DstrShr. *je* appears here as an adverbial preceding an indefinite NP which is quantificationally *dependent*, i.e. an NP_2 in the above sense. Since the notion of dependence is relational, *je* refers back, as it were, to some distributional domain NP_1. In Choe's words, its function is "to mark DstrShr for distributive dependency just as the function of *-self* in English is to mark an anaphor for anaphoric dependency." That *je* in this use is part of the NP and not some kind of verb phrase adverbial is shown by a test where *je* shows up inside a prepositional phrase, as under (2). Another test is topicalization, like in (3): the floated quantifier *alle* in (3a) stays behind, whereas *je* in (3b) had to go with the topicalized NP *einen* Lei.[1]

(2) a. Die Männer sprangen [$_{PP}$ in je ein Auto].
'The men jumped into one car each.'

b. Er überraschte die Kinder [$_{PP}$ mit je zwei Äpfeln].
'He surprised the children with two apples each.'

c. Die Kandidaten sollten [$_{PP}$ aus je zwei Urnen] Kugeln herausgreifen.
'The candidates were to pick (??balls/?a ball) from two urns each.'

(3) a. Die Touristen bekamen alle [einen Lei] umgehängt ⇒
[Einen Lei] bekamen die Touristen alle ungehängt.

[1] Here and in the following examples I have tried to give English paraphrases, most of the time translating *je* by postnominal or "shifted" *each*. Grammatical judgments are given for German and English sentences alike, using the familiar symbols '*' (out) and '?' (marginal, questionable). About most English sentences, however, I did not have any intuitions; special thanks go to Craige Roberts, her husband and to Peter Sells for providing the missing judgments.

'The tourists all received a lei' ⇒ 'A LEI the tourists all received.'

b. Die Touristen bekamen [je einen Lei] umgehängt ⇒
[Je einen Lei] bekamen die Touristen ungehängt.
'The tourists received a lei each.' ⇒ '*A lei each the tourists received.'

2. *je* can also appear as **key on the distributional domain**. Consider (4) below; in (4a) every person is supposed to get at most four tickets, the domain being the collection of all persons. Identifying the domain both in (4b) and (4c) is a little bit less straightforward; it is not just the collection of all strokes or students but a suitable partition over those collections; in fact, we have to form the collection of mutually disjoint groups of strokes and students, respectively: for every group of 1000 strokes you'll be paid 10 marks, and for every group of two students there was a microscope available. (This quantification over groups is what I called "genuine plural quantification" in Link (1987b), with the qualification that the plural sums quantified over here have to be made up by mutually disjoint sets of atoms.)

(4) a. Das Limit war vier Karten (je/pro) Person.
'The limit was four tickets per person.'

b. Sie zahlen 10 Mark (für je(weils) 1000 Anschläge / pro 1000 Anschläge).
'They pay 10 marks (for every 1000 strokes/per 1000 strokes).'

c. Je(weils) zwei Schüler teilten sich ein Mikroskop.
'Every two students shared a microscope.'

3. Finally, there also seems to be a **floated quantifier** use of *je*, which can be considered as marginal, however. An example is (5).

(5) Die Männer rannten (alle/jeder/je) in eine andere Richtung los.
'The men (all/each) headed off in a different direction.'

5.2 The Distributional Domain

In what follows I shall concentrate on *je* in its anti-quantifier position. Let us first look at the distributional domain NP_1. It has to be semantically plural and definite; this places a restriction on the determiners that can occur in NP_1, as the following sentences show.

(6) a. Die Kinder bekamen je drei Äpfel. (definite PNP)[2]
'The children got three apples each.'

b. (Alle/?beide) Kinder bekamen je drei Äpfel.
'(All the/*both) children got three apples each.'

c. *(Jedes/ein/kein) Kind bekam je drei Äpfel.
'*(Every/a/no) child got three apples each.'

d. (?Einige/*viele/*die meisten) Kinder bekamen je drei Äpfel.
'(?Some/*many/*most) children got three apples each.'

e. *Kinder bekamen je drei Äpfel.
'*Children got three apples each.'

f. Sie bekamen je drei Äpfel. (plural pronoun)
'They got three apples each.'

It seems, however, that an indefinite PNP, if it is specific enough, may also form the distributional domain; so (7a) is at least marginally possible, and with numerals like in (7b) the construction is just fine.

(7) a. ?Maria gab (einigen) Kindern je drei Äpfel.
'Mary gave (some/*∅) children three apples each.'

b. Maria gab zwei Kindern je drei Äpfel.
'Mary gave two children three apples each.'

On the other hand, group-denoting terms are both definite and semantically plural, and yet they do not readily serve as distributional domain:

(8) a. ?Die Mannschaft deckte je einen Gegner.
'*The team guarded an opponent each.'

b. ?Das Kommitee stimmte für je einen Kandidaten.
'*The committee voted for a candidate each.'

c. ?Die Mannschaft trank je drei Bier.
'*The team drank three beers each.'

The domain can be a conjunction, however. (9) gives examples of NP conjunctions, and under (10) we find some interesting cases of VP conjunctions filling the role of the distributional domain. In (10a) we have altogether four children who, in groups of two, are distributed over two events: one is a transport to the ball game, the other a transport to the hospital. (10b) means that one clerk processed my application, and another handled my damage claim; finally, three cooks are involved in the scene reported in (10c), one for each of the three jobs of preparing, roasting, and stirring.

---

[2]'PNP' abbreviates "plural noun phrase".

(9) a. Hans und Maria probierten je zwei Weine.
'John and Mary tasted two wines each.'

b. Dem Präsidenten, dem Kanzler und dem Außenminister wurde je ein Dolmetscher zugeteilt.
'An interpreter each was assigned to the president, the chancellor, and the foreign minister.'

(10) a. Je zwei Kinder wurden zum Stadion gebracht und ins Krankenhaus eingeliefert.
'*Two children each were taken to the ball game and sent to the hospital.'

b. Je ein Bediensteter bearbeitete meinen Antrag und nahm meine Schadensmeldung entgegen.
'*(A/one) clerk each handled my application and took care of my damage claim.'

c. Je ein Koch präparierte die Vorspeise, briet das Fleisch und rührte die Soße an.
'*(A/one) cook each prepared the hors d'oeuvre, roasted the beef, and stirred the gravy.'

An interesting question arises in connection with "multi-layered" group terms in NP_1 position: to which extent is the key able to discriminate different levels in those terms? Consider the sentences under (11).

(11) a. Die blauen Wildlederschuhe (kosten je 150 Mark/haben je einen Fleck).
'The blue suede shoes (cost 150 marks each/have a stain each).'

b. Die Meiers und die Müllers (testeten/teilten sich) je ein Auto.
'The Meiers and the Müllers (tested/shared) (one/a) car each.'

c. ?Napoleon und Blücher und Wellington befehligten je eine Armee.
'?Napoleon and Blücher and Wellington commanded an army each.'

d. *Kasparow und die Großmeister spielten je eine Schachpartie.
'*Kasparov and the grandmasters played a game of chess each.'

e. Kasparow und jeder Großmeister spielten eine Schachpartie.
'Kasparov and every grandmaster played a game of chess.'

e. Der Unterausschuß des Senats und der Unterausschuß des Repräsentantenhauses votierten für je eine Gesetzesvorlage des Präsidenten.
'The senate subcommittee and the house subcommittee voted for (one/a) presidential bill each.'

The German (11a), with its first VP, refers to pairs of shoes, and to single shoes with its second VP. The switch is governed by the descriptive content of the VP involved: the price typically refers to pairs of shoes, whereas stains affect shoes individually. A similar diagnosis can be given for (11b). The sentence with *sich teilen* (*share*) unambiguously distributes the DstrShr over the two families because of the collective predicate involved, while the verb *testen* (*test*) in the sentence admits of two readings, involving distribution over both families and family members. (11c), however, cannot possibly be read in its intended historical sense, viz., that Napoleon commanded one army, and Blücher and Wellington together commanded another army. Likewise, (11e) has a meaning which (11d) does not have, viz., that Kasparov played every grandmaster; that is, *je* cannot build asymmetric pairs of the form $\langle$Kasparov,$x\rangle$ where x is a grandmaster, and distribute over them. Finally, (11f) has only one reading, in which there is just one bill for each of the two subcommittees; the committee members cannot serve as distributional domain. Incidentally, this casts doubt on Landman's view in (1989) that conglomerates like *the working class* are built up from subgroups at various levels of a hierarchy that resembles the transitive closure of a set. Even if you think that is what the working class consists of "ontologically" (say a collection of labor unions each of which comprises local committees of union members) then that structure might be world knowledge, but it is linguistically inert: From a linguistic point of view, at least, the working class is one and cannot be split up into a host of individual consumers by any distributional key, or so my theory of plurals claims.

5.3 The Distributive Share

While it does not seem to be a historical accident that the German universal quantifier *jeder* is a contraction of *je* and *der*, the German definite article,[3] modern German requires an indefinite DstrShr after the anti-quantifier *je*

[3]This has been stressed by Dietmar Zaefferer (personal communication), who also insists that a historical explanation of this fact is called for.

(see the examples under (12)). From the semantic point of view this is obvious since the a scopeless definite NP is not really dependent quantificationally. There are marginal cases, though, in which a definite NP does make sense as DstrShr (even if it is not the most elegant prose), in particular when the DstrShr is a relational term like *father*; see (13) below. Note, however, that in (13b,c) we might as well say that these sentences derive their acceptability from the fact that *je* can also be regarded as floated rather than anti-quantificational. The same remark applies to sentence (14) where the DstrShr is a concrete disjunction rather than an NP with an indefinite article. Note also that the *jeweils* variants are always o.k. since *jeweils* may distribute externally (see below).

(12) a. Die Kinder bekamen (*je/jeweils) (viele/wenige) Äpfel.
'The children got (many/few) apples (*each/each time).'

b. Die Kinder bekamen (?je/jeweils) (mindestens/höchstens) dreiÄpfel.
'The children got (at least/at most) three apples (?each/each time).'

c. Die Kinder bekamen (*je/jeweils) (den/jeden) Ball.
'The children got (the/every) ball (*each/each time).'

(13) a. ?Die Mädchen nannten je ihren Vater (als) ihren besten Freund.
'*The girls called their father each their best friend.'

b. ?Die Schüler hatten je den Bus verpaßt.
'*The students had missed the bus each.'

c. Die Männer wählten je das kleinere Übel.
'The men chose the lesser evil each.'

(14) ?Die Abgeordneten gaben ihre Stimme je Engholm, Barschel oder Ronneburger.
'*The delegates voted for Engholm, Barschel or Ronneburger each.'

5.4 Accessability

1. The next issue is accessability. Choe (1987) observes quite correctly that, in principle, the domain and the DstrShr have to be clause mates for the distribution to be triggered. To a certain extent, both the domain and the DstrShr can be embedded in an NP which is a clause mate of its companion

NP. Examples for distribution across NP boundary are listed under (15): in (15a) the domain is situated under a preposition, in (15b) under another NP inside a PP, and in (15c) in a relative clause whose head is a clause mate of the DstrShr; while (15b) is impeccable, (15a) and (15c) are marginally acceptable. (15d) has the DstrShr as a conjunct of a conjunctive object phrase which is clause mate of the domain. This sentence is fine again, but then perhaps we have to analyze it as a straightforward case of conjunction reduction, which would account for the easy accessability of *je*.

(15) a. ?[$_{PP}$ An [den Messeständen]] prüfte Hans je ein Angebot.
'?At the fair booths John examined (an/one) offer each.'

b. [$_{PP}$ Für [die Sicherheit [der Minister]]] stand je ein Beamter zur Verfügung.
'?A guard each was available for the security of the ministers.'

c. ?[Der Mann [der die Weine verkaufte]] öffnete je eine Flasche.
'?The man who was selling the wines opened a bottle each.'

d. Die Jungen bekamen [den Pokal und [je einen Fußball]].
'The boys received the cup and a football each.'

2. The following sentences contain two potential domains or two potential DstrShr's. When there are two potential domains like in (16) the key seems to pick the definite PNP closest to it for distribution, e.g., *die Kinder* in (16a); but distribution over the subject NP *die Mütter* is fine as well. The latter reading is even predominant if the NP closer to the key is only a pronoun (*sie* in (16b)). But linear distance is not the whole story, as one might expect: in (16c), the genetive NP *der Kinder* is subordinated to *Mütter* and cannot serve as domain for the distributive key at all.

(16) a. Die Mütter überraschten die Kinder mit je zwei Äpfeln.
'The mothers surprised the children with two apples each.'

b. Die Mütter überraschten sie mit je zwei Äpfeln.
'The mothers surprised them with two apples each.'

c. Die Mütter der Kinder brachten je einen Kuchen.
'The mothers of the children contributed a cake each.'

Now consider the sentences in (17).

(17) a. Die Mütter erzählten zwei Kindern je ein Märchen.
'The mothers told two children a fairy tale each.'

b. Die Mütter erzählten je zwei Kindern ein Märchen.
'The mothers told two children each a fairy tale.'

c. Die Mütter erzählten je zwei Kindern je ein Märchen.
'*The mothers told two children each a fairy tale each.'

(17a) has at least two readings; the first is like the one in (7b) and also comports with the principle observed in (16a): there is a group G of mothers and a group G' of two children, and G collectively told each child in G' a fairy tale; the second reading relates the key to the subject phrase. Now suppose there are three mothers; how many listeners are there in the second reading? The answer is two, which means that the indefinite PNP *zwei Kindern* takes wide scope over the rest of the sentence. How are we to produce a reading in which every mother told two different children a tale? Well, the key has to go into the NP *zwei Kindern*, like in (17b). But the indefinite NP *ein Märchen* is now up for distribution: were there three or six tales told? It seems to me that the natural reading of (17b) has one tale for each group G' of children. But double distribution is perfectly possible, and can indeed be expressed in German using the anti-quantifier *je* twice: witness (17c). In fact, this sentence gives rise to the quantifier structure $\forall\exists\forall\exists$, which is quite complex. It will be shown below how such a construction can be given a formal treatment in the theory of plurals, by repeatedly using the distributivity operator 'D' introduced in Link (1987b).

3. Distribution works in passive sentences as well, irrespective of the linear order of domain and DstrShr; see sentences (18a,b) and (18c,d). This leads to forward anaphora in the distribution relation where the key precedes the domain; although more natural in passive constructions, forward anaphora are also found in active sentences, witness (19).

(18) a. Von den Müttern wurde je eine Kleinigkeit beigesteuert.
'A little something each was contributed by the mothers.'

b. Je eine Kleinigkeit wurde von den Müttern beigesteuert.
'A little something each was contributed by the mothers.'

c. Den drei Besuchern wurden je zwei Dolmetscher zugeteilt.
'To the three visitors two interpreters each were assigned.'

d. Je zwei Dolmetscher wurden den drei Besuchern zugeteilt.
'Two interpreters each were assigned to the three visitors.'

(19) a. Je ein Professor las die eingereichten Dissertationen.
'?A professor each read the submitted dissertations.'

b. Je ein Lehrer war für die Eltern da.
'(A/one) teacher each was available for the parents.'

4. An important issue which will be addressed now is the question to which extent the distribution works across clause boundaries and in control contexts. It is not for me to relate the data to current syntactic theory, so I shall only present them and hope they are of interest to the syntactician. In (20a) forward anaphora into a lower *daß*-sentence fails completely; turned around as in (20b) the sentence becomes better, but is still quite awkward. In (20c) also the subordination relation is changed (the domain now being in the higher clause), and this sentence is not so bad. Similarly, with a *verbum dicendi* like *sagen* in sentence (21a), some distributional transfer across clause boundary seems possible if only because it is taken as elliptic for (21b). Notice that replacing *je* by *jeweils* here produces a different meaning; with *jeweils* like in (21c), distribution over the PNP in the higher clause is in fact excluded.

(20) a. *Je ein Test ergab, daß die Mittel krebserzeugend sind.
'*A test each showed that the drugs produce cancer.'

b. ??Daß die Mittel krebserzeugend sind, hat je ein Test zweifelsfrei nachgewiesen.
'*That the drugs produce cancer a test each proved beyond doubt.'

c. ?Die Mittel sind krebserzeugend, weil je ein Test dies zweifelsfrei nachgewiesen hat.
'*The drugs produce cancer since a test each proved that beyond doubt.'

(21) a. ?Die Sekretärinnen sagten übereinstimmend aus, daß Oliver North je einen Shredder angefordert habe.
'?The secretaries unanimously said that Oliver North had asked for a shredder each.'

b. Die Sekretärinnen sagten übereinstimmend aus, daß Oliver North bei ihnen je einen Shredder angefordert habe.
'The secretaries unanimously said that Oliver North had asked them for a shredder each.'

c. Die Sekretärinnen sagten aus, daß Oliver North jeweils einen Shredder angefordert habe.
'The secretaries said that Oliver North had asked for a shredder each time.'

In infinitival constructions there seems to be a considerable difference in acceptability between the German *je* and its supposed English counterpart, the shifted *each*. This lack of fit is, of course, partly due to the fact that the infinitive constructions in German and English don't quite work the same way. A number of relevant examples is listed here.

(22) a. Die Wachen (sahen/hörten) je einen Häftling weglaufen. (*Control*)
'*The guards (saw/heard) a prisoner each run away.'

b. Hans und Maria wollten je einen Yoga-Kurs machen. (*Subject Control*)
'?John and Mary wanted to join a yoga class each.'

c. Die Männer beschlossen, je ein Auto zu mieten.
'The men decided to rent a car each.'

d. Die Parteien versprachen, je einen Vertreter zu schicken.
'The parties promised to send a spokesman each.'

e. Die Wachen zwangen je einen Häftling, sich hinzuknien. (*Object Control*)
'?The guards forced a prisoner each to kneel down.'

(23) a. Die Wachen schienen je einen Häftling zu beobachten. (*Raising*)
'The guards seemed to watch a prisoner each.'

b. Die Linienrichter wähnten je einen Spieler im Abseits. (*no full infinitive*)
'*The linesmen believed a player each to be off-side.' (*Raising*)

c. Hans und Maria ließen je ein Fahrrad reparieren.
'John and Mary had a bike each repaired.'

d. Die Wachen ließen je einen Häftling entkommen.
'The guards let a prisoner each escape.'

e. Die Lehrer ließen je ein Kind ein Gedicht aufsagen.
'?The teachers let a child each recite a poem.'

f. Die Lehrer ließen das Kind je ein Gedicht aufsagen.
'*The teachers let the child recite a poem each.'

5.5 *je* vs *jeweils*: Events as Distributional Domain

Since German is more liberal with respect to the clause mate constraint, as we saw, it seems natural to ask whether the distributive key needs a domain in the sentence at all. I for one have no problems with sentences like those in (24),[4] uttered in a suitable context which supplies the distributional domain, e.g., a number of baskets in a market place, or mothers at a PTA meeting.

(24) a. Je drei Äpfel waren faul.
'*Three apples each were rotten.'

b. Je(weils) eine Kleinigkeit wurde beigesteuert.
'A little something (each/each time) was contributed.'

Note again that the sentences with *jeweils* instead of *je* give only an incomplete paraphrase. *jeweils* introduces a much stronger temporal connotation. (25a) for instance suggests that John bought two papers on Monday, on Tuesday, and so on, and (25b) might have been uttered with reference, say, to Bob Dylan's recent concert tour through Germany. Similarly, in (25c) *je* distributes over the Marshall Islands, whereas *jeweils* more likely distributes over a set of occasions, e.g. the yearly Nuclear Free Pacific Conferences that the islands (collectively) sent a representative to.

(25) a. Hans kaufte jeweils zwei Zeitungen.
'John bought two news papers each time.'

b. Jeweils ungefähr hundert Karten wurden verkauft.
'About a hundred tickets were sold each time.'

c. Die Marshall-Inseln schickten (je/jeweils) einen Abgesandten.
'The Marshall Islands sent a representative (each/each time).'

In a framework that has the notion of *event* available the difference between *je* and *jeweils* can be captured formally. The model structure for events presented in Link (1987a), for instance, allows one to distinguish between sums of events that are parametrized along different "dimensions". Thus, the sentence (11c) with *je* would be represented by a sum of events parametrized along the agent dimension: the elements of the sum consist of events of the form 'x sent a representative' where the parameter x runs over all agents involved (viz., the Marshall Islands). The sentence with *jeweils*, however, can also be interpreted as a different sum of events: here the individual events have the form 'the Marshall Islands sent a representative at occasion y', where the parameter y runs over the dimension of appropriate occasions (e.g. the yearly Nuclear Free Pacific Conferences). Thus, the distributional domain is formed *externally* as a certain class of events provided by the context.

[4]Choe (1987) reports that informant Angelika Kratzer disagrees.

5.6 The Semantics of *je* Constructions

It is not the purpose of this chapter to give detailed semantic representations of all the *je* constructions presented above. I'd like to indicate here at least in principle, however, what my present views are about the semantics of those constructions. In Link (1987b) I used the distributivity operator 'D' to achieve the semantical effect of *je* which I identified as a distribution over cases, internal of external. I still think that the D operator is the right tool as long as no events are introduced into the semantics. The basic idea is that a VP containing *je* translates into a VP′ headed by the D operator. If VP′ is the schema $\phi \equiv \phi[u]$ we have the following definition ((Link, 1987b, 171); see also Link 1983 for the basic concepts of the theory of plurals that are used below).

(26) $$^D\phi[u] \quad \equiv \quad \lambda x \forall u\, [\, u \,{}^{\cdot}\Pi\, x \,\rightarrow\, \phi[u]\,]$$

That is, $^D\phi \equiv {}^D\phi[u]$ denotes the set of all individual sums (i-sums) x such that every atomic part u of x has the property ϕ. For example, let $\phi \equiv \phi[u]$ stand for $\exists y[\text{Apfel}'(y) \wedge \text{bekam}'(u,y)]$ ("u received an apple"); then sentence (27a) has the representation (27b).

(27) a. Die Kinder bekamen je einen Apfel.
'The children received an apple each.'

b. $^D\phi[\sigma z\, \text{Kind}'(z)] \;\equiv\; \lambda x \forall u\, [\, u \,{}^{\cdot}\Pi\, x \rightarrow \phi[u]\,](\sigma z\, \text{Kind}'(z))$
$\equiv\; \forall u\, [\, u \,{}^{\cdot}\Pi\, (\sigma z\, \text{Kind}'(z)) \rightarrow \phi[u]\,]$
$\equiv\; \forall u\, [\, u \,{}^{\cdot}\Pi\, (\sigma z\, \text{Kind}'(z)) \rightarrow \exists y[\text{Apfel}'(y) \wedge \text{bekam}'(u,y)]\,]$

As desired, that means that for every child u there is an apple y such that u receives y . Now Choe (1987a):145-148 thinks there is a problem for this approach when the sentence contains more than one potential DstrShr, his example being the Korean translation of (17a,b) (the sentences under (17) are repeated here for convenience).

(28) a. Die Mütter erzählten zwei Kindern je ein Märchen.
'The mothers told two children a fairy tale each.'

b. Die Mütter erzählten je zwei Kindern ein Märchen.
'The mothers told two children each a fairy tale.'

c. Die Mütter erzählten je zwei Kindern je ein Märchen.
'*The mothers told two children each a fairy tale each.'

But the familiar technique of quantifying-in not only yields correct representations for the first two sentences but also for the double anti-quantifier case (17c). To begin with, (17a) has two most salient readings according to what serves as distributional domain. The preferred reading LF(i) in (29) below is the one in which the dative NP *zwei Kindern*, which is contiguous to the DstrShr, is distributed over; in this case, the subject NP *die Mütter* is most naturally read collectively and can be considered as scopeless. Thus, under (i) there are two children each of whom was told a fairy tale by the mothers' joint effort. Let $\phi[u, v] \equiv \exists w\,[\,\text{Märchen}'(w) \wedge \text{erzählte}'(u, v, w)\,]$ be the schema which is more ore less common to the representations to be given for the sentences in (17), viz., "u told v a fairy tale." From it we derive various logical phrases ψ_i on which the D-operator operates; in each case such a phrase ψ_i is thereby asserted to hold of the individual sum in the distributional domain. The appropriate phrase for the reading (i) of sentence (17a) is the schema $\psi_1 \equiv \psi_1[v] \equiv \phi[\sigma x\, \text{Mutter}'(x), v]$ ("the mothers told v a fairy tale"). Then we have:

(29)
$$\begin{aligned}&\exists y\,[\,(\mathbf{2}\ \text{Kinder}')(y) \wedge {}^{D}\psi_1(y)\,] \equiv\\ &\exists y\,[\,(\mathbf{2}\ \text{Kinder}')(y) \wedge \forall v\,[\,v\,{}^{\cdot}\Pi\, y \rightarrow\\ &\exists w\,[\,\text{Märchen}'(w) \wedge \text{erzählte}'(\sigma x\, \text{Mutter}'(x), v, w)\,]\,]\,]\\ &\qquad\qquad\qquad = \text{LF (i) of (17a)}\end{aligned}$$

("there is a sum y of two children such that for all atomic parts v of y [i.e., for every child v] there is a fairy tale w such that the sum of all mothers [i.e., the mothers collectively] told w to v").

However, (17a) can also be read to mean that the DstrShr distributes over the subject NP. In the sentence at hand the dative NP *den Kindern* still has wide scope; but this is not necessary for the distribution to work over the remote phrase: when both the subject NP and the indirect object NP are indefinite phrases carrying a numeral (as in *drei Mütter erzählten zwei Kindern je ein Märchen* — 'three mothers told two children a fairy tale each'), then scope order in the preferred reading is from left to right, but the NP *drei Mütter* can still function as distributional domain. In the second reading (ii) for (17a), then, the suitable schema is $\psi_2 \equiv \psi_2[u] \equiv \phi[u, y]$ ("u told y a fairy tale"), and ${}^{D}\psi_2[u]$ is applied to the i-sum of the mothers; thus

(30)
$$\begin{aligned}&\exists y\,[\,(\mathbf{2}\ \text{Kinder}')(y) \wedge {}^{D}\psi_2(\sigma x\, \text{Mutter}'(x))\,] \equiv\\ &\exists y\,[\,(\mathbf{2}\ \text{Kinder}')(y) \wedge \forall u\,[\,u\,{}^{\cdot}\Pi\,(\sigma x\, \text{Mutter}'(x)) \rightarrow\\ &\exists w\,[\,\text{Märchen}'(w) \wedge \text{erzählte}'(u, y, w)\,]\,]\,] \qquad = \text{LF (ii) of (17a)}\end{aligned}$$

("there is a sum y of two children such that for all atomic parts u of the sum of all mothers [i.e., for all mothers u] there is a fairy tale w such that u told w to y [i.e., to the children collectively]").

In (17b) the situation is unambiguous: the subject NP *die Mütter* is the distributional domain for the DstrShr *je zwei Kindern*; the appropriate schema here is $\psi_3 \equiv \psi_3[u] \equiv \exists y\,[(\mathbf{2}\text{ Kinder}')(y) \wedge \phi[u,y]]$ ("there is an i-sum y consisting of two children such that u told y a fairy tale"). The D-term applied to the i-sum $\sigma x\,\text{Mutter}'(x)$ yields

(31) $$^{D}\psi_3(\sigma x\,\text{Mutter}'(x)) \equiv \forall u\,[\,u\,{}^{\cdot}\Pi\,\sigma x\,\text{Mutter}'(x) \rightarrow \exists y\,[\,(\mathbf{2}\text{ Kinder}')(y) \wedge \exists w\,[\,\text{Märchen}'(w) \wedge \text{erzählte}'(\sigma x\,\text{Mutter}'(x), y, w)\,]\,]\,] \quad = \text{LF of (17b)}$$

("for all atomic parts u of the sum of all mothers [i.e., for all mothers u] there is a sum y of two children and a fairy tale w such that u told w to y [i.e., to the children collectively]").

Finally, in (17c) the distribution procedure is iterated: first, the direct object NP *je ein Mädchen* is distributed over the indirect object NP *je zwei Kindern* which, in turn, distributes over the subject NP *die Mütter*. With $\psi \equiv \psi_4[v]$ as $\phi[u,v]$ and $\psi_5 \equiv \psi_5[u]$ as $\exists y\,[(\mathbf{2}\text{ Kinder}')(y) \wedge ({}^{D}\psi_4)(y)]$ we have

(32) $$^{D}\psi_5(\sigma x\,\text{Mutter}'(x)) \equiv \forall u\,[\,u\,{}^{\cdot}\Pi\,(\sigma x\,\text{Mutter}'(x)) \rightarrow \exists y\,[\,(\mathbf{2}\text{ Kinder}')(y) \wedge ({}^{D}\psi_4)(y)]\,] \equiv \forall u\,[\,u\,{}^{\cdot}\Pi\,(\sigma x\,\text{Mutter}'(x)) \rightarrow \exists y\,[\,(\mathbf{2}\text{ Kinder}')(y) \wedge \forall v\,[\,v\,{}^{\cdot}\Pi\,y \rightarrow \exists w\,[\,\text{Märchen}'(w) \wedge \text{erzählte}'(u,v,w)\,]\,]\,]\,] \quad = \text{LF of (17c)}$$

("for all atomic parts u of the sum of all mothers [i.e., for all mothers u] there is a sum y of two children such that for all atomic parts v of y [i.e., for every child v] there is a fairy tale w such that u told w to v").

I did not show here, of course, what an explicit translation algorithm of turning *je* sentences into semantic representations might look like. For instance, the compositionality principle would require the DstrShr *je ein Märchen* to receive a separate translation. Obviously this can hardly be done without a lot of abstracting if the desired output above is to be reached. Furthermore, the quantifying-in device seems to be unusual in that it looks like it has to be mandatory in the *je* cases rather than optional. As soon as some mechanism has made out the distributional domain for the key, the domain is "taken out" of the appropriate context, leaving a schema for the D-operator to operate on. Such a procedure is necessary since semantically, the domain has to gain scope over the DstrShr. This is admittedly vague, but any translation algorithm has to create the effect one way or the other. My purpose here was to show which additional requirements for the overall shape of the grammar come from the existence of anti-quantifier constructions in language.

Things look somewhat different, however, when the event structures in Link (1987a) are adopted. Then a distributive reading receives a representation as a sum of events which is parametrized by the elements of the distributional domain. The sentences in (17) differ in the domain over which the summing takes place; for instance, (17b) has a representation as a sum of events whose atoms are of the form "mother u told two children a fairy tale", and the summing runs over all mothers. The anti-quantifier *je*, then, is seen to obligatorily trigger a representation in terms of sums of events.

Chapter 6

First-Order Axioms for the Logic of Plurality

In the previous chapters the concept of plurality was investigated from a structural point of view. That led to the lattice-theoretic approach which introduced structured individuals, viz., the individual sums, into the domain of discourse. Since the main purpose has been to model plural phenomena in language and to contribute to their philosophical understanding the emphasis was put on the *semantic aspect* of plurality; no metalogical treatment was given so far. In particular, I have not been too specific about the precise nature of the axioms that govern the use of plural objects. In Chapter 1, for instance, I just assumed the models interpreting the plural part of the theory LPM to be Boolean algebras because they are rather familiar structures. But then we saw that not all the features of a Boolean algebra were actually needed to describe the behavior of plurals. For instance, we didn't have to speak about meets and complements most of the time since there are no ready-made expressions in language denoting such operations. That suggests a more "genetic" approach that introduces the necessary axioms one by one and tries to justify each of them both on linguistic and philosophical grounds.

In this chapter I am going to present such an axiom system, which I call LP again. It characterizes a certain semilattice structure which is seemingly weaker than a Boolean algebra since it never mentions meets or complements; but it turns out that in the presence of additional axioms (non-zero) meets and complements can be defined. In fact, the given axioms already suffice to yield the full complete atomic Boolean structure minus its zero element after all. This is a well-known mathematical fact, but it will be described here for completeness of exposition. Some of the relevant proofs will be given in the Appendix.

I will also relate these axioms to other systems in the literature that deal with structured domains. One prominent example in the philosophical tradition are the *mereological systems* of various kinds that characterize the notion of part-whole relation in a general way.[1] In fact, the plural lattice amounts to a "classical" mereology although its motivation is completely different. That makes one wonder why the structure is basically the same. I will come back to this question below.

Among the systems that deal specifically with plurality I shall discuss the work of Jan Tore Lønning (1989, 1997). It takes a second-order approach to plurality but also addresses the question, which parts of genuine plural discourse, if any, can be described within the bounds of first-order logic.

That brings us to metatheory proper. There is a particular result that I will make use of in the final chapters of this book. Let us call $ML2^-$ that version of monadic second-order logic which quantifies only over non-empty sets. $ML2^-$, just like regular monadic second-order logic ML2, doesn't admit of a complete axiomatization (with respect to the usual standard[2] models). Now it can be shown that $ML2^-$ can be embedded into LP, which means that LP is incomplete, too. The reason for the incompleteness is the fact that we want to allow the formation of sums from arbitrary collections of objects, and this is a second-order device. However, if we restrict the collections to the ones that are *definable in the language* then one does have a completeness result. The price to be paid is that the notion of validity has now to be understood with respect to definable structures that are not the full models anymore.

I won't discuss the question here whether pluralic discourse in language refers to full models or not. I do think, however, that the full models are the "intended" ones, and that we have to accept incompleteness as a fact of life, so to speak. Is LP a second-order logic is disguise, then? Regarding its logical strength, the answer is Yes, when we stick to standard models.[3] But there is still a point here, I submit, in keeping to the first-order format: The higher-order objects that second-order logic quantifies over are best viewed as *properties* and not pluralities. It is true that in extensional semantics properties are *modeled* by sets, and that non-empty sets can be mapped onto the pluralities in a one-to-one fashion, but that doesn't turn pluralities into properties *conceptually*. This is argued more fully in Chapter 13 where

[1] Eberle (1970) and Simons (1987) are excellent sources on mereology.

[2] The distinction is here between *standard* and *generalized* models in the sense of Henkin (1950); for an account, see Shapiro (1991).

[3] We will see some of the power of LP in Chapter 14 where David Lewis's "nominalistic" reconstruction of set theory is carried out within LP.

I discuss the relation of LP to George Boolos's nominalistic interpretation of monadic second-order logic.

What pluralities are, and what the kind of discourse is that they are intended to model, has been amply described in the previous chapters from a linguistic point of view. With the focus now on the logical side, I'll shortly review the basic features of LP and its application in linguistic semantics.

6.1 The original LP arsenal and its use

The logic of plurality LP is couched in a first-order language with equality and λ-abstraction. The most important special symbols are: '$\oplus$' for (finite) *plural conjunction*, forming i-sums; '$\leq_i$' for the *individual part-of relation* (also called the "below" relation in what follows); 'σ' for *pluralic definite description*; and '*' for the *plural operator* on 1-place predicates P. The star produces predicates of the form $^*P$ that are true of exactly those complex objects which are built up from the P-elements by means of a general summing operation on the domain. The meta-variable 'P' is taken to range not only over basic predicates but also over complex ones that are formed by λ-abstraction.

Defined notions are the predicate 'At', which singles out the *atomic elements* of the domain, and the *distributivity operator* 'D' that marks the distributive reading of a predicate.

Such a system is able to deal with, among other things,

- indefinite and definite plural NPs
- collective and distributive readings
- plural anaphora.

It can be added to other representation languages to cover plural phenomena (e.g., discourse representation theory, DRT). Moreover, recent proposals for an event-based algebraic semantics (Bach (1986), Chapter 11 of this book, and Krifka (1989a)) build on the structural approach to objects in the individual domain that is put forth in LP, and extend it to the verbal domain.

A few typical examples illustrating the treatment in LP of various plural constructions are given below.

(1) a. Some delegates filed a resolution. (*indefinite plural NP*)

b. $\exists x(^*Px \wedge Qx)$ (*collective reading*)

(2) a. The delegates filed a resolution. (*definite plural NP*)

b. $Q(\sigma x Px)$ (*collective reading*)

(3) a. The delegates voted for the resolution. (*definite plural NP*)

b. ${}^{D}Q(\sigma x Px) \Leftrightarrow \forall u(u \leq_i \sigma x Px \rightarrow Qu)$ (*distributive reading*)

(4) a. The delegates met in the lobby; they discussed the resolution.

b. $\exists x\,(\,x = \sigma x Px \,\wedge\, Qx \,\wedge\, Rx\,)$ (*anaphora*)

(5) a. John and Mary are delegates and share an office. (*conjoined plural NP*)

b. $\exists x(x = j \oplus m \wedge {}^{*}Px \wedge Qx)$
$\Leftrightarrow\ {}^{*}P(j \oplus m) \wedge Q(j \oplus m)$
$\Leftrightarrow\ Pj \wedge Pm \wedge Q(j \oplus m)$
(*mixed distributive and collective predication*)

As can be seen from the logical representations given, existential quantification runs over i-sums, not only over atomic individuals. Thus in (1) it is actually some i-sum or collection of individuals that is said to jointly file the resolution. Since only the *members* of this collection x (and not x itself) have the property of being a delegate the predicate P for 'delegate' has to be starred: the effect is that all atomic individuals below x are asserted to be delegates. Note that this move avoids the possible "redundancy option" of granting i-sums of P-elements the property P, too. For problems with this option see Chapter 1.

In sentence (2) we have again the collective reading in which the property in question is predicated of an i-sum, this time of the definite collection of all the delegates present in the context. By contrast, when the assertion is that each individual delegate did something, e.g., vote for the resolution as in (3), the distributivity operator D has to be applied. The representation is then equivalent to a universal quantification as shown on the right-hand side of the formula (3b). The original translation ${}^{D}Q(\sigma x Px)$ allows for a rather natural compositional treatment of mixed distributive and collective predication as shown in (5). For an extensive coverage of distributivity phenomena see Roberts (1987b).

Sentence (4) shows a simple text with an anaphoric plural pronoun together with its first-order representation. It goes without saying that a more adequate format (like DRT, for instance) is needed to take care of the dynamic nature of anaphoric processes, the more so when we deal with the quite complex data of plural anaphora.

Let me now turn to a logical issue that touches the core of the representations in LP. The examples given above show apparent instances of plural quantification, but one might try to paraphrase them in such a way that they involve quantification over atomic, i.e., singular, individuals only. If one succeeds in doing so in special cases, then the more general question can be raised, whether there is any need in language to quantify irreducibly over pluralities. That brings up the problem of what I called *genuine plural quantification.* It was discussed to some extent in Chapter 4, but I'd like to add some comments here. Are there linguistic contexts in which quantification is not only expressed syntactically in terms of pluralic expressions (like *All men are mortal*) but where at the same time the objects quantified over are genuine i-sums? *All men are mortal* is not of this kind since it is completely equivalent to the distributive quantification *Everything that is human is mortal.* Now a sentence like *some delegates filed a resolution* seems to be an instance in point: it is true in a model of LP if there is an i-sum of delegates that collectively filed a resolution; so the existential quantifier runs over the full domain of i-sums here, not only over the atomic objects. By contrast, however, genuine *universal* plural quantification is harder to come by. One example is *Any three men will be able to lift this piano.* Another example that seems to come close is *All competing companies have common interests*; the intended interpretation here is that whenever there is an i-sum of companies whose members are in competition with each other then these members will show common interests. But change *all* to *most*, and this interpretation pattern gets into trouble. It is not at all the case that the sentence *most competing companies have common interests* means that most i-sums x of competing companies have the property that the members of x show common interests; but this would be the proper reading if there were quantification over sums involved. Rather, it is a clear fact of language that sums are not counted and weighed. Thus, the sentence seems to mean something like this: in *most of the situations* where we have a group of competing companies, the members of this group will have common interests. But we have no means in LP to express quantification over situations.

That genuine plural quantification is a rather unstable matter is shown by an important example of Lønning's (1989:56) where a sentence with an indefinite plural NP and a collective verb is *negated*, viz., (6a):

(6) a. Some boys did not share a pizza.

b. Somebody did not buy a book.

Consider (6b) first. What it means is that there is a person u such that no book was bought by u. By analogy, we would expect (6a) to mean that there is an i-sum of boys x such that no pizza was shared by x. But this reading seems to almost invariably come out true as soon as the number of boys exceeds 3, say. Lønning has a model with 6 boys which makes for $2^6 - 6 - 1 = 57$ pluralic i-sums, and chances are that at least one such i-sum has the property of not sharing a pizza. The most salient reading of (6a), however, is this, I think: there are some boys who did not *take part in* the activity of sharing a pizza (they were not hungry or they were too hungry to share). Thus we see that when a sentence with a collective VP is negated there is a strong tendency to interpret the VP distributively. If that is not possible or only marginally possible the sentence sounds odd, e.g., *some demonstrators didn't disperse*. For more on this issue see again Lønning (1989) and also Verkuyl (1988).

This is not the place for rehearsing all the arguments for providing a formal framework in which genuine plural quantification can be expressed. Suffice it to say that there is some indication that language has this device available. I am now going to explore the logical features of that framework.

6.2 LP axioms

6.2.1 The logical basis of free logic

Let us now turn to the question of axiomatics for the formal language LP. LP is formulated as a first-order theory with identity '$=$' and operators 'λ' and 'ι'. Thus the quantifiers of LP just bind first-order variables that range over individuals; syntactically, there is no quantification over higher-order entities. However, the axioms will impose a certain structure on the individuals which is strong enough to simulate monadic second-order logic within LP. The λ-operator forms abstracts of the form $(\lambda x\phi)$ for arbitrary formulas ϕ, yielding 1-place predicates which serve as names for complex properties expressible in the language. With the help of the ι-operator we can form the familiar definite descriptions $(\iota x\phi)$. The presence of definite descriptions immediately raises the question, how to deal with denotation gaps. The natural move here, short of giving up bivalence, is to adopt a *free logic* which allows for having singular terms in the language that do not always denote. Various systems of this kind have been proposed in the literature[4] which behave quite differently, so I cannot just refer to "free logic" if I want to be precise about the axioms of LP. For explicitness, then, I give the axioms of a particular free logic, called *FL*, which is *not* an

[4]Here are some standard references to the literature on free logic: Leonard (1956); Hintikka (1959); Lambert (1991b); van Fraassen (1966a,b); Scott (1967); Eberle (1969); Lambert and van Fraassen (1972)., An important source where many classical papers on free logic and its philosophical applications are collected is available with Lambert (1991a). For free logic techniques in mathematical logic, see also Beeson (1985); Feferman (1995) and references cited therein.

extension of classical first-order logic. Its main features are the following:[5] (i) FL allows for empty domains. (ii) The self-identity of terms is not a theorem. (iii) Argument places in predicates have *existential import*, that is, a term t can only be an argument in a true predication if t denotes (note that open formulas are not predicates in the present sense). (iv) A distinction can be made in FL between *weak* and *strong negation* by using appropriate λ-terms.

FL forms the logical basis of LP, and LP will be a proper extension of FL.

The language of FL is a first-order language $\mathcal{L}_1$ with identity and special symbols 'λ', 'ι', and '$\mathbf{E}$'. $\mathbf{E}$ is the 1-place *existence predicate*; thus, '$\mathbf{E}\,t$' reads "t exists" or "t denotes." The general syntactic variable for predicates is 'π', that for function symbols 'δ'. All predicates of arity $n \neq 1$ are basic predicate symbols, but for $n = 1$ predicates can be complex λ-terms. Thus, *1-place predicates* of FL are $\mathbf{E}$, 1-place predicate symbols or λ-terms. *Terms* of FL are variables or expressions of the form $(\delta t_0 \ldots t_{m-1})$ and $(\iota x \phi)$. *Atomic formulas* of FL are expressions of the form $\mathbf{E}\,t$, $s = t$, and $\pi t_0 \ldots t_{n-1}$ where s and t are terms. The set of *formulas* of FL is the closure of the set of atomic formulas under the usual logical operations.[6] A *well-formed expression (wfe)* of FL is a term, a predicate or a formula of FL. Syntactic variables for 1-place predicates are 'P' and 'Q'; for formulas 'ϕ' and 'ψ'; for variables 'x', 'y', 'z'; for individual terms 's', 't', 'r'; for well-formed expressions 'ζ' and 'η'.

The usual notions of *bound* and *free* occurrence of a variable, and that of being *free for* an occurrence in a wfe, now apply to expressions built with all variable-binding operators, the quantifiers, λ and ι. If ζ is a wfe, x a variable and t a term, then ζ^x_t is the *proper substitution*[7] of t for x in ζ. Again, if ζ is a wfe and t a term, then $\zeta[t]$ is ζ, indicating that t occurs in ζ at some designated place(s) for which t is free in ζ. For a wfe ζ, let $FR(\zeta)$ be the set of variables occurring free in ζ.

The *axioms* of FL are the following:

(Ax.1) ϕ (if ϕ is a *Tautological Schema*)

(Ax.2) $\forall x\,(\phi \to \psi) \to (\forall x \phi \to \forall x \psi)$ (*Quantifier Distribution*)

(Ax.3) $\exists x \phi \leftrightarrow \neg\forall x \neg\phi$ (*Quantifier Negation*)

(Ax.4) $\mathbf{E}\,t \to (\forall x \phi \to \phi^x_t)$ (*Instantiation*)

[5] Cf. Eberle (1969) whose system of free logic comes closest to the present one.

[6] Actually, one has to give a simultaneous recursive definition of 'term', '1-place predicate', and 'formula'.

[7] See, for instance, Kalish *et al.* (1980).

(Ax.5) $\forall x(x = x)$ (*Self-Identity*)

(Ax.6) $s = t \to (\phi[s] \to \phi[t])$ (*Leibniz' Law*)

(Ax.7) $s = t \to \mathbf{E}\, s \wedge \mathbf{E}\, t$ (*ExId*)

(Ax.8) $\pi t_0 \ldots t_{n-1} \to \mathbf{E}\, t_0 \wedge \ldots \wedge \mathbf{E}\, t_{n-1}$ $(n > 0)$ (*ExPr*)

(Ax.9) $\mathbf{E}\,(\delta t_0 \ldots t_{m-1}) \to \mathbf{E}\, t_0 \wedge \ldots \wedge \mathbf{E}\, t_{m-1}$ $(m > 0)$ (*ExOp*)

(Ax.10) $\mathbf{E}\, t \to [(\lambda x \phi)t \leftrightarrow \phi^x_t]$ (λ-*Conversion*)

(Ax.11) $Q(\iota x P x) \leftrightarrow \exists y\,(\forall x(Px \leftrightarrow x = y) \wedge Qy)$ (*Iota-Elimination*)

A *theorem of FL* is everything that is deducible from these axioms by means of the classical first-order inference rules, *Modus Ponens* (from $\phi, \phi \to \psi$ to infer ψ) and *Generalization* (from $\phi \to \psi$ to infer $\phi \to \forall x \psi$ if $x \notin FR(\phi)$). Notation: $\vdash_{FL} \phi$.

As an example of a theorem I mention the principle of existential generalization which reads in FL:

(T.33) $\vdash_{FL}\ \mathbf{E}\, t \wedge \phi^x_t \to \exists x \phi$ (*Existential Generalization*)

Note that the schematic self-identity $t = t$ cannot be derived anymore in the usual way by instantiation from (Axiom 5). In fact, it is not a theorem of FL, and thus, FL is not an extension of classical first-order logic. Rather $t = t$ amounts to the existence of a denotation for t, i.e., it says that t exists. This is expressed in the following two simple FL-theorems:

(T.34) $\vdash_{FL}\ \mathbf{E}\, t \leftrightarrow t = t$

(T.35) $\vdash_{FL}\ \mathbf{E}\, t \leftrightarrow \exists x(x = t)$ $(x \notin FR(t))$

Thus we see that the existence predicate is definable in terms of the identity sign. But I want to keep it for perspicuity.

Recall that the *uniqueness condition* for predicates ("there is exactly one P") is defined as

(D.45) $\exists! x P x \leftrightarrow \exists x P x \wedge \forall x \forall y\,(Px \wedge Py \to x = y)$

With this notation we can prove the *existence condition* for definite descriptions, by iota-elimination (set $Q \doteq P$) and (ExPr):

(T.36) $\vdash_{FL}\ \mathbf{E}\,(\iota x P x) \leftrightarrow \exists! x P x$

(T.37) $\vdash_{FL}\ \mathbf{E}\,(\iota x P x) \leftrightarrow P(\iota x P x)$

Let me list a few *non-theorems* in FL:[8]

(T.38) $\nvdash_{FL} \quad t = t$

(T.39) $\nvdash_{FL} \quad \neg\mathbf{E}\, s \wedge \neg\mathbf{E}\, t \to s = t$

(T.40) $\nvdash_{FL} \quad \phi[t] \to \mathbf{E}\, t$

6.2.2 Proper axioms for LP

Let us now proceed to LP proper. In the examples of Section 6.1 we had the special symbols '$\oplus$', '$\leq_i$', 'σ', '*', and '^{D}'. Let me start with the most basic symbol of plural theory, $\oplus$. It is a 2-place operation symbol and satisfies the following three axioms of a *semilattice* in the algebraic sense (see Appendix):

(Ax.12) $\forall x \forall y\, (\, x \oplus y \;=\; y \oplus x\,)$ (*Symmetry*)

(Ax.13) $\forall x \forall y \forall z\, (\, x \oplus (y \oplus z) \;=\; (x \oplus y) \oplus z\,)$ (*Associativity*)

(Ax.14) $\forall x\, (\, x \oplus x \;=\; x\,)$ (*Idempotence*)

Plural conjunction observes these axioms, that is, it is symmetric, associative, and idempotent. For example, if John and Mary share a room, Mary and John share it; if John and (Bill and Sue) meet, then (John and Bill) and Sue meet. Idempotence is more of a logical nature since, although trivially John and John is John, nobody would ever refer to John by 'John and John'. Indeed, there is a principle governing the pragmatics of reference involving conjunctive plural terms to the effect that a speaker using the term '*A* and *B*' (where '*A*' and '*B*' are names) is bound to refer to an i-sum consisting of two different objects named by '*A*' and '*B*'; thus the proposition that the author of *De re publica* denounced Catiline cannot be expressed by the sentence *Cicero and Tully denounced Catiline.* The same applies to conjunctive phrases with more than two conjuncts.

Doubts have been raised in the literature also against the other two axioms (Axiom 12), (Axiom 13), again on empirical grounds. For instance, while *John and Mary are husband and wife* is fine, *Mary and John are husband and wife* is not, seemingly violating symmetry. But then, not even *Mary and John are wife and husband* is o.k., which hints at a socially codified order of terms here. More of a problem is a sentence like *The first to cross the finish line were Mary and Sue, arriving in that order.*

[8](T.38) and (T.39), for instance, are endorsed by the popular free logic FD2 in Lambert and van Fraassen (1972).

But in such a sentence the conjunctive plural term is not only *used* but also *mentioned* by means of the phrase *in that order*, very much in the same way as in Quine's example *Giorgione was so-called because of his size*, where the name 'Giorgione' is also both used and mentioned at the same time. In general, my assumption here is that when symmetry seems to be violated there is an extra effect superimposed on the plain plural structure, so we can keep the axiom. Associativity has been contested with examples like *Napoleon and Blücher and Wellington fought against each other at Waterloo*, where historical knowledge leads us to interpret the sentence in a non-associative way, viz., "Napoleon against Blücher and Wellington." This problem relates to the issue of a putative "intermediate group level," a discussion of which can be found in Chapter 7. Again, I follow the line here that the basic meaning of plural conjunction is associative.

One more general question in connection with modeling pluralities has to be settled: Is it really necessary to admit i-sums consisting of any two objects under the sun, say, the pope and the prime number 37? While such a combination sounds like outright nonsense at first blush, it certainly doesn't do any harm; what is more, it might occur in sentences, e.g., *The pope and the prime number 37 are utterly different in their properties, the first being a concrete object, the second abstract etc.*); and arbitrary summing, being a total operation, simplifies the theory. These arguments in favor of admitting the summing of arbitrary entities are in my view sufficient for an ontologically "modest", but structurally adequate natural language semantics. Anybody who is more ambitious ontologically and wishes to put forward a more "contentful" theory of plural objects will be hard pressed to characterize classes of summable objects in a natural way. Note, however, that from a purely formal point of view there would be no obstacle to making plural conjunction a partial operation since we already operate in a free logic setting.

It is well-known that (Axioms 12–14) induce a partial order that guarantees the existence of finite least upper bounds (suprema) with respect to this ordering, making it into a *(join) semilattice* in the order-theoretic sense.[9] This partial ordering is the i-part relation '$\leq_i$', defined by

(D.46) $$t \leq_i s \;\leftrightarrow t \oplus s = s$$

With this definition and the axioms for $\oplus$ we can prove that $\leq_i$ is a partial order, viz., it is reflexive, transitive, and antisymmetric.

[9]A discussion of the correspondence between the algebraic and the order-theoretic perspective in lattice theory can be found in the Appendix.

However, there are a number of reasons to reverse the logical order between $\oplus$ and $\leq_i$. First of all, $\oplus$ is going to be taken below as but a special case of the general summation operation, so it is definable in terms of σ. Note also that as far as intuitive justification goes, the partial order properties seem much less controversial than their algebraic counterparts. (I conclude that this fact lends additional support to the axioms for $\oplus$ as given above.) Finally, taking $\leq_i$ as primitive brings the exposition more in line with mereological approaches.

Thus, I define the *language of LP* as $\mathcal{L}_1$, extended by the new primitive symbol $\leq_i$. Every well-formed expression of FL is one of LP, but now there are more well-formed expressions in virtue of new atomic formulas of the kind '$s \leq_i t$'. We can then list as the first three proper axioms of LP:

(Ax.15) $\quad \forall x\,(\,x \leq_i x\,)$ $\qquad$ (*Reflexivity*)

(Ax.16) $\quad \forall x \forall y \forall z\,(\,s \leq_i t \wedge t \leq_i r \rightarrow s \leq_i r\,)$ $\qquad$ (*Transitivity*)

(Ax.17) $\quad \forall x \forall y\,(\,x \leq_i y \wedge y \leq_i x \rightarrow x = y\,)$ $\qquad$ (*Antisymmetry*)

Now just reading Definition (D.46) from right to left will define the $\oplus$ operation in terms of $\leq_i$. Axioms 12–14 then come out as LP theorems (the proofs drawing precisely on the partial order properties; see Appendix).

By means of the i-part relation we can give yet another characterization of the existence of a term t: t exists iff there is an i-part below it.

(T.41) $\quad \vdash_{LP}\ \mathbf{E}t \leftrightarrow \exists x\,(\,x \leq_i t\,)$ $\qquad$ ($x \notin FR(t)$)

6.2.3 Structuring the plural semilattice

The above axioms are too weak for a representation of the intuitions about the structure of the plural lattice. To begin with, the domain should be non-empty. We have to postulate this now since free logic, unlike classical logic, admits empty universes.

(Ax.18) $\quad \exists x \mathbf{E}\,x$ $\qquad$ (*Existence*)

Call this axiom *(Ex)*, for short. Furthermore, it should be clear that there cannot be a "real" object which is part of every existing object; so there should be no entity in the universe that is below all the other objects. Therefore we want to ban a zero or bottom element from the plural lattice. The corresponding axiom is *(NZ)*:

(Ax.19) $\quad \neg\exists x \forall y\, x \leq_i y$ $\qquad$ (*Non-Existence of Zero*)

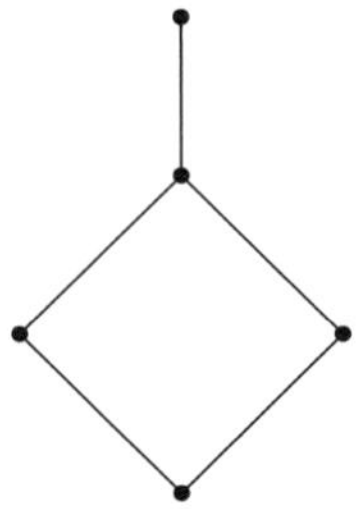

Figure 6.1: A semilattice with a bottom element

For instance, Figure 6.1 shows a simple structure with a bottom element that is excluded by the No-Zero principle. Let us dub structures observing this principle $^{-}$-*semilattices.*

Observe that, fittingly, Axiom 19 excludes one-element domains. Thus it makes sure that we deal not only with at least something, but actually with a plurality of things.

In Chapter 1 the domain was a full Boolean algebra, but its zero element was assigned to serve the role of a dummy object in a free logic set-up, and as such it was no proper element of the domain. But then, by the No-Zero principle, we cannot start from the two-element Boolean algebra with the bottom element removed since that leaves us again with the trivial structure whose single element is both top and bottom.

A schematic picture that I find useful in connection with semilattices is depicted in Figure 6.2. It illustrates the typical bottomless structures that the theory of plurals deals with. The base of the pyramid visualizes the atoms of the plural lattice, whereas the points higher up signify proper i-sums.

That there are atoms at all has not yet been established by the axioms so far. In a Boolean algebra atoms are smallest non-zero elements. Here the definition of an atom reads:

(D.47) $\quad At\,t \leftrightarrow \mathbf{E}\,t \wedge \forall x\,[\,x \leq_i t \rightarrow x = t\,]$

The following axiom says that every object x has an i-part u which is atomic. It is illustrated in Figure 6.3.

(Ax.20) $\quad \forall x \exists u [At\,u \wedge u \leq_i x]$ *(Atomicity)*

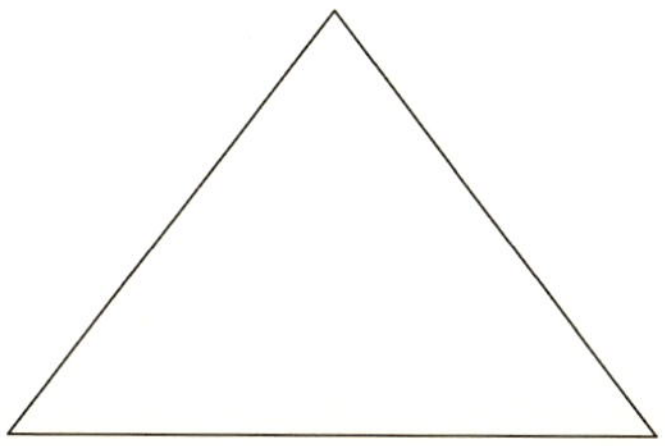

Figure 6.2: Schematic 'pyramid' picture of bottomless semilattices with atoms

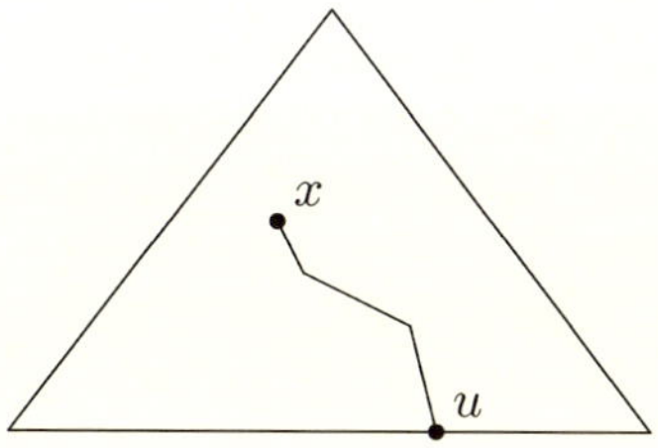

Figure 6.3: Illustration of Ax.20

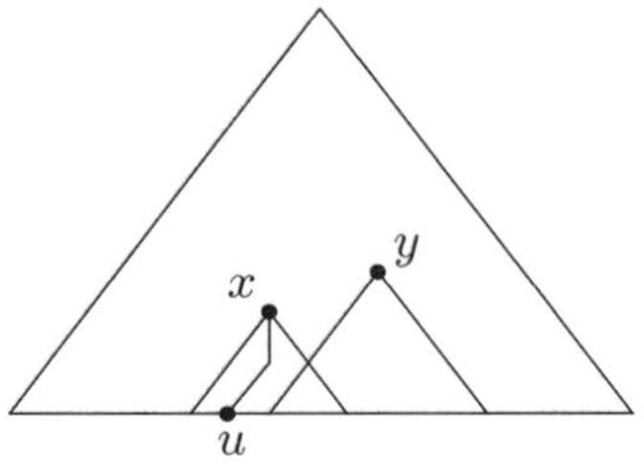

Figure 6.4: Illustration of Ax.21

Actually we don't have to adopt Atomicity as an axiom since it follows from a stronger axiom that is needed anyway. This one says that for every pair of objects x, y such that x is not below y there is an atom u below x which is not below y; in other words: if x is not below y then there is an atom below x that "separates" x from y. Let us call the axiom *Atomic Separation*; Figure 6.4 contains an illustration of it. The corresponding structures will be called *S^a-semilattices.*

(Ax.21) $\forall x \forall y[\neg x \leq_i y \rightarrow \exists u[At\, u \wedge u \leq_i x \wedge \neg u \leq_i y]]$ (*Atomic Separation*)

In addition to there being at least two different elements (by Axiom 19), Axiom 20 entails that there are at least two *incomparable* elements, i.e., there are x, y such that $\neg x \leq_i y$ and $\neg y \leq_i x$. For suppose not; then we would have Connectivity, and $\leq_i$ would be a linear order. By Axiom 19, its domain doesn't have a smallest element, so it must be an infinitely descending chain. But that chain doesn't have atoms, contradicting Axiom 20. Thus, Axiom 20 excludes descending chains of ever smaller i-parts like the one in Figure 6.5, structures that certainly don't qualify intuitively as models for a plural lattice.[10]

Atomic Separation does even more. It makes sure that that the nodes in the structure always *branch*. For instance, take the semilattice in Fig-

[10]This is not to say, however, that the plural lattices we are about to characterize may not *contain* such chains; indeed, the power set 2^ω of the set ω of natural numbers, minus the empty set, is a semilattice with respect to set-theoretic inclusion; it satisfies all the axioms listed in this chapter, and yet it contains infinite descending chains, e.g., $\omega, \omega \setminus \{0\}, \omega \setminus \{0,1\}, \omega \setminus \{0,1,2\}, \ldots$. (Thanks to Jan Tore Lønning for giving this example.)

Figure 6.5: Infinite descending chain blocked by Ax.20

ure 6.1, with the bottom element removed. This structure satisfies Axioms 19 and 20, but the two upper points cannot be separated by the two atoms since the uppermost node doesn't branch. The edge between them couldn't stand for the i-part relation since for a plural sum to "grow bigger," it would have gain at least one atom, which graphically means branching. Thus the structure is not an intended one and is properly blocked by Axiom 21. In fact, Atomic Separation turns out to be a crucial axiom that allows the representation of any object as the i-sum of all its atomic parts (see Axiom 30 below). A semilattice that doesn't qualify in this respect and violates Atomic Separation is displayed in Figure 6.6: its "slim waist" prevents a characterization of the two points on the third level in terms of the atoms below them. It is a model of Axiom 20, though, since every element has an atom below it. That Atomic Separation implies Atomicity in the presence of Axiom 19 is shown in the Appendix; see Proposition 15 there.

While atomicity is a most natural requirement in the present theory, matters are different in the case of *mass terms*; there, the semantics of language suggests that there be no smallest parts, or at least not necessarily so (see Chapter 1, and Pelletier and Schubert (1989); Landman (1991)).

We now add the σ-operator to the language of LP and include among its terms expressions of the form $\sigma x P x$, standing for the sum of all elements that have the property P. The first axiom we have to give for σ is an *existence condition*. Axiom 22 guarantees the existence of $\sigma x P x$, provided P is non-empty. This is a kind of completeness property but it falls short

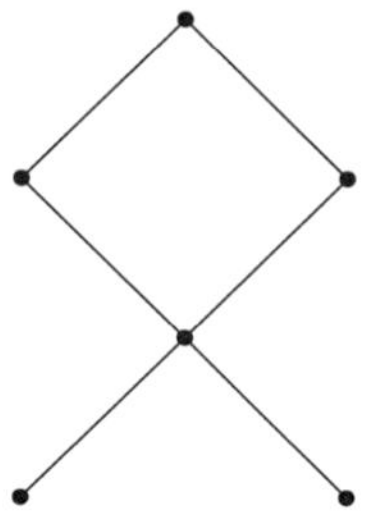

Figure 6.6: A model for Ax.20 blocked by Ax.21

of characterizing the full notion of completeness in a semilattice (existence of suprema of arbitrary non-empty sets), which is not first-order definable (see Lønning (1989)). The present axiom will be called "Definable Completeness" as it postulates a supremum for each non-empty subset that has a name (viz., P) in the object language. An illustration can be found in Figure 6.7. As the shape of the P-area is meant to indicate the extension of a predicate P may contain individuals whose atomic parts fail to belong to it (think of a property like owning something, which can be truly collective: if John and Mary own a house as common property, neither of them owns it alone in the legal sense).

(Ax.22) $\exists x Px \rightarrow \mathbf{E}(\sigma x Px)$ (*Definable Completeness*)

(If there is an object with property P then the i-sum $\sigma x Px$ of all objects that are P exists.) The semilattice structures satisfying this axiom are called *C^d-semilattices.*

The following two axioms give the σ-operator its intended meaning as the supremum (with respect to the i-part relation) of all individuals that have the property P.

(Ax.23) $\mathbf{E}(\sigma x Px) \rightarrow \forall y[Py \rightarrow y \leq_i \sigma x Px]$ (*σ-properties*)

(Ax.24) $\mathbf{E}(\sigma x Px) \rightarrow \forall y[\forall x[Px \rightarrow x \leq_i y] \rightarrow \sigma x Px \leq_i y]$

(Every object y which is P is below the i-sum $\sigma x Px$; and whenever some object y is above all x that are P it is also above $\sigma x Px$.) Thus $\sigma x Px$ is an *upper bound* for P, and it is the *least upper bound* for P. Axioms 23 and 24

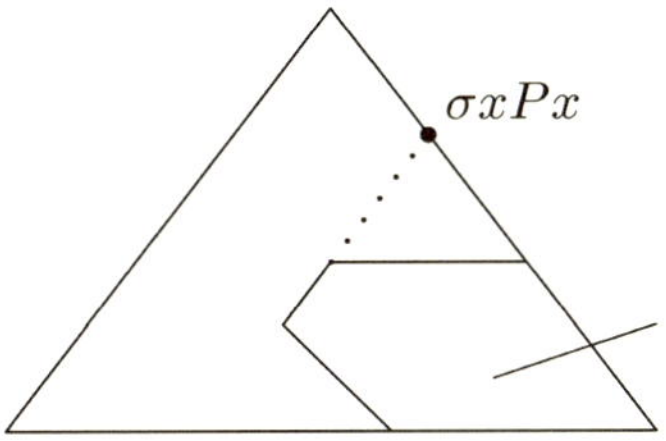

Figure 6.7: Illustration of Ax.22

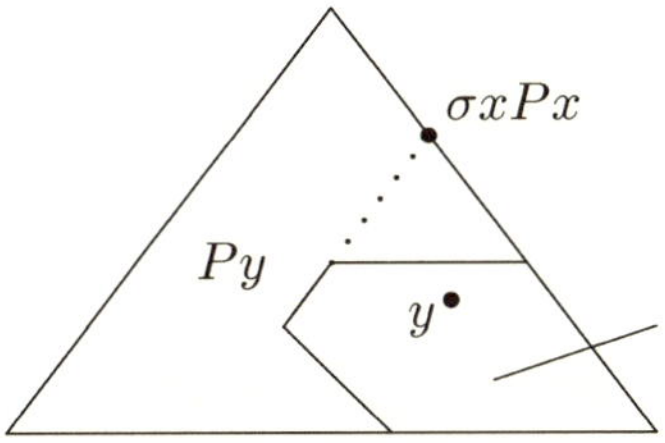

Figure 6.8: Illustration of Ax.23

say, then, that $\sigma x P x$ denotes the supremum of all objects falling under P whenever it exists. Illustrations for these axioms are found in Figures 6.8 and 6.9.

Since the least upper bound is taken with respect to a partial order it is uniquely determined. We could therefore *define* $\sigma x P x$ in terms of ι as *the* unique i-sum having the σ-properties with respect to P. This will be done in the next section where the "official" version of the LP axioms are given.

We now give a characterization of the plural operator *. Intuitively, an element t falls under the predicate $^*P$ if t is "made up" exclusively from elements that are P (see Chapters 1 and 2). It turns out that the following condition is adequate in expressing this idea; it comprises the case where the objects falling under P are not necessarily atomic.

(Ax.25) $\quad {}^*Pt \;\leftrightarrow\; t = \sigma x(Px \wedge x \leq_i t)$

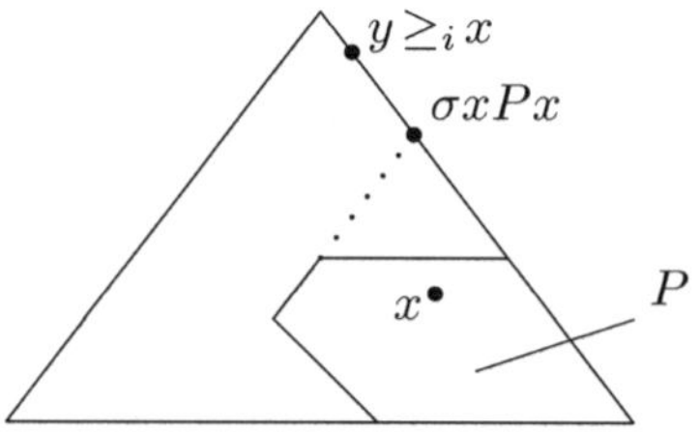

Figure 6.9: Illustration of Ax.24

(t is a $^*P$ iff t is the supremum of all P-elements below it.) With this axiom the following theorems, which describe the connection between P and $^*P$ more closely, can be shown to hold. They say in turn: that all P are $^*P$; that $\sigma x Px$ is a $^*P$ for non-empty P; that if P is non-empty so is $^*P$ (in fact, with Axiom 26 below the converse also holds); that the suprema of P and $^*P$ are the same if P is non-empty; and that $\sigma x Px$, if it exists, is maximal in $^*P$.

(T.42) $\vdash_{LP} \forall x(Px \rightarrow {}^*Px)$

(T.43) $\vdash_{LP} \exists x Px \rightarrow {}^*P(\sigma x Px)$

(T.44) $\vdash_{LP} \exists x Px \rightarrow \exists x\, {}^*Px$

(T.45) $\vdash_{LP} \exists x Px \rightarrow \sigma x Px = \sigma x {}^*Px$

(T.46) $\vdash_{LP} \forall y({}^*Py \wedge \sigma x Px \leq_i y \rightarrow \sigma x Px = y)$

For a sample proof, consider T.42. Given Px we have to show that $^*Px$, i.e., by Axiom 25, $x = \sigma y(Py \wedge y \leq_i x)$. Note first that $\sigma y(Py \wedge y \leq_i x)$ exists: by Px and $x \leq_i x$ we see that $P' := \lambda y(Py \wedge y \leq_i x)$ is non-empty; thus we have $\exists y P'y$ or $\exists y(Py \wedge y \leq_i x)$, and $\mathbf{E}\, \sigma y(Py \wedge y \leq_i x)$ with Definable Completeness. Now we have $Px \wedge x \leq_i x$, hence $x \leq_i \sigma y(Py \wedge y \leq_i x)$. To prove this inequality in the other direction it suffices to show, by Axiom 24, that $\forall z(Pz \wedge z \leq_i x \rightarrow z \leq_i x)$; but this is trivial. By antisymmetry of $\leq_i$ we arrive at the above equality, q.e.d.

The idea behind the last axiom, Axiom 26 below, is the intuitive requirement that two different collections of atomic objects should not have identical i-sums. For instance, let P denote the set { John, Mary } , its

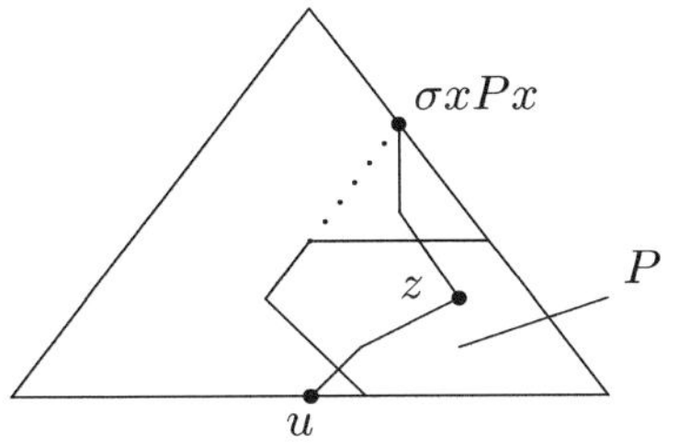

Figure 6.10: Illustration of Ax.26

sum being John ⊕ Mary. When Sue is a third person, the i-sum John ⊕ Mary should differ from the i-sum John ⊕ Mary ⊕ Sue; for otherwise, Sue will invariably share all common activities of John and Mary, a situation that might not always be welcomed by them. However, unless Axiom 26 or something like it is fulfilled the four-element structure { John, Mary, Sue, John ⊕ Mary ⊕ Sue } with atoms $a =$ John, Mary, Sue and $a \leq_i$ John ⊕ Mary ⊕ Sue for all atoms a would be an admissible semilattice. The above requirement is automatically met in Boolean lattices; but when we start from the considerably weaker semilattice structures an extra axiom has to be added to guarantee that.[11]

The following formulation of the axiom, illustrated in Figure 6.10, guarantees the desired property (see Figures 6.11 and 6.16, and Proposition 3 of the Appendix). Semilattices satisfying it are said to have *σ-prime atoms* and will be called *P-semilattices.*

(Ax.26) $\forall u[At\, u \wedge u \leq_i \sigma x Px \rightarrow \exists z[Pz \wedge u \leq_i z]]$

(For all atoms u below a sum $\sigma x Px$ there is an object z which is P and is above u.)[12] Figures 6.11 and 6.12 show semilattice structures that are excluded by Axiom 26. Note that the ladder structure in Figure 6.12 is not blocked by Atomic Separation in the sense of Axiom 21.

[11]This was first pointed out explicitly in the literature on plurals by Landman (1989); see also his (1991).

[12]Recall that the condition $u \leq_i \sigma x Px$ in the antecedent implies the existence of $\sigma x Px$.

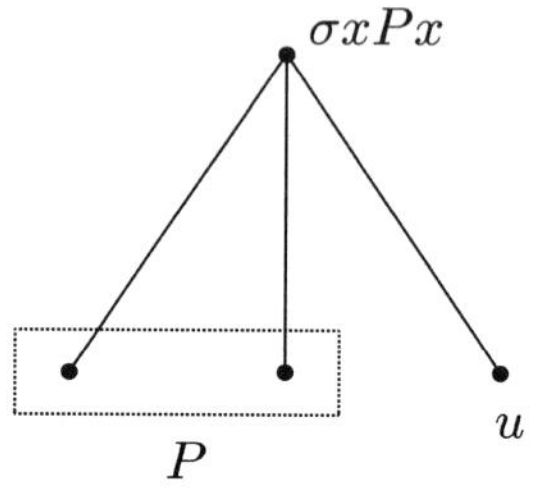

Figure 6.11: A semilattice structure blocked by Ax.26

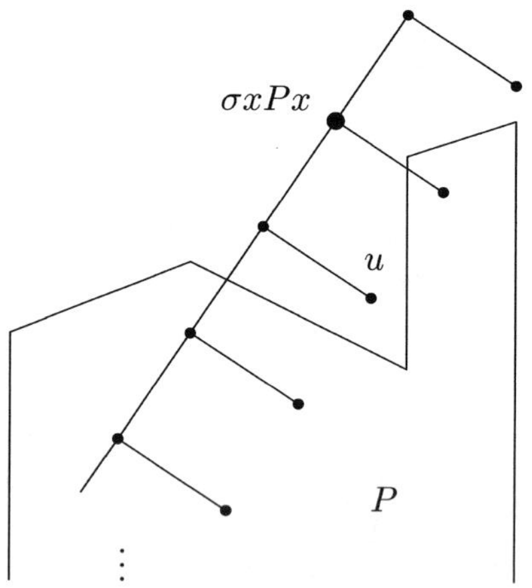

Figure 6.12: Another structure blocked by Ax.26

As will be shown in the Appendix, an analogue of Axiom 26 produces the "most general" semilattice $L(A)$ over a given set of atoms A in the following sense: there is no way to build up a more detailed semilattice than $L(A)$ from the arsenal A. Accordingly, other semilattices over A will be either isomorphic to $L(A)$ or will "fuse" some elements somewhere that are distinguished in $L(A)$. Call such a semilattice T a *test semilattice*. Then any mapping f from A into T can be uniquely extended to a complete semilattice homomorphism g from $L(A)$ into T.[13] Semilattices which satisfy Axiom 26 have this extension property. Structures with the extension property are called "free" in the mathematical literature, which explains why sometimes Axiom 26 is also said to express the property of freedom.

6.2.4 Lønning's system rephrased in LP

Lønning (1989) presents a logical framework for the treatment of plurality which is formulated within a generalized quantifier language, so the axioms assume a different shape there. A comparison with the system LP shows that Lønning retains the full strength of the λ-operator but introduces a second kind of quantification which is restricted to atomic individuals only. Call the corresponding quantifiers $\forall_a$, $\exists_a$, and the corresponding σ-operator σ_a. Then we have

(D.48) $\quad \forall_a x Px \leftrightarrow \forall x(At\, x \rightarrow Px)$

(D.49) $\quad \exists_a x Px \leftrightarrow \exists x(At\, x \wedge Px)$

(D.50) $\quad \sigma_a x Px = \sigma x(At\, x \wedge Px)$

Let 'u','v','w' be variables that range over atoms only; then we can drop the subscript 'a' and write '$\forall u Pu$' for '$\forall_a u Pu$', and similarly with '$\exists u Pu$' and '$\sigma u Pu$'. With this convention it is possible to write down the following of Lønning's axioms in terms of restricted quantification (it is assumed that P and Q apply to atomic individuals only).

(Ax.27) $\quad \exists x Px \rightarrow \forall u[Pu \rightarrow u \leq_i \sigma v Pv]$

(Ax.28) $\quad \exists x Px \rightarrow \forall y[\forall u[Pu \rightarrow u \leq_i y] \rightarrow \sigma v Pv \leq_i y]$

($\sigma v Pv$ is an upper bound for the (atomic) elements of P, and it is the least upper bound.) Except for the restricted quantification these axioms express the above σ-properties.

(Ax.29) $\quad {}^D Pt \leftrightarrow \neg\forall u\, t \leq_i u \wedge t = \sigma u(u \leq_i t \wedge Pu)$

[13] An illustration is given in the Appendix.

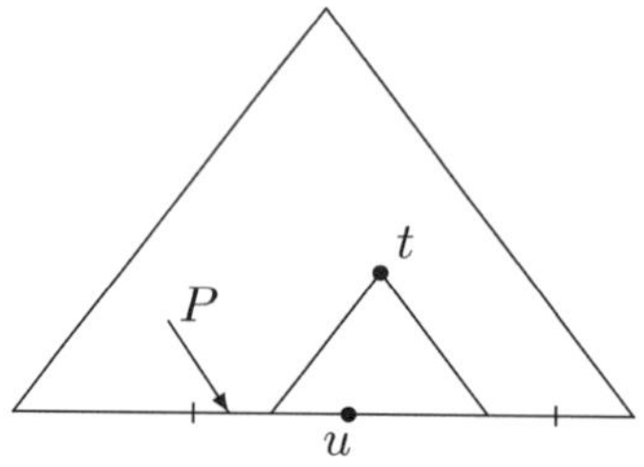

Figure 6.13: Illustration of Ax.29,30

(The distributive version ^{D}P of P applies to t just in case t is not the zero element and t can be written as the sum of all atoms u below t that are P.) In Lønning's restricted set-up the star operator and the D operator coincide, so * is dropped in favor of D here. From Axiom 29, together with Axioms 30 and 31 below, the definition of the distributivity operator given, e.g., in Chapter 4 can be formally derived (see Appendix):

(D.51) $\quad {}^{D}Pt \leftrightarrow \forall u[u \leq_i t \rightarrow Pu]$

(Ax.30) $\quad \mathbf{E}t \rightarrow t = \sigma u(u \leq_i t)$ $\quad$ (*σ-generation*)

(A term t equals the sum of its atomic parts.) Call this axiom *(SG)*. Axioms 29 and 30 are illustrated in Figure 6.13. The sectioned part of the pyramid base consists of all the atoms that are P, and the base of the small pyramid consists of all atoms below t. σ-generation, then, means that every element in the semilattice is uniquely determined by the base section below it, i.e., by its atomic i-parts. A structure like the one in Figure 6.14 is therefore blocked by Axiom 30.

(Ax.31) $\quad \sigma uPu = \sigma uQu \rightarrow \forall u[Pu \leftrightarrow Qu]$ $\quad$ (*Injectivity of σ*)

(When the sums over two atomic predicates P and Q are equal then P and Q are coextensive.) This axiom expresses the *injectivity of the summing operation*, which takes a predicate and produces a sum term. It can be seen in Figure 6.15 that a situation blocked by Axiom 26 is also excluded by Axiom 31.

Figure 6.14: A structure blocked by Ax.30

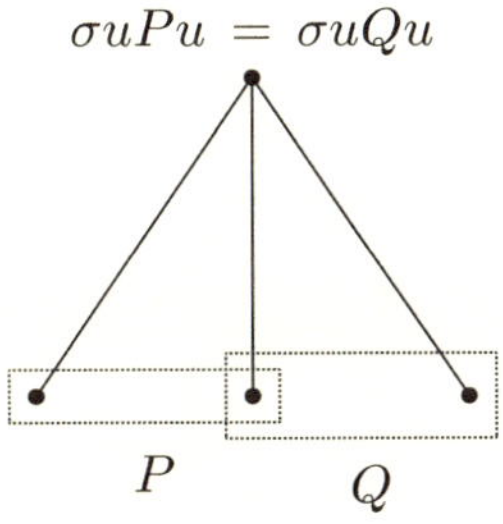

Figure 6.15: Structure violating Ax.31

It can be shown now that Axioms 27–31 are theorems of Axioms 12–26. Axioms 27,28 obviously follow from Axioms 23,24; Axiom 29 follows from Axiom 19 under Definition 51; Axioms 21,22,27,28 entail Axiom 30, and Axioms 23,26 entail Axiom 31. Conversely, Axiom 30 implies Axiom 21, and Axiom 26 is entailed by Axioms 30,31. All this will be shown in the Appendix. A consequence of these results is that Lønning's axioms are basically equivalent to the axiom system given here when we ignore the question of full completeness which cannot be settled within first-order logic anyway. The structure of the plural lattice can thus be described either as a *bottomless, separating semilattice with a supremum term-operator* σ *and with* σ*-prime atoms* (that is, as a $C^dS^aP^-$ *semilattice*) or else as a *bottomless* σ*-generated semilattice with an injective supremum term-operator.*

Now it is a mathematical fact that when we take the resulting model structures and add the property of (full) completeness we find ourselves already within the class of *complete atomic Boolean algebras minus zero.* The corresponding representation theorem will be formulated in the Appendix.

6.3 Metatheory

The purpose of giving various equivalent formulations of the axioms of LP was to shed some light on their content from different angles. I will now streamline the discussion and settle with a particular set of axioms for LP which does three things: (i) It minimizes the number of primitive symbols of LP over FL, that is, it just adds the part-whole relation $\leq$ (I omit the index now for brevity and generality; see the next point.) (ii) It collects the reference to the atomic structure into one axiom, so as to facilitate comparison with non-atomic lattices as they are used in modeling homogeneous (e.g., mass) objects. (iii) It brings the discussion in line with the literature on mereology; this connection will become particularly important in the final chapters of the book.

Let us note, to begin with, that we keep Axioms 15 – 17 (dropping the index), which express the partial order properties of $\leq$; I will refer to these properties collectively with *(PO).* Then the central mereological notion of *overlap* is defined thus:

(D.52) $s \circ t \leftrightarrow \exists x(x \leq s \wedge x \leq t)$ (*overlap*)

It follows immediately from T.41 above that self-overlap is another existence criterion:

(T.47) $\mathbf{E}t \leftrightarrow t \circ t$

I add two simple definitions.

(D.53) $s < t \leftrightarrow s \leq t \wedge s \neq t$ (*strictly less than*)

(D.54) $s \mid t \leftrightarrow \neg s \circ t$ (*disjointness*)

We take over the Axioms of Existence, (Ex), and Non-Existence of Zero, (NZ) (Axioms 18 and 19). Also, the axioms involving the σ-operator and will basically be retained. A slight reformulation is due to the fact that σ will now be defined in terms of the description operator. Let me start by defining an "up-arrow" ("down-arrow") operator on predicates, turning a P into a predicate ${}^{\uparrow}P$ (${}^{\downarrow}P$) which is true of all the upper (lower) bounds of P-objects:

(D.55) $\quad {}^{\uparrow}P \; := \; \lambda x \forall y (Py \rightarrow y \leq x)$

(D.56) $\quad {}^{\downarrow}P \; := \; \lambda x \forall y (Py \rightarrow x \leq y)$

Obviously we have

(T.48) $\quad \forall x \, ({}^{\downarrow\uparrow}P \, x \; \leftrightarrow \; \forall z \, (\forall y (Py \rightarrow y \leq z) \rightarrow x \leq z \,))$

Thus, the down-up-arrow yields the lower bounds of the upper bounds of a predicate P. We want to define $\sigma x Px$ as the least upper bound of the P-objects in case P is non-empty; using the arrow notation we can define:

(D.57) $\quad \sigma x Px \; := \; \iota x \, ({}^{\uparrow}P \, x \wedge {}^{\downarrow\uparrow}P \, x \,)$

If $\sigma x Px$ exists we get the following biconditional which expresses the property of an upper bound, when read from left to right (choosing $\sigma x Px$ for y), and that of a least upper bound when read from right to left.

(T.49) $\quad \mathbf{E} \, (\sigma x Px) \; \rightarrow \; \forall y \, (\sigma x Px \leq y \; \leftrightarrow \; \forall z (Pz \; \rightarrow \; z \leq y) \,)$

The existence of least upper bounds for non-empty predicates is secured by the axiom of Definable Completeness (DC, for short), which now reads:

(Ax.32) $\quad \exists x Px \; \rightarrow \; \exists x \forall y \, (\, x \leq y \; \leftrightarrow \; \forall z (Pz \; \rightarrow \; z \leq y) \,)$ $\qquad (DC)$

Uniqueness of the least upper bound follows from antisymmetry of $\leq$, whence the theorem:

(T.50) $\quad \exists x Px \; \rightarrow \; \mathbf{E} \, (\sigma x Px)$

With the general summing operation at our disposal we can define the circled plus which is the simple join operation on terms:

(D.58) $\quad s \oplus t \; := \; \sigma x \, (\, x = s \vee x = t \,)$

It is easily verified, using T.49, that $s \oplus t$ has the properties of a join for existing s, t:

(T.51) $\quad \mathbf{E} \, s \wedge \mathbf{E} \, t \; \rightarrow \; s \leq s \oplus t \wedge t \leq s \oplus t$

(T.52) $\quad \mathbf{E} \, s \wedge \mathbf{E} \, t \; \rightarrow \; \forall y \, (\, s \leq y \wedge t \leq y \rightarrow s \oplus t \leq y \,)$

The axioms that turn the structures described so far in this section into a proper plural lattice are Atomic Separation (Ax.21) and σ-prime Atomicity (Ax.26).[14] Our goal is to confine the concept of atom to one single axiom. What we will do is to replace Axiom 21 by a more general principle of separation which doesn't mention atoms, and to replace Axiom 20 by the Axiom of σ-generation. That is all we need. The general separation axiom (S for short) reads:

[14]Recall that Atomicity (Ax.20) was derivable.

(Ax.33) $\forall x \forall y (\neg x \leq y \rightarrow \exists z(z \leq x \wedge z \mid y))$ (*Separation*)

(If x is not below y then there is a z below x which is disjoint from y.) Separation is called the *Strong Supplementation Principle (SSP)* in Simons (1987); it is shown there that in the presence of full summation it follows from the *Weak Supplementation Principle (WSP)*, which is called here *Weak Separation (WS)* to conform with the names of the other separation principles.[15]

(Ax.34) $\forall x \forall y (x < y \rightarrow \exists z(z \leq y \wedge z \mid x))$ (*Weak Separation*)

Thus we have, with $L = PO + DC$:

(T.53) $WS \vdash_L S$

The reason for this theorem is that with σ we can define the greatest lower bound of two terms whenever they overlap; in this case the predicate $\lambda x(x \leq s \wedge x \leq t)$ is non-empty, and with

(D.59) $s \odot t := \sigma x(x \leq s \wedge x \leq t)$

we have:

(T.54) $s \circ t \rightarrow \mathbf{E}(s \odot t)$

Suppose now that x is not below y, and x and y overlap. Then $s \odot t$ provides a term strictly less than x to which (WS) can be applied to produce a separating z.[16]

The original set of LP axioms contained one more principle, that of σ-prime atomicity (Axiom 26). An "atom-free" formulation of this principle is what I will call *σ-overlap, (SO)*:

(Ax.35) $\forall y (y \leq \sigma x P x \rightarrow \exists z(P z \wedge z \circ y))$ (*σ-overlap*)

Axiom 26 is obviously a special case of Axiom 35. But the latter also entails a principle of *distributivity* that Landman (1991) takes as constitutive for the mass term lattice of homogeneous objects.

(Ax.36) $\forall x \forall y \forall z (x \leq y \oplus z \rightarrow x \leq y \vee x \leq z \vee \exists y' \exists z' (y' \leq y \wedge z' \leq z \wedge x = y' \oplus z'))$ (*Distributivity*)

[15]Still another name for it is *Witness* (see Landman 1991), which is also used in our Chapter 12.

[16]For an explicit derivation, see Simons (1987):31.

This principle guarantees the law of distributivity with respect to the operations $\oplus$ and $\odot$, which is an essential ingredient of the intended Boolean structure. What is still missing for a full Boolean algebra is the unit element, the zero element, and the operation of complementation satisfying the appropriate laws. Now we explicitly banned a zero element from our ontology;[17] however, the unit, or universal, element as well as complementation can be defined in terms of σ:

(D.60) $\mathcal{U} \;:=\; \sigma x\,(x = x)$ (*Universal Element*)

(D.61) $\bar{s} \;:=\; \sigma x\,(x \mid s)$ (*Complement of s*)

$\bar{s}$ exists iff $s \neq \mathcal{U}$. This operation satisfies the complementation laws, suitably modified to fit the zero-free structures.

Let us define now the *logic of mass terms, LM*: it is the free logic FL, together with the axioms (PO), (Ex), (NZ), (DC), (WS), and (SO). It can be shown that LM describes definably complete Boolean algebras minus zero.[18]

The system LM is, in fact, basically equivalent to the system CM of *Classical Mereology* as it is described in Simons (1987):37. Ignoring minor differences, the systems share (WS) and a principle which amounts to the existence of general sums for non-empty predicates. In LM this is (DC); CM features a version of it, call it *ES*, which is equivalent in the present setting:

(Ax.37) $\exists x Px \;\rightarrow\; \exists x \forall y\,(x \circ y \;\leftrightarrow\; \exists z\,(Pz \wedge z \circ y))$ (*Existence of Sums*)

There is one single axiom that brings us back to plural lattices; it is the above principle of σ-generation, (SG) (Axiom 30) that every object can be built up uniquely from a set of atoms. It can be shown that (SG) entails (SO), so we can drop (SO) from our axiom set in the case of atomic structures.

We have arrived at our final set of axioms for LP: The *Logic of Plurality* is FL, together with the axioms (PO), (Ex), (NZ), (DC), (WS), and (SG). The models for LP are definably complete atomic Boolean algebras minus the zero element ($C^d AB^-$ structures, for short; see Appendix). Atomicity, Atomic Separation, and σ-injectivity are then theorems of LP.

---

[17] A lattice is of course Boolean only if it has a zero element; in the present context, however, we allow ourselves to speak of 'Boolean' also when the zero element of the lattice is removed.

[18] See again Landman (1991).

There are four metalogical issues that I want to note concerning LP. First, the system is a *conservative extension of FL*; that is, any LP theorem that doesn't involve proper LP symbols is derivable in FL. Next, LP is *incomplete* with respect to the class of (fully complete) CAB^- structures. One way to see this is to note the fact mentioned above that $ML2^-$, the monadic second-order logic with quantification over non-empty sets, can be embedded into LP (see Lønning (1989, 1997)); thus the incompleteness of $ML2^-$ is imported into LP. Here is an explicit translation $*$ between the languages of the systems, similar to Lønning's, which is basically a *syntactic interpretation* from $ML2^-$ into LP.[19]

Let $X_0, X_1, \ldots$ be an fixed enumeration of the second-order variables in $ML2^-$, and $x_0, x_1, \ldots$ a fixed enumeration of the variables for i-sums in LP. Let $ML2^-$ and LP share their first-order variables $u, v, w, \ldots$ and their first-order predicates $P, Q, \ldots$; all individual terms $t, s, \ldots$ are translated identically. Then we have the following translation rules:

Syntactic Interpretation $*$: $ML2^- \rightarrow$ LP:

1. $(X_i)^* = x_i$, $P^* = \sigma u P u$;
2. $(Tt)^* = t^* \leq T^*$ if T is some X_i or some (first-order) predicate P;
3. $(\neg\phi)^* = \neg\phi^*$
4. $(\phi J \psi)^* = \phi^* J \psi^*$ for some 2-place connective J
5. $(\forall u \phi)^* = \forall u \phi^*$
6. $(\forall X_i \phi)^* = \forall x_i \phi^*$
7. $ML2^- \vdash \phi \Rightarrow LP \vdash \phi^*$

The last condition is in need of verification; that is, we have to prove by induction on the length of derivations that whenever $ML2^-$ proves a formula ϕ then its translation ϕ^* can be proved in LP. The basic step in this inductive proof involves the axioms of $ML2^-$ whose translations must come out as LP-theorems (they don't have to be axioms in LP). Now the main axiom schemata for $ML2^-$ are universal instantiation and comprehension:

Universal Instantiation:

$\exists u T u \rightarrow (\forall X \phi \rightarrow \phi^X_T)$ if T is some X_i or some (first-order) predicate P, and T free for X in ϕ;

[19]The standard notion of syntactic interpretation for classical logic can be found, e.g., in Tarski *et al.* (1953). It has to suitably adapted to fit into the current context of free logic.

Comprehension:

$$\exists uPu \to \exists X\forall u(Xu \leftrightarrow Pu) \qquad \text{if } X \text{ is not free in } \phi$$

The first axiom translates to the restricted scheme of universal instantiation in free logic. Call the comprehension scheme ψ; in order to see that the translation ψ^* is derivable in LP we can appeal to (Ax.37), the Existence of Sums, ES, which is an LP-theorem. Let the i-sum variable x be the translation of the second-order variable X; then ψ^* is the LP-formula

$$\exists u.u \leq \sigma uPu \to \exists x\forall u\,(\,u \leq x \leftrightarrow Pu\,)$$

Now given the antecedent of ψ^* we can infer $\exists uPu$, whence, by ES, we get the formula $\exists x\forall u\,(\,x \circ u \leftrightarrow \exists z(Pz \wedge z \circ u)\,)$. But since u is atomic, $x \circ u$ is equivalent to $u \leq x$, and Pu is equivalent to $\exists z(Pz \wedge z \circ u)$. The latter is the case because P is a first-order or distributive predicate and hence can be true of atomic individuals only. That yields the consequent of ψ^*, and by the deduction theorem, ψ^* itself is a theorem of LP.

A further metalogical property of LP is its *generalized completeness.* While LP is incomplete with respect to the standard models, it is possible to give a completeness proof for generalized Henkin models of definably complete AB^- structures. This was basically done in Eberle (1970).

The final issue is, how much of the expressive power of LP is actually needed to represent pluralic discourse of natural language. Lønning (1989, 1997) addresses this question and shows, among other things, that *definite plural logic* (with definite plural expressions only) is complete, and when indefinite pluralic expressions are admitted (an obvious desideratum, of course), completeness can be attained in a certain subclass of sentences called *persistent.* For instance, arbitrary quantification over *atomic* individuals only will be persistent. As for genuine plural quantification, positive existential sentences are persistent, too. Suppose we managed to "explain away" linguistic data like the *competing companies* case mentioned above that seem to hint at genuine universal plural quantification in language. Then this strategy will only succeed if the *negation* of indefinite plural NPs does not bring in universal plural quantification through the backdoor. This is the reason why it would be important to give an in-depth linguistic analysis of the interplay between indefinite NPs and negation in plural sentences, like the above example (6). Lønning's discussions contains an important step toward this goal.

Chapter 7

Ten Years of Research on Plurals – Where Do We Stand?

7.1 Introduction

The ten years mentioned in the title basically refer to the decade of the eighties and the early nineties which witnessed a tremendous amount of research activity on plurals in linguistic semantics and philosophy. This activity grew out of the realization that plurals are all-pervasive in language and hence cannot be regarded as an exotic topic by anyone who wants to give a reasonably complete account of the structure of language.

The pluralic idiom of natural language was typically neglected in the development of formal logic. This discipline evolved out of serious problems in the foundations of mathematics. In the context of mathematics, plural expressions occur only in the informal mathematical *argot*, the metalanguage that the mathematicians use to talk about their subject. It is a technical subdialect of natural language that is used to communicate in an imprecise but very efficient way mathematical ideas that if one were to take the trouble could be written down in the classical formal language of logic which only contains "singular" quantification. Thus a statement like *All non-negative real numbers have real square roots* could be more clumsily rendered as *For every non-negative real number x there is at least one real number y such that the square of y is x.* And in cases where the plural seems more essential to the statement, like in *almost all elements of the sequence lie in this neighborhood*, which means that all but a finite number of elements have the property in question, then this plurality is rephrased in the singular mode by speaking of sets: *with the exception of a finite set of elements of the sequence every member of the sequence lies in this neighborhood.* So it is fair to say that in mathematics, plurals serve no theoretical purpose and can basically be ignored.[1]

Outside the realm of mathematics natural language was not represented in a formal language. It was the philosophers, under the lead of Quine, who took up the Fregean exercises of formalizing natural language as a means for becoming clear about the ontological commitments hidden in linguistic locutions. Now Quine, in his logical regimentation program he prescribed to language (Quine (1960)) dealt explicitly with mass terms; but he did not say much about plurals. He did mention, though, certain peculiar plural sentences that proved recalcitrant to a straightforward first-order representation (Quine (1972)). So it is perhaps surprising that in spite of a rather long tradition of logical analysis of natural language one of the first philosophical papers explicitly dealing with valid arguments involving plurals is probably Massey (1976). In linguistic quarters at that time, Montague Grammar was already in full swing, but people there considered the incorporation of plurals into the Montagovian framework just a routine exercise (see e.g., Bartsch (1973); Hausser (1974b); Bennett (1975)).

Scha (1981) marks the starting point of systematic semantic research on plurals. My paper Link (1983a) (Chapter 1 of this volume) tried to bring together the philosophical tradition and the work in linguistic semantics. In it I layed out a different conception of dealing with semantic matters: Instead of set-theoretic modeling I proposed an algebraic approach to the study of plurals and mass terms. The main linguistic motivation was the striking structural analogy between the plural domain and the domain of mass expressions that can best be captured in an algebraic setting. But there was also a rather firm philosophical intuition behind this approach on which I will shortly comment in section 7.2.

When the study of plurals grew into an independent subject of research it had to situate itself with respect to the leading overall paradigms in linguistic semantics, like Montague Grammar (MG), Generalized Quantifier Theory (GQT), Discourse Representation Theory (DRT), or Situation Theory (ST). There are demands on the systems that go in both directions here: On the one hand, the theory of plurals (PT) has to be flexible enough to meet the formal needs of any one of these systems. One the other hand, such a general framework should not only be able to accommodate plurals in some way or other; it should moreover be structured in such a way as

[1]From a linguistic point of view it would be worthwhile to do some systematic field work with mathematicians to find out exactly how they translate their informal talk, which is full of pluralic locutions, into singular mathematical statements.

to lend itself to a natural and theoretically coherent incorporation of PT. The kind of plural theory I tried to advance was meant to fit smoothly into any of those frameworks. Among them, DRT has been particularly successful in incorporating plural theory. In fact, some of the basic claims of PT proved to be most valuable in the context of plural anaphora and have led to important developments there (see Kamp and Reyle (1993)). While I will not comment on ST here, GQT has to be mentioned, which can be considered as a kind of successor framework of MG. The classical paper on GQT, Barwise and Cooper (1981), remains silent on genuine pluralic issues; they are discussed to a certain degree within the framework in Chapter 4, with emphasis on their linguistic aspects. A more systematic GQ version of plural theory is used in Lønning (1989) where also fundamental metalogical results on PT are arrived at. Finally, van der Does's important study (1992) provides an abstract GQ setting for pluralic determiners and modifiers.

Over the years it has become evident that a typical difficulty in studying plurals is the fact that plural terms are notoriously vague in their reference; in this way they serve the overall efficiency of language in a remarkable way. Formal representations, on the other hand, are typically calibrated for a high degree of precision. The problem here is come up with representations that are optimally tuned to this empirical level of accuracy. Thus the question is not only how plural phenomena can be represented at all, but also how to avoid overprecision. Let me illustrate this point with a few examples.

(1) The Romans built the aqueduct. They were excellent architects.

(2) [George Bush during the gulf war:] The Germans are awful. They have provided the Iraquis with those horrible chemical weapons, but now they are hiding behind their paychecks.

(3) [A martian:] These earthlings are strange. They build those wonderful structures they call cities only to level them off again every other decade.

Obviously, not all of the Romans built the aqueduct, nor were all of them excellent architects; and presumably, the ones who actually erected the aqueduct were not identical with the architects. Or take the Germans; certain German business men profited from the chemical weapons sale, and certain German politicians hid behind their paychecks, perhaps. But even if they entertain a covert or open relationship these groups can hardly be called identical. The same applies to the various groups of earthlings. Thus the anaphoric plural pronoun can apparently be used (and functions well!) without there being a fixed entity which it refers back to at every occurrence.

Here is another example: Do the sentences under (4) mean that *all* doors were opened, that *all* tools were used at every occasion?

(4) a. The burglars opened the doors with their special tools.

b. The burglars used their special tools to open the doors.

To use the universal quantifier $\forall$ for the representation of *all* the definite plural noun phrases involved would give their interpretation a degree of precision that is not matched by the empirical semantic facts concerning these data. It is clear then that considerable care has to be exercised to avoid the fallacy of over-representation.

This chapter consists of three parts. In the next section I will give a broad overview over some of the current research activities in the field. Among them, four issues are singled out here for special consideration. Section 7.3 discusses what in my eyes constitutes a prominent case of over-representation. Section 7.4 finally is concerned with the issue of distributivity. Here I shall mainly comment on recent work by Jaap van der Does and Jan Tore Lønning.

Fixing terminology. Before I start I'd like to settle with some terminology just to give a precise frame for discussion. In general I shall presuppose here some acquaintance with the basic framework of plural logic LP as laid out for instance, in Link (1983a).

A *plural term* is a syntactic expression of natural language (mostly a plural NP) or an expression in the formal representation language (for instance LP). A *plural object* or *plurality* is a semantic entity, the denotation of a plural term. Sometimes a plural term may fail to refer to a plurality: (i) when the plural term is of a purely syntactic nature, like the English word 'scissors'; (ii) when a plural term is used in a spurious way; an example is *all men are mortal* which is totally equivalent to *every man is mortal*; (iii) when a plural term is a *general* NP in the sense that is genuinely quantificational; an example is *most Germans*; (iv) when the plural term is a bare plural in its generic use, e.g., in *dinosaurs are extinct*. In the latter case there is evidence that *dinosaurs* refers to a *natural kind* and not to a concrete collection of entities.[2]

The concept of a *plural character* is a functional notion: a plural character is a plurality that enters a collective relation like *meet* or *share*. It

[2]For a report on the state of the art in the linguistic study of generics see Carlson and Pelletier (1995).

seems that pluralities do not always serve as plural characters; sometimes there are only there to delineate a certain group that is being referred to in a given context of utterance; but the relation they are involved in are purely distributive so that it is only the atomic individuals making up the plurality that have the property in question.

The terms *group* and *collection* will be used in a loose and informal way and often interchangeably; only in Section 7.3 *group* has a technical meaning. By contrast, technical terms are: *set*, *class*, *fusion*, *individual sum (i-sum)*; in set-theoretic usage a set is a class which is a member of some other class; a class that is not a set is called a proper class; the ordinary usage is such that when a class is known to be a set it is called a set. In philosophical usage, the term *class* is often indiscriminately applied to both sets and classes, in the sense that classes are abstract collections of individuals; he who recognizes classes believes in universals and goes beyond the nominalist position which maintains that there are only particulars/individuals.

7.2 Current Areas of Research

Cross-linguistic Research. When plurals attracted the attention of the research community in linguistic semantics people soon became curious to learn what kind of cross-linguistic data could possibly be uncovered that have a bearing on the theoretical distinctions to be drawn in the field of semantics. As to the conceptual side intriguing data had already been accumulated in various centers of universal linguistics around the world (see, e.g., Biermann (1981); Zaefferer (1991) and Chapter 9 of this book for relevant information). Here I want to focus on a particular issue, the question of *distributivity.* As it turns out, there is an enormous amount of data in various typologically unrelated languages to the effect that the distributive mode of predication tends to be specifically marked in language. Let us consider a type of sentence where there is a plural subject NP together with a predicate VP which is unspecified with respect to the collective/distributive distinction. In English, for instance, the word *each*, either in a floated quantifier or a postnominal position in the VP, can then be inserted to force the distributive reading of the sentence. A typical example is

(5) a. The men each had a beer.

b. The men had a beer each.

A particle with a similar function is *je* in German, which behaves much like the postnominal *each* (see Chapter 5). But it is much more flexible and powerful. It can occur repeatedly in a sentence, creating quite intricate quantifier structures (6a), and it can occur even if there is no plural NP that can serve as distributional domain for it (then the missing antecedent has to be constructed from context); see (6b).

(6) a. Die Mütter erzählten je zwei Kindern je ein Märchen.
    b. [*The mothers told two children each a fairy tale each.*]

(7) a. Je drei Äpfel waren faul.
    b. [*Three apples each were rotten.*]

Choe (1987a) discovered a similar particle in Korean, and he proposed the fitting term *anti-quantifier* for it. Gil's data on Georgian (Gil (1988)) as well as long-standing facts about Pashto[3] which can be traced back at least to Lorimer (1915) show that distributivity can also be marked by reduplication. Thus it appears that language uses a wide range of possibilities to specify the distributive reading of a sentence.

Algebraic Semantics for Natural Language. The lattice-theoretic approach to plurals is to be seen as part of a more general program of an algebraic semantics for natural language. Apart from plurals, considerable headway has been made here in the study of mass terms; see Bunt (1979); Roeper (1983); Lønning (1987b); Krifka (1987, 1989b); Landman (1991), and Chapter 1 of this book. Of even greater importance is the semantic research on *events*, including the question whether semantic representations should generally be couched in a language of events. Algebraic work on this subject include Hinrichs (1985); Bach (1986); Krifka (1987); Lasersohn (1988); Krifka (1989b); Kamp and Reyle (1993), and Chapters 10–12 of this book.

Metatheory. In the discussion of plurality the focus has of course been on the problem how the plural semantics is able to account for the linguistic data. On the other hand, there are natural metatheoretical questions that arise in this context. For instance, the logic of plurality LP is couched in a first-order language; does that mean that its logical strength is the same as that of pure first-order predicate calculus (with identity)? In particular, is there a complete axiomatization for LP? It is these and related questions that are addressed in Lønning's important study Lønning (1989). Lønning does not discuss the system LP but a similar one, called *plural logic, PL,* that is built on GQ theory. By embedding monadic second-order logic into PL and from the fact that already monadic second-order logic is incomplete he obtains the result that PL is not complete with respect

[3]Thanks to Dietmar Zaefferer for calling my attention to this.

to classes of semantic structures that display the closure under arbitrary joins (in particular, complete lattices and complete atomic Boolean algebras (CAB)). To delineate the "first-order part" of plural theory, Lønning discusses various subsystems; thus, *Definite Plural Logic* (with definite plural terms like *the boys* and quantification over atoms, but no quantification over proper i-sums) is complete with respect to the class of complete atomic Boolean algebras. Finally, drawing on a classical paper in the study of second order logic (Orey (1959)), so-called *persistent* formulas are considered; such formulas do not distinguish between full CAB structures and *definably complete* atomic Boolean algebras (DCAB), that is, suitably "thinned-out" structures that admit of a generalized completeness proof in the sense of Henkin's general models. Typically, simple existential sentences with no negated collective predicate in their matrix are persistent. Apart from persistent formulas, a larger class of "standard" formulas is considered which, however, still leaves out general universal quantification over arbitrary sums. The question then is whether the pluralic locutions in language make full use of the expressive power of PL or whether their logical representations can be kept within the class of persistent (or standard) formulas. In Lønning (1989), this issue is discussed at length. For the status of universal plural quantification in language, see also Chapter 4. The discussion shows that examples for this quantificational mode are hard to find. In particular, it appears that from a linguistic point of view, the interaction between existential plural quantification and negation is as yet purely understood.[4] The same holds for the question to which extent the use of pronouns anaphoric to proper sums transcend the realm of "firstorderizable" sentences of English; see the intriguing examples in Boolos (1984b,a) and Lønning's discussion of them. Future work will have to pay due attention to these issues.

Plural Anaphora. It can be said that there is a good compatibility between plural theory and DRT. In fact, some of the basic claims of PT proved to be particularly valuable in the context of anaphora. As far as the linguistic aspects of plural anaphora and their semantic representation are concerned, an enormous amount of progress has been made within the framework of DRT; see Kamp and Reyle (1993). In this work, a wealth of new relevant linguistic data are discussed, and a number of novel techniques of anaphora resolution are developed to account for them. Here is a list of

[4]Thus, for instance, while the sentence *Some boys shared a pizza* can be represented by a simple existential plural quantification involving a positively occurring collective predicate (which guarantees its persistence) the negated sentence *Some boys did not share a pizza* doesn't seem to amount to a universal quantification over sums of boys. Rather, under negation proper i-sums tend to "dissolve" and lead to distributive predication. For problems with the role of negation in defining plural determiners, see below.

major themes in the study of plural anaphora, together with some of the techniques just mentioned, and typical illustrative sentences.

- Pick-up of suitable antecedents
  - a maximality constraint: no reference to subsums of plural terms
- Construal of antecedents
  - generic use: *Mary killed a wasp. She hates them.*
  - summation: *John took Mary for a ride. They had a lot of fun.*
  - Σ-abstraction: *Every classmate of John's took a girl to Acapulco. They had a lousy time.*
- Dependent plural pronouns
  - λ-abstraction: *Sarah and Mike sent a card to their mother, and Alice did, too.*
- Pluralic donkey sentences: *Few people who own two cars make equal use of them.*

There is a recent critique of the treatment of plural anaphora in DRT by Krifka (1996). This paper takes a different approach which is based on "parameterized sum individuals", an idea going back to Rooth (1987).

Formal Ontology and Philosophy of Mathematics. A basic tenet of plural theory has been a *denotational view* towards definite plural terms like *the men*: such a term denotes a plural object in LP. In linguistic semantics there has been some controversy about the nature of those plural objects. First they were basically assumed to be sets without much of an argument. When it was realized that *mixed predicates* like *lift the piano* subsume individuals and collections alike, and hence that there has to be an extension for them of uniform type, the typical move was to have individual terms denote singleton sets. Thus the denotations of predicates were pushed one level up in the set-theoretic hierarchy. While this is fine from a purely representational point of view, where all that matters is getting the truth conditions right, that practice leaves something to be desired for those who are also concerned with the methodological and philosophical questions arising in this context. As for methodology, my argument has been, as I mentioned above, that set-theoretical modeling in the realm of pluralities misses an important generalization when it comes to incorporating mass terms in one uniform theory of semantics. The philosophical problem is *ontological*: Does the mere fact that people use plural expressions commit them to abstract entities like sets? My answer is of course no, and part of the original motivation for the algebraic approach was to show a way how to avoid that commitment. If you are given a domain of first-order individuals then pluralities of those individuals, conceived as mereological fusions, have the same ontological status as the individuals you start with. As (Lewis, 1991, p. 81) says: "But given a prior commitment to cats, say, a commitment to cat-fusions is not a *further* commitment. The fusion is nothing over and above the cats that compose it."

Now there are actually two different issues involved here: one concerns the set-theoretic vs algebraic approach, where there might not even be a real philosophical disagreement between the two camps subscribing to those approaches; the linguists that stick to their set-theoretic representations typically show a "don't care" attitude towards philosophical issues like ontological commitment. Both sides, however, embrace the denotational view. On the other hand, the position has also been taken that the denotational view has to be given up; in fact, Schein (1993) claims that the view is outright paradoxical, and that plural theory is inconsistent. However, this claim is easily disposed of, since, obviously, LP has a model, viz. atomic Boolean algebras minus the zero element; for a more detailed rebuttal, see Chapter 13.

In this context, an interesting line of research has emerged in the philosophy of mathematics that uses the device of plural quantification to give a nominalistic interpretation of (monadic) second-order logic which steers clear of any commitment to classes; see Boolos (1984b,a, 1985a). The basic linguistic locution that Boolos uses is the relation ... *is one of them* between a first order object and several things ("them"), which he insists are *plurally referred to* and do not form a single set-like collection. Although this relation is a partitive construction and mereology is the theory of parts and wholes, Boolos doesn't give a mereological account; in fact, he never mentions mereology. Such an account, however, is developed in the final chapters of this book. There it is argued that the distinction that has been made in the philosophy of mathematics between the *logical* and the *iterative* conception of set can be understood in such a way as to equate logical sets with mereological fusions (actually i-sums in LP). That opens up the possibility to develop *within LP* the nominalistic reconstruction of set theory given by Lewis (1991, 1993), which is based on a separate mereology of the iterative sets plus Boolos's plural quantification. This is carried out in Chapter 14.

AI-related research. Finally, reference should be made to the growing interest of the NL processing community in plural phenomena. Relevant work includes Schütze (1989); Allgayer and Reddig-Siekmann (1990); Link and Schütze (1991); Aone (1991) and papers in Guarino and Poli (1995). This volume also shows how mereology in general, of which plural theory is but a particular instance, can be put to use for both structuring and unifying the "ontologies" of knowledge-based systems.

7.3 Groups and the Problem of Over-Representation

Several authors have felt the need to introduce multi-level plural objects, which have come to be called "groups". That move was prompted by examples like the following (Landman (1989)).

(8) The boys and the girls had to sleep in different dorms, met in the morning at breakfast, and were then wearing their blue uniforms.

Here the NP *the boys and the girls* has at the same time to stand for a uniform i-sum (to fit the VP 'meet'), to distribute one level down (to fit 'sleep in different dorms'), and to distribute two levels down (for the VP 'wearing'). There are notorious problems for the theory of plurals with such examples since the summing operation obliterates structure: once the sum of the boys and the girls is formed there is no unique way to retrieve the relevant intermediate parts that were used to build that sum; only its atomic parts can always be regained. The question is how to react to this situation. It could be argued, to begin with, that (8) is not representative for the general linguistic phenomenon of conjoining VPs with collective and intermediate level readings, but rather hinges essentially on the word *different* which admits of a special treatment to dispose of Landman's sentence. One reason for trying to avoid the combination of collective and intermediate level readings could be that there is a way to treat the conjunction of two VPs Q and Q' *without* resorting to groups when Q, Q' are of one of the following types: (i) Q, Q' have the same distributivity type; (ii) Q is collective, Q' distributive; (iii) Q is intermediate, Q' distributive (see Lønning (1989)). The trick is here to combine term conjunction in LP with generalized quantifier conjunction. For example, consider sentence (9), adapted from Lønning, which describes a situation with a mixed double tennis match.

(9) Steffi and Michael and Arantxa and Emilio got \$ 10,000 for the match.

Under its distributive reading (9) implies that \$ 40,000 were handed out; in the collective case the money totals \$ 10,000, but it is also possible that \$ 20,000 were paid (intermediate level reading). Now let $\lambda PP(s \oplus m)$

denote the set of all properties applying to the i-sum of Steffi and Michael, $\lambda PP(a \oplus e)$ the set of all properties applying to the i-sum of Arantxa and Emilio. Then $\lambda P[P(s \oplus m) \wedge P(a \oplus e)]$ (the set of all properties that apply to the sum of Steffi and Michael as well as to the sum of Arantxa and Emilio) is an appropriate representation for the subject NP in (9) when the intended reading is the intermediate level one. Now conjoining two VPs Q, Q' poses no problem in the cases (i) – (iii) above: while (i) is obvious, the appropriate term for case (ii) is $\lambda x(Qx \wedge^D Q'x)(s \oplus m \oplus a \oplus e)$, for case (iii) $\lambda P[P(s \oplus m) \wedge P(a \oplus e)](\lambda x(Qx \wedge^D Q'x))$. But there is no way to use the same method when a collective and an intermediate level VP are conjoined. This gets the above proposal into trouble empirically since at least in German there are clear cases of such VP conjunctions that do not use the word *different*:

(10) a. Die norwegischen Delegierten und die schwedischen Delegierten trafen sich in der Lobby des Bayerischen Hofs und erhielten je einen Dolmetscher.

b. [*The Norwegian delegates and the Swedish delegates met in the lobby of the "Bayerischer Hof" and were assigned an interpreter each.*]

(11) a. Im Landheim haben die Jungen und die Mädchen je einen Schlafsaal zur Verfügung, können sich aber zu gemeinsamen Aktivitäten im Aufenthaltsraum treffen.

b. [*In summer camp the boys and the girls sleep in a dormitory each, but they are allowed to meet in the hall for common activities.*]

In the face of data like these a number of people, including the present author, have concluded that the introduction of intermediate level entities, called *groups*, is inevitable (see, e.g. Hoeksema (1983); Link (1984); Landman (1989)). There are two main avenues to proceed here: one is to give some kind of minor modification of the theory just enough to accommodate the data, without touching its core; the other is to give up the theory altogether. While I myself gave, more or less reluctantly, an obvious patch-up to the basic system LP just for that one extra level, Landman thought that this kind of evidence is reason enough to revise the whole of LP in such a way that its mereological character is obliterated. In fact he brings set-theoretic comprehension in again through the back door. Thus, in order to keep the intermediate i-sums in the above NPs separate, the sums (or unions) of atomic individuals are closed off by forming their singletons. For instance, the subject NP in sentence (9) in its intermediate group reading is represented as the set of rank 2,

$$\{\{s, m\}, \{a, e\}\}$$

where s stands again for Steffi, m for Michael, a for Arantxa and e for Emilio. Nor does Landman stop at this rank; rather, he basically erects the full cumulative hierarchy of order ω over any given domain of discourse. Thus he defines for a given set A of basic individuals:[5]

$$\begin{aligned} V_0^A &= A \\ V_{n+1}^A &= 2^{V_n^A} \setminus \{\emptyset\} \quad (n \in \omega) \\ V_\omega^A &= \bigcup_{n \in \omega} V_n^A \end{aligned}$$

Now while the cumulative hierarchy of pure sets has 16 elements of rank 3 Landman's has 127 elements of this rank if A contains just 2 members, and approximately $2^{33\,000}$ if A has 4 elements. Here is an overhead if there ever was one. Set-theoretic modeling has led Landman astray. What's more, there were never, I think, convincing empirical reasons to deal with more than one intermediate level.

What, then, should be done with that remaining level? Recently, intriguing arguments have been advanced to the effect that there is no need to introduce a new kind of entities over and above the mereological i-sums. I shall briefly reproduce the main idea of a paper by Schwarzschild [1990], but see also Schwarzschild (1991); Krifka (1991b).

Assume that the animals on a farm are just cows and pigs, with young animals and old animals among them. Thus we have the following equality of extensions: $\|$ *animal* $\|$ $=$ $\|$ *cow* $\|$ $\cup$ $\|$ *pig* $\|$ $=$ $\|$ *young animal*$\|$ $\cup$ $\|$ *old animal*$\|$. Assume further that the farmer separates the young animals from the old animals. Then the situation can be described by either one of the following five sentences, albeit with varying accuracy:

(12) a. The young animals and the old animals were separated.
b. The animals were separated. $\downarrow$
c. The animals were separated by age. $\downarrow$
d. The cows and the pigs were separated by age. $\downarrow\uparrow$
e. The cows and the pigs were separated. $\downarrow\uparrow$

The down and up arrows hint at two generalizations that Schwarzschild abstracts from these data: The *mereological generalization* says that whenever P is true of a group G then P is true of the sum of the entities of level one in the group, denoted by $\downarrow G$; examples for P are *were separated*, *talked*

[5]Landman's definition is slightly different in that he generously throws in another power set operator at the ω level.

to each other, were given different foods, etc. According to the second principle, the *upward closure condition*, when P is true of a sum G of entities of level one P is true of any group $\uparrow G$ formed from G. The claim is then that while natural language predication does not differentiate between levels it is still important which way the i-sum in question is referred to by the plural NP: it determines the *operative subsums* (Schwarzschild) according to which the predicate is to be understood unless there is some other clue that does this job, either in the sentence itself (e.g. an adverbial like *by age*) or in the context of utterance. In (12a), then, the operative subsums are the ones mentioned in the subject NP. (12b) holds because if (12a) is true it is also true that the animals were separated according to some criterion not made explicit here; this is done in (12c) through the adverbial *by age.* Now since according to our assumption the animals are coextensive with the cows and the pigs the NP *the cows and the pigs* can be substituted salva veritate for the NP *the animals* in (12c), whence the truth of (12d). Mereological generalization is applied here, followed by the upward closure condition. Finally, the criterion *by age* is dropped, yielding what might out of context be a misleading way of expressing the situation at hand, but it is nonetheless true. This shows that the NP used to refer to the group under consideration does by no means always contain the information as to what the operative criterion is according to which the group is split up.

Similar observations apply to the *Waterloo* example given in Hoeksema (1983). When the sentence *Napoleon and Blücher and Wellington fought against each other at Waterloo* is uttered without a pause after *Napoleon* it is hard for a hearer who missed the relevant chapter of history to grasp the intended reading. So it is actually intonation that determines the appropriate operative subsums here. In the last example I'd like to mention, which is again taken from Schwarzschild (1990), the operative subsums are given by cross-sentential deixis.

(13) a. The pictures that came from Bill's parents and the pictures that came from Sheila's parents were separated.

b. The books were separated that way, too.

Examples like this undermine the conception that linguistic evidence forces the semanticist to introduce yet another layer of entities (the group level entities) on top of the sum objects of LP theory. It seems that proliferation of entities according to this conception is rather a case of over-representation than a piece of reasonable semantic modeling. Following the suggestions of Schwarzschild (1990); Krifka (1991b) we conclude that a more careful analysis should put greater emphasis on the interplay between semantics proper and discourse phenomena to be treated in a format like DRT.

7.4 Distributivity

Ever since Scha's seminal paper Scha (1981) there has been a dispute over the question of how many readings there are for a sentence involving (one or more) plural NPs. In Chapter 2 I expressed my doubts that Scha's fairly complex classification of readings is the right one, and I proposed a rather simple scheme instead which was generated by the basic collective/distributive distinction (C/D distinction, for short). In particular, I came up with seven readings for sentences like (14).

(14) Four men lifted three tables.

Since there are two pluralic argument places, $2 \times 2 = 4$ readings came from the C/D distinction; a possible difference in scope still multiplied the readings by 2 minus 1 (no scope in the double collective case). Now this was a logician's view working in the Montagovian tradition; hence its main goal was to display the various quantifier structures that a plural sentence like the above gives rise to when treated rigorously in a formal framework like LP. Now there were two kinds of critical reactions to this. One was that there is no scope distinction operative here. This objection came from quarters which had for some time taken exception to the Montagovian view that the paradigm sentence *every woman loves a man* has to readings due to scope. While I do appreciate the point made here I will not discuss this issue but rather ignore possible scope distinctions altogether. That leaves us with four readings. For some, even that is too much; thus Lønning (1991) says that sentence (15a) is not ambiguous contrary to what my scheme predicts.

(15) a. John ate three apples.

b. John lifted three tables.

c. John juggled with 6 plates.

In his eyes it is a matter of indeterminacy rather than ambiguity whether those three apples were eaten one by one or swallowed in one fell swoop (pragmatic star). Today, again, I feel sympathetic to his view, although not all questions are answered here. In special contexts the C/D distinction might still matter, witness (15b,c). More importantly, when event variables are introduced into the representation language the question inevitably arises as to what the temporal relations are between the various actions expressed by a plural sentence.

Let us accept the diagnosis of indeterminacy for the sake of exposition. Then the above scheme has been stripped down to two readings only coming from the C/D ambiguity in the pluralic subject NP. This is what Lønning calls the *standard proposal.*[6]

The standard proposal has been objected to in the literature as being too simplistic. Two basic arguments can be distinguished that support this criticism:

- the argument from undergeneration of readings
- the argument from a missed generalization

Simplifying the historical record somewhat, the argument from undergeneration of readings takes up Scha's view that in addition to the collective and the distributive reading we have to admit a *mixed reading* which looks into the subcollections of a given plurality. Over the years a host of intriguing linguistic examples have come up that were designed to substantiate this claim. A "neoclassical" example derives from Gillon (1987):[7]

(16) a. Hammerstein, Rodgers and Hart wrote musicals.

b. Hammerstein, Rodgers and Hart wrote a musical.

The argument goes like this. None of the three wrote a musical all by himself; so the sentence is not distributive. But also there is no musical that was written by all three of them together; so the sentence cannot be read collectively, either. Rather, there is a certain *cover* Y of the set consisting of the three composers such that every element of Y is a collection satisfying the VP property. (The various accounts differ from one another with respect to the specific nature of those covers: they can be minimal covers, proper partitions or again overlapping "pseudo-partitions"; see Lønning (1991) for references). In our case, musical history says that, for instance, the collection consisting of the two pair sets {Hammerstein,Rodgers} and {Rodgers,Hart} is a verifying cover for the sentence.

However, I agree with Lønning's analysis that the argument is not as cogent as it might appear. To begin with, the b) sentence cannot have the

[6] I am following here his lucid exposition in Lønning (1991).

[7] For the benefit of the uneducated reader (like myself) here are some basic facts about Hammerstein, Hart and Rodgers: Richard Rodgers and Oscar Hammerstein II together wrote the musical *Oklahoma!* whereas Rodgers and Lorenz Hart wrote *On Your Toes* together, for instance. Again, Rodgers, Hart and Herbert Field wrote *A Connecticut Yankee*, Hammerstein and Jerome Kern *Show Boat*, Rodgers, Hammerstein, and Joshua Logan *South Pacific*. But it seems indeed that neither did the three of them write any musical together, nor did any of them write one all by himself. (This is taken from Gorton Carruth, *The Encyclopedia of American Facts and Dates*. Eighth edition, Harper and Row, New York 1987.) Special thanks go to Brandon Gillon for drawing the attention of the community to American culture.

cover interpretation; it simply doesn't mean that. Either it is read collectively, or the indefinite NP *a musical* is distributed over the subject NP which means that the three individuals each have the property in question. There seems to be no way to get at the intermediate subcollections. Now the a) sentence might invite a mixed interpretation more easily but even here serious doubts remain that have to do with the meaning of bare plural phrases like *writing musicals.* Such properties are *homogeneous* in character and do not imply that a person satisfying them bear the verb relation to a *complete* object falling under the bare plural concept. In the case at hand, each of our three composers can truly be said to have written musicals even if no one wrote a single musical all by himself; their *involvement* in musical writing is sufficient (cf. Link and Schütze (1991)).

Observe that this is of course an empirical argument pertaining to a certain class of typical examples. Our theory should perhaps be flexible enough to cover the mixed readings, too, just in case future evidence is given to support them. So for the benefit of further discussion let me give representations of them within the algebraic theory of plurals.[8] Even if pluralities are no higher-order objects covers certainly are; so we have to extend the first-order theory LP to include second-order variables ranging over sets of i-sums. For the present purpose it would be pointless to give a rigorous representation in the object language; rather, the notation will be "semantic", i.e., metalinguistic, but with free use of both the suggestive λ-notation and the circled plus to build finite i-sums. Let E be the domain of individuals (i-sums) with variables x, y, z, and $\sqsubseteq$ the i-part relation on E; A is the set of atomic i-sums, with variables u, v, w, and 2^E the set of sets of i-sums with variables X, Y, Z. The principal ideal of an i-sum x is defined by

(17) $x^{\downarrow} := \{y \mid y \sqsubseteq x\}$ the principal ideal of x

When we intersect this set with A we get the set of atoms below x; by the *sup-generation property* of the plural lattice (see Chapter 6) we have:

(18) $x = \sup(A \cap x^{\downarrow})$

Now let us follow van der Does (1992) and introduce certain transformations on 2^E, δ for "distributive" and μ for "mixed". Let $\mathcal{C}$ stand for the various classes of covers like the ones mentioned above.

[8] In fact those representations will look pretty much the same as the ones in van der Does (1992). The reason is that van der Does has as atomic individuals singleton sets over a basic set X, such that set-theoretic union on the power set 2^X models the summing operation. As Lønning [1991] observes, the basic set X serves no purpose whatsoever. This is a typical occasion for a mathematician to abstract away from that artifact of the representation. It is here that the lattices come in.

(19) $\delta \;=\; \lambda X\, \lambda y\,.A \cap y^{\downarrow} \;\subseteq\; X$

(20) $\mu \;=\; \lambda X\, \lambda y\,.\exists Y \in \mathcal{C}(\, y = \sup Y \;\&\; Y \;\subseteq\; X\,)$

Let X be the property $[\![\mathit{wrote.musicals}]\!]$ and $oh \oplus rr \oplus lh$ the i-sum consisting of Oscar Hammerstein, Richard Rodgers and Lorenz Hart. Then we have:

$$\text{(21)} \qquad oh \oplus rr \oplus lh \in \delta([\![\mathit{wrote.musicals}]\!]) \iff$$
$$\forall u(\, u \in A \,\&\, u \sqsubseteq (oh \oplus rr \oplus lh) \;\Rightarrow\; u \in [\![\mathit{wrote.musicals}]\!]\,) \iff$$
$$oh \in [\![\mathit{wrote.musicals}]\!] \;\&\; rr \in [\![\mathit{wrote.musicals}]\!] \;\&\; lh \in [\![\mathit{wrote.musicals}]\!]$$

Thus this gives the desired distributive reading. By contrast, the μ-operator formalizes the mixed case, as our example illustrates:

$$\text{(22)} \qquad oh \oplus rr \oplus lh \in \mu([\![\mathit{wrote.musicals}]\!])$$
$$\iff \;\; \exists Y \in \mathcal{C}(\,(oh \oplus rr \oplus lh) \;=\; \sup Y \;\&\; Y \;\subseteq\; [\![\mathit{wrote.musicals}]\!]\,)$$

With $Y \;=\; \{oh \oplus rr, rr \oplus lh\}$ we get the intended meaning:

$$\text{(23)} \qquad oh \oplus rr \;\in\; [\![\mathit{wrote.musicals}]\!] \;\&\; rr \oplus lh \;\in\; [\![\mathit{wrote.musicals}]\!]$$

Since $\sup(A \cap x^{\downarrow}) \;=\; x$ we see that δ is a special case of μ, in fact the one in which the cover Y is the finest partition of x. Another limiting case is the cover $Y \;=\; \{x\}$ which yields the collective reading.

This observation brings us to the second argument against the standard proposal, viz. the argument from a missed generalization. If both the collective and the distributive reading are but special cases of the mixed reading, so the argument goes, then there is no ambiguity there after all. What we have is one representation with two opposite limiting cases. This is what van der Does (1992) calls the No Ambiguity Strategy (NAS).

While van der Does, in his joint work with Henk Verkuyl (Verkuyl and van der Does (1996)), seems to support this strategy he rejects it explicitly in his dissertation. I concur with his later assessment on this topic. Empirical work across languages has established beyond doubt, I think, that the distributive mode of predication is highly marked in language, as already the few data mentioned in section 7.2 show. The reason seems to be obvious: distributive predication has universal quantificational force and is thus equipped with a precise logical interpretation. By contrast, the collective mode is mostly vague and indeterminate, even when made explicit by means of adverbs like *together.* But predictably, no one has come up yet with a language where there is a special marker for, say, the pseudo-partitional reading.

Thus the empirical line is drawn between the distributive vs the non-distributive (the rest). The rest can be split up in *collective in a narrow sense* (one collective relation with a plural character) or *collective in a broad sense* where the mixed cases are included (lots of collective and/or distributive relations with corresponding plural and/or singular characters). Among the latter, only the *cumulative* case of Scha (1981) seems to have a special linguistic status in that both the subject NP and the object NP involved in cumulative constructions are *autonomously referring* plural terms. Note, however, that the cumulative "reading" is particularly prominent when both NPs are specified numerically, like in Scha's famous sentence about the Dutch firms and their American computers. Van der Does [1992: p. 55] claims he can also read sentence (24a) cumulatively, where the determiner of the object phrase is no numeral, but pluralic *some.*

(24) a. Hammerstein, Rodgers and Hart wrote some musicals.

b. Some musicals were written by Hammerstein, Rodgers and Hart.

I disagree because that would amount to giving the object phrase a degree of specificity that seems hard to come by. Things look better, however, when the sentence is passivized, as in (24b). Even so, the defender of the standard proposal will always say that it is the indeterminacy in the relation of writing music, and not a linguistic ambiguity, that gives room for the cumulative interpretation here.

In summary, I look at the non-distributive domain as a matter of indeterminacy, not ambiguity (see, again, Link and Schütze (1991)). Among the non-distributive phenomena only the narrowly understood collective interpretation and the cumulative interpretation stand out as having a fairly context-independent status. By contrast, the basic collective/distributive ambiguity is well-entrenched in language even if mathematically, both the collective and the distributive reading are but special cases of a more general cover interpretation.[9]

A related issue of long standing is the question where exactly the C/D ambiguity arises, in the NP or in the VP. Well, it's the VP, I think, and I have always been convinced of that. So I embrace van der Does's Verb Phrase Strategy (VPS), which he rejects, but which is in my opinion corroborated by ample empirical evidence. Prominent among this evidence are

[9]The analysis given raises an interesting methodological point. Maintaining the ambiguity seems to be at variance with a common methodological principle in science. Galileo, for instance, criticized Aristotle on the grounds that he draws a categorical, in fact metaphysical, dividing line between the state of motion and the state of rest. The failure to view the state of rest as the limiting case of the state of motion at velocity zero constituted a major stumbling block for the progress of science, according to Galileo. Thus Aristotle plainly missed a crucial generalization. Ironically, I find myself siding with empirical linguistics against abstract scientific methodology.

Craige Roberts's arguments from anaphora (Roberts (1987b),1987a) and sentences involving mixed collective and distributive predicates, like (25a).

(25) a. Four men went to the bar together and each had a beer.

b. $\exists x(x \in [\![\mathit{4\ men}]\!]\ \&\ x \in [\![\mathit{went.together}]\!] \cap \delta([\![\mathit{had.a.beer}]\!]))$

The representation (25b) shows the ease by which such sentences are handled under VPS. Furthermore, data from the German distributivity operator *je* show that the pluralic domain of distribution that *je* operates on is not always unique and can often be only determined by context. For instance, example (7a) above, *Je drei Äpfel waren faul*, doesn't contain a plural NP that could serve as the distributional domain for *je*, but the sentence is perfectly fine in German. It can be felicitously uttered in a situation where there is a fruit stand displaying apples in several baskets each of which contains three rotten apples.[10] It is hard to imagine how a pluralic nominal domain that in some cases has to be reconstructed from context could "know" that it should serve as distributional domain for an antiquantifier like *je*.

These observations by no means imply, of course, that the NP (the determiner, that is) does not contribute to the meaning of sentences in an essential way. I even suspect that the Determiner Strategy (DETS), which is endorsed by van der Does, is not really that different from the VPS after all. But how could that be in view of van der Does's claim that the DETS in effect supersedes the VPS in that it can handle data on which the latter fails? The crucial case that is adduced in favor of DETS is the behavior of non-MON$\uparrow$ determiners.

To answer this question, let us start with the basic picture of GQ Theory. The determiner of the subject NP essentially controls the quantificational force of the sentence. No discourse phenomena are taken into account. When the NP contains a numeral, for instance, all that counts is the *number of objects* involved in the relation expressed by the sentence. Thus consider sentence (25a) above. It is true in GQ theory under the usual interpretation of unmodified numerals if the number of men that went to the bar together and had a beer each is at least four. This truth condition is obviously upwardly monotone. However, it doesn't bring out the fact that the plural NP can and in most cases will be used *referentially* when the speaker has a particular group of four men in mind. This i-sum is then the topic of the rest of the discourse, never mind how many men ordered a beer besides that. In DRT there would be a plural reference marker for it. The referential use of indefinite NPs was acknowledged by researchers in the GQ

---

[10]For more on *je* see Chapter 5.

tradition when Barwise (1987) compromised with discourse theories by distinguishing between *singular* and *general* NPs: indefinite NPs are typically singular in Barwise's sense[11] in that they are open to a referential interpretation. Unlike the original GQ representation, the LP representation given in (25b), which a simple existential statement, can easily be amended to a quantifier-free discourse format that captures the referential use. That is why I keep the existential quantifier in front. One has to bear in mind, of course, that this entails upward monotonicity.

Van der Does argues that this approach cannot be sustained in the non-MON↑ cases. Those involve determiners like *exactly three* or *at most four*. They are genuinely quantificational in that they give rise to a *general* noun phrase in Barwise's sense: a generalized quantifier which does not provide a reference object for the discourse. DRT typically sets up a duplex condition for them. There is nothing referential involved here. But on the other hand, we do have sentences like the following, in which the subject NP seems to be both referential and quantificational.

(26) At most four men went to the bar together and had a beer each.

Let us consider various possible eye-witness reports:

1. To the best of my recollection, there were no more than 4 men in the room; they went to the bar together and had a beer each, etc.

2. At most 4 men went to the bar together; there were lots of men in the room but it was those men – I am sure they were not more than four – that apparently tried to pick a fight with the stranger at the bar. First they had a beer each etc.

3. The bar keeper: There were at most 4 men that went to the bar together and had a beer each; I can tell it from the number of broken glasses they threw at the stranger: no more than four are missing. etc.

What this shows is that the non-MON↑ determiners are special in that they take scope over the material that follows. The scope does not necessarily extend to the sentence boundary, but can rather vary form context to context. Thus sentences involving such determiners cannot be evaluated *unless the material is specified that goes under their scope.* According to the first report there were no more than four men in the room; the scope is *'men'.* According to the second report there were no more than four men

[11]Including pluralic indefinite NPs! The opposition here is "singular term" vs "general term" in the logical sense; it has nothing to do with the number distinction.

that went to the bar together; here the scope is *'men that went to the bar together'*. And the third report says that there were no more than four men that went to the bar together and had a beer each; here the scope extends over all of *'men that went to the bar together and had a beer each'*. According to this story there might well have been more than four men that went to the bar together; but only a group of no more than four men was such that its members had a beer each.

Now it is plain that it would lead to inadequate results if one were to give representations for the referentially used non-MON↑ determiners, starting with an existential quantifier, *and adding no extra clause* that specifies the scope of the determiner. Since van der Does [1992] does not consider the possibility of adding a scope indicator he thinks that the standard approach is bound to fail here; he concludes from this that the proper representation has to be the one which just applies the cardinality condition to the collection satisfying the property under the scope of the determiner, which he takes to extend to the sentence boundary. That means that the non-MON↑ determiners produce *general NPs*, that is, genuine generalized quantifiers which do not provide a reference object for the discourse. In DRT, this corresponds to a duplex condition without reference marker. But the different context-dependent interpretations of sentence (26) above show clearly that in some way or other an antecedent for later anaphora can nevertheless be created. How can a representation account for this? Well, I think, DR theorists would resort to *antecedent construal by abstraction* (see Kamp and Reyle (1993)). Informally, for instance, a GQ scheme like *at most n X are Y* would be taken up by the pluralic abstraction term *those X that are Y*. In the first order framework of LP a referential plural NP has to be expressed by an existential quantification. Does that mean that we are bound to produce a MON↑ quantificational force, as van der Does (1992) seems to suggest? I think not. We do have to add, however, a universally quantified clause to prevent this effect.

Let us consider the *exactly* case first and illustrate it by the following example, which contains the collective predicate *meet*.[12]

(27) a. Exactly five boys met.

b. $\exists x(\,|x| = 5 \;\&\; x \in [\![boys]\!] \cap [\![met]\!]\,)$

c. $\exists x(\,|x| = 5 \;\&\; x \in [\![boys]\!] \cap [\![met]\!]$
$\qquad \& \; \forall y(\,|y| = 5 \;\&\; y \in [\![boys]\!] \cap [\![met]\!] \;\Rightarrow\; y = x\,))$

d. $\exists x(\,|x| = 5 \;\&\; x \in [\![boys]\!] \cap [\![met]\!]$
$\qquad \& \; \forall y(\,y \in [\![boys]\!] \cap [\![met]\!] \;\Rightarrow\; y = x\,))$

[12]In the following the "cardinality" operator $|\cdot|$ is meant to take individual terms and return the number of atoms in the denotation of that term, that is, $|x| = \mathrm{card}(x^{\downarrow} \cap A)$.

e. $\exists x(\,|x| = 5 \;\&\; x \in [\![boys]\!] \cap [\![met]\!]$
$\&\; \forall y(\,y \in [\![boys]\!] \cap [\![met]\!] \;\Rightarrow\; y \sqsubseteq x\,))$

Let $\phi[x]$ be the property of being an i-sum x that consists of boys that met. It will certainly not do to interpret sentence (27a) as (27b); the reason is that (27b) is compatible with all kinds of different i-sums having property ϕ, the cardinalities of which do not even have to equal 5. A fairly natural idea here is to give a representation which is modeled after the uniqueness condition for singular definite descriptions. When we add a universally quantified clause expressing this condition, we still have two options, (27c) and (27d); the difference is whether or not the cardinality condition is included in the uniqueness clause. (27c) says that there is a certain sum of five boys that met, and that this sum is unique among all the other five-atoms sums in having the property ϕ; but there may still be other sums x such that $\phi[x]$ of different cardinality. However, that doesn't give the right meaning. In (27d), the cardinality condition is dropped; but now we are facing a different problem: The current logical form admits of exactly one i-sum x such that $\phi[x]$, and for this x we have $|x| = 5$; if we change the property ϕ into one that is downward persistent (i.e., one that if true of x is also true of all y below x) then this contradicts the uniqueness of x for cardinalities of x greater than one. Downward persistent properties abound; for instance, all starred distributive predicates in plural theory like being a sum of boys[13] are of this kind. Now in the plural domain two concepts for expressing the notion of exactness have to be distinguished that coincide in the singular case: *uniqueness* and *maximality.* While we have to give up the former, we can still postulate the latter: (27e) displays the maximality condition in its extra clause. Thus, exactness does not mean that there is only a single individual in the domain with the property in question, but rather that there is a maximal one among them, and this unique element is picked out.

Note that from this representation we get back the familiar condition of GQ theory for *exactly five* when the predicate *met* is replaced by a distributive predicate, say *laughed*:

(28) a. Exactly five boys laughed.

b. $\exists x(\,|x| = 5 \;\&\; x \in [\![boys]\!] \cap [\![{}^{*}laughed]\!]$
$\&\; \forall y(\,y \in [\![boys]\!] \cap [\![{}^{*}laughed]\!] \;\Rightarrow\; y \sqsubseteq x\,))$

c. $\mathrm{card}([\![boy]\!] \cap [\![laughed]\!]) = 5$

[13]If P is a distributive predicate (i.e., one which is true of atomic individuals only) then ${}^{*}P$ is true of all i-sums whose atomic parts each have the property P. Nouns like *boy* are distributive; in the text I often use *boys* instead of **boy*.

The transition from (28b) to (28c) can be shown as follows.[14] Since $[\![boys]\!] = [\![{}^*boy]\!]$ the formula (28b) yields $x \in [\![{}^*boy]\!] \cap [\![{}^*laughed]\!]$ and $[\![{}^*boy]\!] \cap [\![{}^*laughed]\!] \subseteq x^{\downarrow}$., hence $x = \sup([\![{}^*boy]\!] \cap [\![{}^*laughed]\!])$. Now observe that $[\![{}^*boy]\!] \cap [\![{}^*laughed]\!] = [\![{}^*boy.that.laughed]\!]$, and therefore, by a property of starred predicates (Theorem (T. 4) in Chapter 6), $x = \sup([\![{}^*boy.that.laughed]\!]) = \sup([\![boy.that.laughed]\!])$. By sup-generation, we have also $x = \sup(A \cap x^{\downarrow})$. From the injectivity of the sup-operator we get $A \cap x^{\downarrow} = [\![boy.that.laughed]\!] = [\![boy]\!] \cap [\![laughed]\!]$. With $|x| = 5$ and the meaning of $|\cdot|$ we arrive at (28c).

There is further support for the representation of *exactly n* that we have given here. *Exactly n* should mean the same as *at least n and not more than n*, and this equivalence also comes out formally when the latter determiner is regarded as a conjunction of *at least n* and the negation of *more than n* under their natural interpretation. Thus consider (29):

(29) a. At least five and not more than five boys laughed.

b. $\exists x(\,|x| \geq 5\ \&\ x \in [\![boys]\!] \cap [\![{}^*laughed]\!]\,)$
$\&\ \forall y(\,y \in [\![boys]\!] \cap [\![{}^*laughed]\!] \Rightarrow |y| \leq 5\,)$

c. $\exists x(\,|x| = 5\ \&\ x \in [\![boys]\!] \cap [\![{}^*laughed]\!]$
$\&\ \forall y(\,y \in [\![boys]\!] \cap [\![{}^*laughed]\!] \Rightarrow y \sqsubseteq x\,))$

The equivalence of (29b) and (29c) is seen as follows. The direction from right to left follows from the isotony of $|\cdot|$. The opposite direction is proved by reductio. Assume that there is a $y \in [\![boys]\!] \cap [\![{}^*laughed]\!]$ such that $y \not\sqsubseteq x$; then, by the separation property of the plural lattice, there is an atom $u \in A$ such that $u \sqsubseteq y$ and $u \not\sqsubseteq x$. By downward persistence of starred predicates, $u \in [\![boys]\!] \cap [\![{}^*laughed]\!]$. Define $x' := x \sqcup u$; then $x \sqsubset x'$. But starred predicates are also cumulative, hence $x' \in [\![boys]\!] \cap [\![{}^*laughed]\!]$. But $|x'| > 5$ which contradicts the second premise in (29b).

Note that the proof fails if genuinely collective predicates are considered; they are in general neither downward persistent nor cumulative. What that means is that it is not enough to restrict the cardinality in the negated determiner *not more than n*; we have to use the $\sqsubseteq$-relation instead. This anticipates a choice that has also to be made when we return to the non-MON↑ determiner scheme *at most n X are Y*. One of the options is to say that there is an i-sum x such that $|x| \leq n$ and $x \in X \cap Y$ while adding the clause that all y with $y \in X \cap Y$ are limited in size by $|y| \leq n$. But that would leave room for the possibility that we are fixing on a discourse referent x with $x \in X \cap Y$, but of smaller size than n, while there are bigger i-sums y around (with $n \geq |y| > |x|$) which also meet the condition $y \in X \cap Y$. To test this option consider (30).

[14]In this argument formal properties of the plural lattice are used; see Chapter 6.

(30) At most four squatters were left in the building. They were dragged out by the police.

I think this two-sentence discourse means that afterwards all squatters are gone. But that is not what our first options predicts; it is compatible with a situation where a group of two squatters was given the treatment by the police while one or two squatters still remain in the house. What is missing here again is the *maximality condition* which guarantees that the anaphoric pronoun *they* refers to all of the squatters that are left in the building. This is the second option which can be represented thus.

(31) a. At most four squatters were left.

b. $\exists x(\,|x| \leq 4 \;\&\; x \in [\![squatters]\!] \cap [\![{}^{*}were.left]\!]$
$\qquad \&\; \forall y(\,y \in [\![squatters]\!] \cap [\![{}^{*}were.left]\!] \;\Rightarrow\; y \sqsubseteq x\,))$

Since our example involves only distributive predicates this again boils down to the usual GQT condition $\text{card}([\![squatter]\!] \cap [\![were.left]\!]) \leq 4$. The only difference is that the latter condition carries no existential import. In order to arrive at a similar cardinality restriction also in the general case of the existence-free MON↓ determiner *at most n X are Y*, we take the supremum of $X \cap Y$ and put the restriction on the number of its atoms: $\text{card}([\sup(X \cap Y)]^{\downarrow} \cap A) = |\sup(X \cap Y)| \leq n$ (if $X \cap Y = \emptyset$ then $[\sup(X \cap Y)]^{\downarrow}$ is considered to be empty, too).

In summary, then, here is a list of some plural determiners with their semantic interpretation in a GQ framework that incorporates plural theory. Lower case variables run over arbitrary i-sums, upper case variables over sets of i-sums.

Pluralic determiners.

(32) $[\![\text{some}_{pl}]\!] = \lambda X \lambda Y. \exists x\,(\,|x| \geq 2 \,\wedge\, x \in X \cap Y\,)$

(33) $[\![n]\!] = \lambda X \lambda Y. \exists x\,(\,|x| = n \,\wedge\, x \in X \cap Y\,)$

(34) $[\![\text{at.least } n]\!] = \lambda X \lambda Y. \exists x\,(\,|x| \geq n \,\wedge\, x \in X \cap Y\,)$

(35) $[\![\text{more.than } n]\!] = \lambda X \lambda Y. \exists x\,(\,|x| > n \,\wedge\, x \in X \cap Y\,)$

(36) $[\![\text{not.more.than } n\,]\!] =$
$\lambda X \lambda Y. \neg \exists x\,(\,|x| > n \,\wedge\, x \in X \cap Y\,) =$
$\lambda X \lambda Y. \forall x\,(\,x \in X \cap Y \;\Rightarrow\; |x| \leq n\,)$

(37) $[\![\text{at.most } n]\!] = \lambda X \lambda Y. |\sup X \cap Y\,| \leq n$

(38) $[\![\,(\text{some.but})\ \text{at.most}\ n\,]\!]\ =$

$$\lambda X \lambda Y.\exists x\,(\,|x| \leq n \ \wedge\ x \in X \cap Y \ \wedge\ \forall y\,(\,y \in X \cap Y \ \Rightarrow\ y \sqsubseteq x\,))$$

(39) $[\![\,\text{exactly}\ n\,]\!]\ =$

$$\lambda X \lambda Y.\exists x\,(\,|x| = n \ \wedge\ x \in X \cap Y \ \wedge\ \forall y\,(\,y \in X \cap Y \ \Rightarrow\ y \sqsubseteq x\,))$$

As argued above, there are two representations for the non-MON↑ determiner *at most*, one with and one without existential import. The latter gives a pure upward cardinality restriction leaving room for zero, while the former lends itself to anaphora in discourse. Furthermore, in the general case of collective predication, *at most n* is not the same as negated *more than n*; genuine plural determiners just don't work that way (even disregarding the existence question). An old theme is recurring here: universal plural quantification has only a highly restricted role to play in language. That discards Equation 16 as suitable representation in the collective case. Observe finally that due the maximality condition the determiners in Equations (38) and (39) are not upwardly monotone anymore. So the empirical advantage over the standard proposal that is claimed by van der Does's account for the non-MON↑ cases evaporates.

Chapter 8

Algebraic Semantics for Natural Language: Some Philosophy, some Applications

8.1 Introduction

The enterprise of intelligent information processing leads together researchers from different fields and backgrounds, e.g., computer science, artificial intelligence, logic, linguistics, and philosophy. Before we proclaim the grand unification of the various goals, methods, and research traditions we should become very clear about the genuine character of the approaches in the different fields.

Information processing, when performed by an intelligent agent, draws on a wide array of knowledge sources. Among them are world knowledge, situational knowledge, conceptual knowledge and linguistic knowledge. The latter in turn subdivides into syntactic, semantic, and pragmatic knowledge. The focus will here be on the semantic knowledge which is part of the general linguistic competence of any speaker of a natural language. Again, semantic information can be lexical or structural. Lexical knowledge heavily interacts with conceptual knowledge, and thus a theory of that knowledge has its roots in psychological as well as philosophical analysis. But structural semantics has more to do with logic than with philosophy. Moreover, there is a very important constraint on any theory of semantics that deals with language proper: it not only has to give the logic of the structural phrases of language but it must also take into account what has come to be called the *dynamics of language use* in greater portions of discourse. Now this constraint dictates a close-to-language analysis which is simply not needed in other forms of information processing. Here is a difference in goals that has to be heeded lest people come to wonder why certain integrated systems of knowledge representation, which combine traditional AI tools with a language understanding component, exhibit a rather poor performance and have a system architecture that is reminiscent of unpleasant zombies.

A formal representation of a certain range of linguistic phenomena has to be *correct* and as *complete* as possible.[1] Correctness means that the formal relation of semantic and/or pragmatic entailment does not contain instances that are intuitively invalid. This is an absolute criterium of adequacy for the representation. But even if the representation system is correct it might, and in general will, be incomplete in that the intuitive entailment relation is not fully captured by the formalism. Moreover, a semantic representation usually gives no more than a *model* of the phenomena to be described, and there is no claim about the "real nature" of the things in the domain of discourse. From a philosophical point of view this might be unsatisfactory, but it is simply different goals that are pursued in the various fields of research: (i) The essence of the semantic enterprise is to give representations according to the criteria of absolute correctness and relative completeness; (ii) An NL understanding system has to satisfy an additional requirement: computational efficiency. That's why first-order representations are preferred: complete axiomatizability makes the valid sentences recursively enumerable and hence more suitable to an algorithmic treatment. (iii) Philosophy, on the other hand, is concerned with ontology, and it puts a different kind of demand on the semantics: formal semantics should provide the means to clarify the ontological status of the entities under consideration. With the help of a formal theory of a certain range of objects it could be stated explicitly which kind of objects somebody embracing this theory is committed to. This seems like a strong requirement since the semantic objects, being not more than modeling devices, cannot be *equated* with the real entities in the domain of discourse. The semantic objects are just designed to achieve the representational goal. But there is some sense in which a formal semantics can "mirror" the ontology; it can characterize its *structural properties.* Structure, however, when expressed mathematically, is abstract algebra. And this is how formal semantics can remain neutral about the question "what the objects really are" and still

[1] Cf. Blau (1978).

serve the needs of the philosopher: by providing an algebraic theory of the domain of entities in question.

The chapter has four parts. Section 8.2 explains the term "algebraic semantics" as it is used in logic. Following Peter Aczel, two senses of "algebraic" are distinguished which I call here *conceptual* and *structural*. These two senses of the algebraic method are applied to NL semantics in Section 8.3. The conceptual part is realized by the method of *structuring the domains of linguistic ontology* in various ways. Thus *plural entities* are recognized along with *mass entities* and *events*. The common outlook here is *mereological*. Section 8.4 gives some applications to the study of plurals that are to show the usefulness of the algebraic approach. The final Section 8.5 addresses the ontology of plurals and comments on some relevant discussion of mereology in recent philosophical work.

8.2 Algebraic Semantics in Logic

The notion of an algebraic semantics has been around for some time in the field of logic. It refers to a certain approach to both classical and non-classical model theory which uses suitable algebraic structures as semantic "mirror images" of the logic at hand. The *locus classicus* here is Rasiowa and Sikorski (1968), but see also Goldblatt (1984).

Consider, for instance, the classical propositional calculus, PC. Let S_0 be the set of propositional variables $p, q, r, \ldots$ and S the set of PC formulas $\phi, \psi, \chi, \ldots$ as usual. We have $S_0 \subseteq S$. Then we can give the following semantics for PC.

Algebraic semantics for PC: Let $\mathcal{B}$ be a Boolean algebra $\mathcal{B} = \langle B, \sqcup, \sqcap, ', \mathbf{0}, \mathbf{1} \rangle$. An *(atomic) $\mathcal{B}$-valuation for PC* is a function $f : S_0 \rightarrow B$. A *$\mathcal{B}$-valuation for PC based on f* is the (unique) function g_f from S to B with $g \supseteq f$ such that

1. $g(\neg\phi) = g(\phi)'$
2. $g(\phi \vee \psi) = g(\phi) \sqcup g(\psi)$
3. $g(\phi \wedge \psi) = g(\phi) \sqcap g(\psi)$
4. $g(\phi \rightarrow \psi) = g(\phi)' \sqcup g(\psi)$

A formula ϕ is *true with respect to f* just in case $g_f(\phi) = \mathbf{1}$. ϕ is *$\mathcal{B}$-valid* iff ϕ is true w. r. t. all $\mathcal{B}$-valuations f. ϕ is *BA-valid* iff ϕ is $\mathcal{B}$-valid for all Boolean algebras $\mathcal{B}$. Now it can be shown (Goldblatt (1984)):

Theorem 1 *$\phi \in S$ is a tautology in the usual sense iff ϕ is BA-valid iff ϕ is $\mathcal{B}$-valid for some particular Boolean algebra $\mathcal{B}$.*

When we pick $\mathcal{B} = \mathbf{2}$, with $\mathbf{2} = \langle\{0,1\}, \max, \min, \lambda x.1-x, 0, 1\rangle$, we have the usual (extensional) semantics. Another prominent Boolean algebra is the *Lindenbaum algebra for PC* which consists of all the equivalence classes of interderivable (i.e., logically equivalent) formulas of PC. Call this algebra $\mathcal{B}_L$. The elements of $\mathcal{B}_L$ can serve as an explication of the notion *the proposition expressed by* ϕ, for a formula ϕ in PC: ϕ denotes

$$\|\phi\| \;:=\; [\phi]_\sim \;=\; \{\psi \mid \psi \sim \phi\}$$

i.e., the set of all formulas that are PC-equivalent to ϕ.

That explication is, however, not well-suited for situations in which there is a need to integrate propositions into the domain of discourse. For this purpose, another standard Boolean algebra has been used, viz. the power set $\wp(W)$ of a set W; indeed, that was the choice of Carnap and the other intensional logicians, who interpreted W as the set of possible worlds.

In view of the one-to-one correspondence between $\wp(W)$ and the set 2^W of characteristic functions on W it can be said that propositions are expressed in terms of truth values and possible worlds. When we add the basic domain E of individuals we get the starting point of the (intensional) Montagovian hierarchy of denotations of finite type over three basic sets: $\{0,1\}$, E and W. Properties and relations are also modeled in terms of these sets, and thereby an explicit interpretation is given to the basic intensional entities.

The *method of set-theoretic modeling* that is used here has the advantage of giving precise explications for concepts that have in the past been notoriously vague (this point is pressed with vigor in Montague (1969)). There are, however, serious problems with this approach, as has become evident over the years. Explicit set-theoretic models come along with a pre-determined criterion of identity which might turn out to be inadequate for certain purposes; regarding propositions, the trouble with the propositional attitudes are a well-known case in point. There is also the conceptual danger of confusing the objects to be modeled with their representations in the model. Under such an identification other important uses of the concept in question that are incompatible with the given model might be ignored.

In philosophy, the recognition of this situation led to dissatisfaction with the very methodology involved. Intensional type logic was given up by George Bealer, for instance, who explicitly embraces an algebraic approach (Bealer (1982)). He takes propositions at face value, as unanalyzed first order objects, and tries to describe their properties through structural axioms that govern their behavior. The same holds for the other intensional objects, viz. properties and relations, which, together with the propositions, are called *PRPs* by Bealer. Thus the PRPs are "flattened out" in a first order domain. This approach leaves room for a more fine-grained notion of proposition that seems to be needed in connection with the attitudes.

Another influential author sharing the algebraic perspective is Peter Aczel. In (Aczel (1989)) he distinguishes two senses of "algebraic":

> First of all I do not feel presupposed to the view that there is only one global notion of proposition that will turn out to be superior to all others. Of course that may be what will happen, but it seems more likely that there will be a need for several notions to serve different purposes. The situation could turn out to be analogous to that of the algebraic treatment of number. There are many notions of number in mathematics, each having their uses. ... Each of this notions determines a ring structure. ... The algebraic notion of a ring can be viewed as an abstraction of the properties of addition and multiplication that are common to many of these different notions. ... So I wish to suggest an analogous approach to the notion of proposition. ...
>
> There is a second sense in which I consider my approach to be an algebraic one. I wish to keep to the guiding principle that the meaning of a compound expression is to be determined by the meaning of its component parts. When there are no variable binding operations, as in the propositional calculus, this principle leads to the familiar view of the expressions of the language as forming a free algebra relative to the signature of operation symbols of the language. Then an interpretation of the language is viewed as an algebra of that signature and the meaning function is the unique homomorphism from the free algebra of expressions to the interpretation algebra. ((Aczel, 1989, p. 21f.))

The two senses of "algebraic" that Aczel speaks about could be called *conceptual* and *structural*, respectively. Aczel's topic is the explication of intensional notions like the notion of proposition. Conceptually, he uses an axiomatic approach that abstracts away from specific set-theoretic models that could be given for this notion in a particular field of application. Here he differs from the classical approach in intensional semantics, where propositions are modeled as sets of possible worlds (or, equivalently, as functions from possible worlds to truth values). On the structural side, however, Aczel follows closely the algebraic methodology that was first explicitly formulated by the leading protagonist of intensional logic, Richard Montague (cf. Montague (1970); so Montague's algebraic attitude in doing semantics was restricted to the structural aspect only). Aczel differs from Bealer in that he defines properties as propositional functions and thus

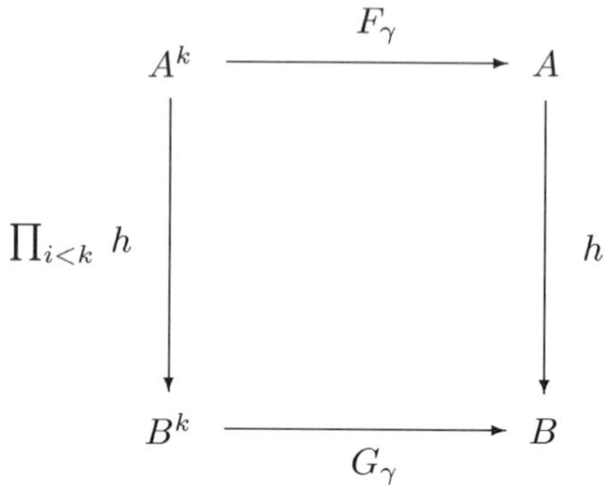

Figure 8.1: Homomorphism between syntactic and semantic algebra

treats them as more intimitely related to propositions. But unlike Montague, he takes the propositions themselves as basic and does not analyze them any further.

Having worked on the algebraization of linguistic semantics for some years I find myself in full accordance with Aczel's outlook. The familiar Tarskian universe of discourse is usually regarded as totally flat. I argue for an algebraically structured ontology that is not given a set-theoretic representation but rather characterized axiomatically. We will of course have to consider (set-theoretic) models for those axioms but there is no commitment as to the "real nature" of the objects considered. Secondly, concepts of universal algebra like homomorphisms do not only play a role in the overall relationship between syntax and semantics; they appear also as a means to describe the various connections between different sorts of entities in the domain of discourse (see below).

The central notion of *homomorphism* can be graphically illustrated by means of *commuting diagrams*. Figure 8.1 is a formal way to express the *principle of compositionality* which relates the *syntactic algebra* $\mathcal{A} = \langle A, F_\gamma \rangle_{\gamma \in \Gamma}$ of a language to the *semantic algebra* $\mathcal{B} = \langle B, G_\gamma \rangle_{\gamma \in \Gamma}$ of its "meanings" (cf. Montague (1970)). The semantic interpretation function h is a homomorphism from $\mathcal{A}$ to $\mathcal{B}$, that is, it "commutes" with the structural operations on the algebras in the following way: the value under h of an application of a k-place syntactic operation F_γ to a family $(x_i)_{i<k}$

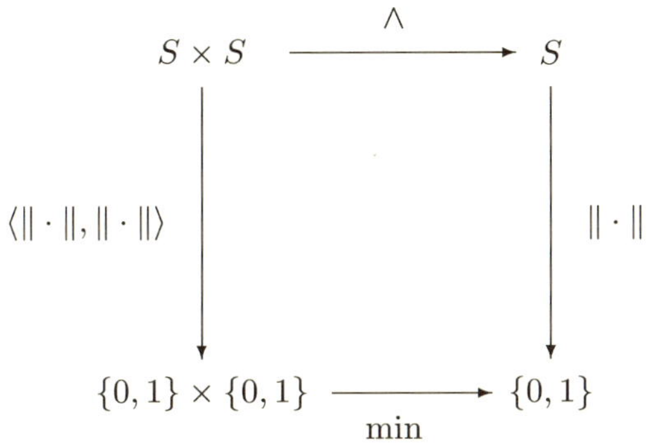

Figure 8.2: Commuting diagram for logical conjunction

of syntactic objects yields to the same result in B as when the semantic counterpart G_γ of F_γ is applied to the family $(h(x_i))_{i<k}$ of h-values of the x_i.

As an example in PC, consider the syntactic operation '$\wedge$' of conjunction. Its semantic counterpart in the set of truth values is the minimum function. Now the valuation function on S, denoted by $\|\cdot\|$ here, is a homomorphism from S to $\{0, 1\}$ in that it commutes with the pair of operations $\langle \wedge, \min \rangle$ (see Figure 8.2), and, analogously, with the other operations on S.

8.3 Algebraic Semantics for Natural Language

Inspired by Montague's *Universal Grammar* Edward Keenan and his collaborators exploited the notion of homomorphism in their analysis of the category structure of natural language (cf. Keenan (1981); Keenan and Faltz (1985)). The central idea is that of lifting the Boolean operations to higher categories. The operations of conjunction, disjunction and negation are not only defined for sentences but also for phrases below the sentence level.

For instance, let X and Y stand for intransitive verb phrases. Then *VP*

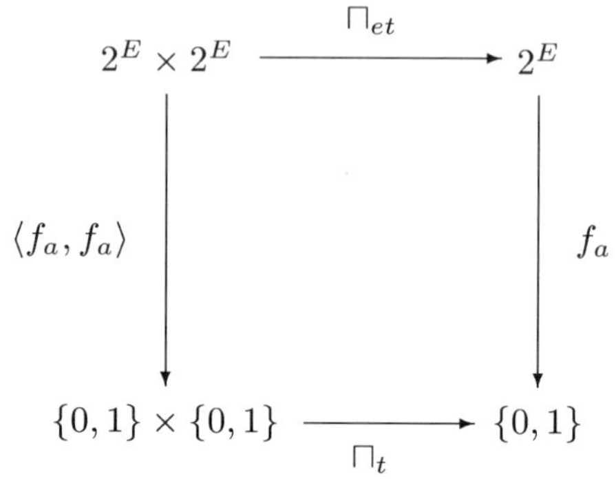

Figure 8.3: Commuting diagram for VP conjunction

conjunction on the algebra 2^E of VP denotations is defined by

$$X \sqcap_{et} Y \ := \ \lambda x.\, X(x) \,\&\, Y(x)$$

For every $a \in E$, the denotation function $f_a : 2^E \to \{0,1\}$, with $f_a(X) = 1$ if $a \in X$, and $f_a(X) = 0$ otherwise, is a Boolean homomorphism, i.e., it commutes with the conjunction operation (see Figure 8.3) and with the other Boolean operations.

A similar consideration applies to NP denotations which are elements of $D_{(et)t} = 2^{2^E}$. For instance, the conjoined NP *John and Mary* denotes the set of all sets containing both John and Mary. The characteristic function f_X for $X \in 2^E$ is a Boolean homomorphism from $D_{(et)t}$ into $\{0,1\}$ as Figure 8.4 illustrates.

In Keenan's Boolean Semantics almost all categories higher up are Boolean algebras. Only the usual Tarskian domain of individuals remains *flat*.[2] While this might be just a formal peculiarity it does hint at a descriptive defect: Boolean semantics cannot fully account for the meaning

[2]In an intensional version of Boolean Semantics dealing with possible worlds one would see that not only the set E of individuals, but also the set W of possible worlds is a flat domain. Thus the Boolean structure is generated exclusively from the Boolean algebra **2** of truth values. A structuring of W has been explored, e.g., in Situation Theory (Barwise and Perry (1983); Cooper *et al.* (1990); Barwise *et al.* (1991)), but also in Angelika Kratzer's highly original work Kratzer (1989).

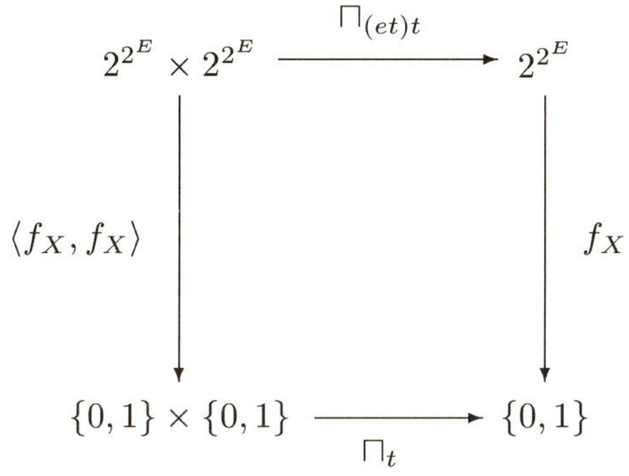

Figure 8.4: Commuting diagram for NP conjunction

of conjoined NPs. It is typically unable to give the proper analysis of *collective predication*, that is, of sentences like (i): *John and Mary meet*. In the conjoined NP *John and Mary* the '*and*' is *not* a Boolean operation; otherwise it would commute with the concatenation operation, yielding the result that the sentence (i) and the sentence (ii): *John meets and Mary meets*, mean the same. But (ii), unlike (i), is nonsense because collective predicates do not apply to singular objects.

8.3.1 Plural lattices

The recognition of the existence of a non-Boolean '*and*' leads back to the *mereological paradigm* in formal philosophy (see Leonard and Goodman (1940); Quine (1960); Eberle (1970); Simons (1987)). The backbone of the logical reconstruction of nominalism in the 1940s is the notion of *fusion* of two individuals a and b, that is, a and b "taken together". This fusion, call it $a \sqcup b$, is just another individual, which is related to a and b by means of a *part-whole relation* $\sqsubseteq$. This relation is usually taken to be a partial order. In fact, under very weak assumptions (Axioms 1 – 3 below) the operations $\sqcup$ and $\sqsubseteq$ are interdefinable through

$$a \sqsubseteq b \;\Leftrightarrow a \sqcup b = b$$

In Link (1983a) I applied this idea to the study of plurals and mass terms in linguistic semantics. I argued for a *structured domain of individuals* that contains not only usual individuals but also their fusions. I insisted that these fusions be different from the *classes* containing the individuals fused. And fusions of fusions do not, as in set theory, raise the level of comprehension but are just fusions of individuals again. For an elaboration on this theme see Section 8.5.

In the mentioned paper I took the domains to be atomic Boolean algebras minus zero outright; a more genetic approach, however, is to formulate postulates for the structure of those domains that are to contain singular and plural objects alike. Let $L \neq \emptyset$ be such a domain and $\sqcup$ the fusion operation on L, which is called the *individual sum operation* (i-sum operation, for short; the function symbol denoting it is the circled plus, $\oplus$). Furthermore, let $\sqsubseteq$ be a part-relation defined by the equivalence above, and A the set of $\sqsubseteq$-minimal elements in L, called the *atoms in L*. Define for a given element a the *principal ideal of a* as the set of all elements "below" a:

Definition 1 $a^{\downarrow} := \{x \in L \mid xa \sqsubseteq\}$

Then the following axioms of the system LP ("the logic of plurality") give a characterization of the intended plural structures; for a detailed motivation see Chapter 6.

(Ax. 1) $\forall x, y\, [\, x \sqcup y = y \sqcup x\,]$ (*Commutativity*)

(Ax. 2) $\forall x, y, z\, [\, x \sqcup (y \sqcup z) = (x \sqcup y) \sqcup z\,]$ (*Associativity*)

(Ax. 3) $\forall x\, [\, x \sqcup x = x\,]$ (*Idempotence*)

(Ax. 4) $\neg\exists x \forall y\; x \sqsubseteq y$ (*Non-Existence of Zero*)

(Ax. 5) $\forall X \subseteq L\, [\, X \neq \emptyset \;\Rightarrow \exists x.\, x = \sup X\,]$ (*Completeness*)

(Ax. 6) $\forall x, y\, [\, x \not\sqsubseteq y \;\Rightarrow \exists u \in A : u \sqsubseteq x \wedge u \not\sqsubseteq y\,]$ (*Separation*)

(Ax. 7) $\forall x \exists u \in A :\; u \sqsubseteq x$ (*Atomicity*)

(Ax. 8) $\forall X \subseteq L,\, u \in A\, [\, u \sqsubseteq \sup X \;\Rightarrow \exists x \in X :\, u \sqsubseteq x\,]$ (*sup-prime atoms*)

(Ax. 9) $\forall x \in L :\; x = \sup(x^{\downarrow} \cap A)$ (*sup-generation*)

(Ax. 10) $\forall X, Y \subseteq A\, [\, \sup X = \sup Y \;\Rightarrow X = Y\,]$ (*sup-injectivity*)

Structures satisfying Axioms 1 – 3 are called *(join) semilattices.* A plural structure is thus a restricted kind of semilattice. Not all axioms are independent; for instance, Axioms 7,9,10 are derivable from the rest (see Chapter 6). Call L *bottomless* if it satisfies Axiom 4. Then we introduce a name for the plural structures.

Definition 2 A *plural lattice* is a bottomless, complete, and separating semilattice with sup-prime atoms.

There is the following representation theorem that justifies the initial introduction of Boolean algebras (for a proof see the Appendix).

Theorem 2 *Plural lattices are, up to isomorphism, complete atomic Boolean algebras minus zero.*

8.3.2 Mass Terms

The denotations of mass terms like *water* are quantities of matter. They differ from pluralities in that they don't have smallest parts. Thus axioms 6 – 10 in Subsection 8.3.1 have to be dropped since they refer to a set of atoms. The remaining axioms are too weak now. One way to restrict the structures in accordance with our intuitions about mass expressions is to postulate the two additional axioms below. Let us first define:

(D. 1) $x \circ y \;:\Leftrightarrow \exists z\,[\,z \sqsubseteq x \wedge z \sqsubseteq y\,]$ (*overlap*)

(D. 2) $x \sqsubset y \;:\Leftrightarrow x \sqsubseteq y \wedge x \neq y$

Then the axioms are (see Landman (1991)):

(Ax. 11) $\forall x, y\,[\,x \sqsubset y \;\Rightarrow \exists z\,[\,z \sqsubseteq y \wedge \neg x \circ z\,]]$ (*Witness*)

(Ax. 12) $\forall x, y, z\,[\,x \sqsubseteq y \sqcup z \;\Rightarrow [\,x \sqsubseteq y \vee x \sqsubseteq z \vee \exists y' \exists z'\,[\,y' \sqsubseteq y \wedge z' \sqsubseteq z \wedge x = y' \sqcup z'\,]]]$
(*Distributivity*)

Theorem 3 (Landman) *Mass term lattices are, up to isomorphism, complete Boolean algebras minus zero.*

A representation language which is to treat both plural and mass terms will have to combine the mass lattices with the plural lattices. Resulting structures are then similar to what was in Link (1983a) called *Boolean model structures with homogeneous kernel*. Basically, we take the set of atoms in the plural lattice to be the regular atomic individuals plus the elements of the mass lattice. This is because we can still form plural objects from two quantities of matter without fusing them in the mass term sense; example: the denotation of the plural term in the sentence *the liquid in this cup and the liquid in that cup have the same color*. This conjoined NP can of course also refer to the quantity of matter that is the mass fusion of the two initial quantities. What is the relation between this quantity and the i-sum of the two quantities? In Link (1983a) I argued that they are systematically related through a *homomorphism* taking i-sums and returning the quantities of matter they are made up from.

There is another algebraic aspect about the theory of mass terms that readily explains the relation between various proposals that have been advanced in the literature. For instance, Lønning (1987a) assigns to a mass term like *water* a single entity in a Boolean algebra, whereas Link (1983a) and Roeper (1983) interpret such a term as a set with certain closure conditions, viz. as a downward closed join subsemilattice of the mass algebra (such a structure is an *ideal*). But we can freely go back and forth between the two interpretations: the ideals are such that they have a greatest element (= the total quantity of the stuff in question); this element, then, is Lønning's denotation of the mass term. Conversely, starting with the latter we can form the principal ideal generated by it, and we are back to the set interpretation.

8.3.3 Events

Algebraic methods find a natural field of application in the theory of events where it is common to speak of parts and wholes. Events are understood here in a broad sense comprising "happenings" as well as states, which is what Bach (1986) calls *eventualities.*

One of the main reasons for semantics to introduce events into the representation language is linguistic evidence showing that there is direct reference to events, as the interpretation of the anaphoric elements in the following sentences shows.

(1) Three men hoisted the flag of the republic. No one had expected *it*.

(2) John dropped his hard disk *which* was stupid / caused quite a damage.

(3) Two people were killed. *It* happened on two consecutive days.

But as Krifka (1991a) observed there is also a characteristic way to refer indirectly to an underlying event structure. The following sentence does not speak about 4000 ships but rather about 4000 passages of an unknown number of ships:

(4) Four thousand ships passed through the lock.

Finally, the distinction between *collective* and *distributive predication* could also be expressed in an event-based framework. Consider the sentence

(5) Denys and Tania own a farm.

Either there is one farm which Denys and Tania own together (collective reading) or there is one farm owned by Denys and one owned by Tania (distributive reading). In the former case there is what I call a *plural character*, Denys $\oplus$ Tania, that enters a singular owning relation with a certain farm; in the latter case we have a *plurality of events*, $e \oplus e'$, where e stands for the owning relation involving Denys and the first farm, and e' stands for the owning relation involving Tania and the second farm. Thus we see that we need the fusion operation both in the domain of individuals and in the domain of events. This suggests the following model structure for an event-based semantics of natural language (see Chapter 11):

Aether models. A model structure for an event-based representation language is a tuple $\mathcal{M} = \langle A, E, T, H, R, \sigma, \tau \rangle$ such that

1. A is a plural lattice (the domain of *individuals*);
2. E is a complete semilattice (the domain of *events*);
3. T is a complete semilattice (the domain of *time stretches*);
4. H is a complete semilattice (the domain of *spatial regions*);
5. $R = \{\rho_1, \ldots, \rho_r\}$ is a finite of set partial functions from E to $A \cup E$ (the set of *thematic roles*);
6. τ is a partial function from E to T and a join homomorphism on its domain (the *temporal trace function*);
7. σ is a partial function from E to H and a join homomorphism on its domain (the *spatial trace function*);

The set A includes all possible denotations of terms in the nominal domain (plural and mass entities, properties, kinds, etc.). E is the set of denotations for the verbal domain; its elements are particular events e that each come along with a characteristic set of "thematic" roles. For instance, if e is an owning event then the owner role and the role of being owned are defined for e; other roles will be undefined. There is also a particular stretch of time $\tau(e)$ during which e holds; I call $\tau(e)$ the *temporal trace of* e. Now while an event like John's running also has a spatial location it is not clear whether that is true of the above e. Thus we see that, in general, the trace functions and the role functions are only partial functions.

The set E also contains fusions or sums of events. The trace functions, if defined, commute with the sum operation: for instance, the time stretch of a sum event is the sum of the time stretches of its parts. Thus the trace functions are (join) homomorphisms.

Rather than motivating every single feature of the above definition let me say that the emphasis is on the perspective, not on the details. A more careful analysis will have to justify, or else modify, the various properties of the model structure as they are formulated here.

Let us apply these tools to the analysis of the collective/distributive distinction exemplified by sentence (5) above. We can give the following DRT[3] style representation for it (t_0 stands for the reference time determined by the time of utterance):

(5) Denys and Tania own a farm.

$$
\begin{array}{rcl}
 & e, x, u, v, y & \\
u & = & \text{Denys} \\
v & = & \text{Tania} \\
x & = & uv\oplus \\
\textsc{own}(e) & \& & \rho_1(e) = x \\
 & \& & \rho_2(e) = y \;\&\; \text{farm}(y) \\
 & \& & t_0 \sqsubseteq \tau(e)
\end{array} \qquad \text{(i)}
$$

$$
\begin{array}{rcl}
 & e, e', e'', u, v, x, y & \\
u & = & \text{Denys} \\
v & = & \text{Tania} \\
e & = & e'e\oplus'' \\
\textsc{own}(e') & \& & \rho_1(e') = u \\
 & \& & \rho_2(e') = x \;\&\; \text{farm}(x) \\
 & \& & t_0 \sqsubseteq \tau(e') \\
\textsc{own}(e'') & \& & \rho_1(e'') = v \\
 & \& & \rho_2(e'') = y \;\&\; \text{farm}(y) \\
 & \& & t_0 \sqsubseteq \tau(e'')
\end{array} \qquad \text{(ii)}
$$

The first scheme (i) represents the collective reading, with one event involving one plural agent. Scheme (ii) expresses the distributive reading, which is about the sum of two singular events, each involving an atomic individual as agent (whence the existence of two separate owners and two different farms owned).

It should be clear that our present notion of event remains a rather general one unless special conditions are formulated to constrain the admissible structures in an suitable way and to allow to make a distinction between states and happenings, for instance. A more detailed classification

[3] Discourse Representation Theory; see Kamp and Reyle (1993).

of events goes back to Vendler (1967c) where four categories are distinguished: (i) *stative events*, expressed by predicates like *resemble, know, believe, love*; (ii) *activities*, expressed by verbs like *run, sleep, practice, eat, read, listen, push a cart*; (iii) *accomplishments*, expressed by phrases like *write a letter, build a house, awaken, run a mile, draw a circle, cross the street*; and (iv) *achievements*, expressed by *die, arrive, discover, recognize, realize, reach the summit*, among others. There are rather stable linguistic tests differentiating each of these *aspectual classes* from the rest (see, for instance, Dowty (1979); Hinrichs (1985); Dowty (1989)). For the present purposes, however, it suffices to draw the Aristotelian distinction between *telic* and *atelic* processes or events, where atelic events cover states and activities, and telic events cover accomplishments and achievements. Now atelic events have the property of *homogeneous reference*: A temporal part of such an event is again an event of the same type. So for instance, if John's running extends from 6 to 7 p. m. then his activity between 6 and 6:30 p. m. is again a running. By contrast, if $\tau(e)$ is the temporal trace of a telic event e, say, Mary's writing a letter, then a proper part t of this time stretch fails to be the trace of a letter-writing event: either some part of the *preparatory phase* or some part of the *culmination phase*[4] of the process e will be missing. This property is called the atomicity of telic events.

We can write down these properties as axioms in our event-based representation language. The property of homogeneous reference is also called the subinterval property. Its formulation below contains the parameter $\gamma(e)$, called the *granularity of* e. It expresses the observation that some time stretches might be too small to be a trace of the event e; thus $\gamma(e)$ fixes the minimal length that a time stretch must have to serve as a trace for e. t_0 is again the time of reference.

Subintervall property of atelic events

(Ax. 13) $$\forall e, t\,[\,V(e) \wedge t \sqsubseteq \tau(e) \wedge |t| \geq \gamma(e) \Rightarrow \exists e'\,(\,e' \sqsubseteq e \wedge V(e') \wedge \tau(e') = t\,)\,]$$

Atomicity of telic events

(Ax. 14) $$\forall e, e'\,[\,V(e) \wedge V(e') \wedge e' \sqsubseteq e \Rightarrow e' = e\,]$$

For more axioms of this kind see Krifka (1989b). Incidentally, this work has also given an formal explanation of a well-known interrelationship between homogeneity and discreteness in the nominal domain and the corresponding properties in the verbal domain. Thus, for instance, while *an apple* stands for a discrete entity, the plural *apples* refers homogeneously. The same transition is observed when the accomplishment of *eating an*

[4]For these concepts see Parsons (1990).

apple (discrete) is changed to the activity of *eating apples* (homogeneous).

8.4 Some Applications to Plural Theory

The fruitfulness of the algebraic approach has been amply demonstrated in the field of linguistic semantics and in the study of plurals in particular. For references regarding plural research, see the bibliography in the present book, but also, for instance, Roberts (1987b); Lasersohn (1988); Krifka (1989b); Lønning (1989); van der Does (1992); Kamp and Reyle (1993); Schein (1993); Verkuyl and van der Does (1996); for a recent and competent overview consult Lønning (1997). In this section I just want to give an idea of how the expressive power of the plural lattice is exploited to deal with plural phenomena in a rather natural way.

One of the main virtues of the present approach is the ease with which plural constructions can be processed in a compositional way. Let me mention two cases in point: (i) the common phenomenon of *number indeterminacy*, and (ii) *mixed collective and distributive predication.* An example for (i) is the following sentence (due to Hans Kamp).

(6) The books that the students had bought were all they could hope to read within a week.

Here it depends on the ability of the student you pick whether he or she bought one or several books. Since the system PL of plural logic has variables that range indiscriminately over individuals and their fusions we can use those variables in a representation of this sentence; thus it is simply left open whether we talk about one or more than one book in each case.

As for mixed predication, consider sentence (7),

(7) Four men went to the bar together and had a beer each.

How this is treated compositionally will be briefly described. The problem is that the subject NP *four men* should on the one hand stand for a collection of individuals such that it can fill the argument place of the collective predicate *go to the bar together*; but on the other hand it should somehow give access to the single individuals talked about since each of them is said to have a beer. The problem is solved by means of a *transformation on verb phrases* that has the effect of letting the argument VP apply to the atoms of the i-sum denoted by the incoming NP. This VP transformation, called the *distributivity operator* D, was first introduced in Chapter 2 (written in 1984) and in Link (1987b), and it was extensively discussed in subsequent work in the field.[5] When the D operator is applied to a predicate α it returns a predicate β which is true of an i-sum x just in case α is true of all the atomic parts of x. It can be succinctly expressed as the operator δ below, which is accompanied by two analogous operators, γ for collectivity[6] and μ for a "mixed reading".[7]

$$(8) \quad \delta \quad = \quad \lambda X\, \lambda y\, .A \cap y^{\downarrow} \subseteq X$$
$$(9) \quad \gamma \quad = \quad \lambda X\, \lambda y\, .y \in E \setminus A \,\&\, y \in X$$
$$(10) \quad \mu \quad = \quad \lambda X\, \lambda y\, .\exists Y \in \mathcal{C}(\, y = \sup Y \,\&\, Y \subseteq X\,)$$

Now it is easy to give a compositional representation of sentence (7):

$$(7') \quad \exists x(\, x \in [\![\mathit{4\ men}]\!] \;\;\&\;\; x \in \gamma([\![\mathit{went.to.the.bar}]\!]) \;\cap\; \delta([\![\mathit{had.a.beer}]\!])\,)$$

When we fill in the definitions for γ and δ we see that the predicate *went.to.the.bar* is applied collectively to the whole i-sum x of the four men in question, whereas the predicate *had.a.beer* distributes down to the atomic parts of x.

Apart from the collective and the distributive reading there is a mixed or "neutral" reading of plural sentences, as some people have proposed. Consider, for instance, the following example due to Gillon (1987).

(11) Hammerstein, Rodgers and Hart wrote musicals.

As a matter of historical fact, this sentence is false under both its collective and its distributive reading, but it is judged true nevertheless. Gillon claims that it is the *cover* $Y = \{\text{Hammerstein} \oplus \text{Rogers}, \text{Rogers} \oplus \text{Hart}\}$ of the total i-sum consisting of all three composers which makes the sentence true: Hammerstein and Rogers wrote a musical together, and so did Rogers and Hart. This reading can be represented by using van der Does's operator μ, as a short calculation shows.[8]

I will not go into the linguistic arguments for or against such a mixed reading here. For references see Gillon (1987); Link and Schütze (1991); Lasersohn (1989); van der Does (1992); Lønning (1991), and Chapter 7.

[5] See, in particular, Roberts (1987b),1987a where convincing linguistic evidence is presented which supports the introduction of such a device. For a formal treatment of this and similar operations see van der Does (1992).

[6] The clause $y \in E \setminus A$ in Equation 9 is really a presupposition rather than part of the defining condition.

[7] Regarding the introduction of γ and μ I follow a suggestion of van der Does (1992); the versions used here are of course "algebraized."

[8] Use the cover Y just mentioned (the $\mathcal{C}$ occurring in the definition of μ stands for a class of appropriate covers). See also Chapter 7 for details.

Suffice it to say that more sophisticated theoretical devices like the VP operators can be easily accommodated within the given framework.

Let me conclude this overview with a few remarks about plurals within the linguistic theory of generalized quantifiers (GQ theory, for short). After a first discussion of the relationship between plurals and GQ theory in Link (1987b)[9] far more thorough treatments of this issue have been given by Lønning (1989) and van der Does (1992). Van der Does, however, gives a set-theoretic account, and as a consequence he has to raise the level of analysis from sets of sets to sets of sets of sets. Thus, when we start out with a basic domain E of individuals then the determiners like *all, some, most* are 2-place relations in 2^E, i.e., of Montagovian type $(et)(et)t$, according to classical GQ theory. To accommodate plurals, van der Does lifts these relations to the type $(et)((et)t)t$. Individuals $a \in E$ become singletons $\{a\} \in 2^E$, and pluralities are arbitrary sets $X \in 2^E$. VP denotations are sets of collections $\mathbf{X}, \mathbf{Y} \subseteq 2^E$. There is a natural part-of relation in 2^E, viz. set inclusion $X \subseteq Y$; then the union of a set of collections $\mathbf{X} \subseteq 2^E$ is the supremum w. r. t. inclusion, i.e., $\bigcup \mathbf{X} \in 2^E$ is the least upper $\subseteq$-bound of $\mathbf{X} \subseteq 2^E$. This arsenal is a perfectly legitimate kit of modeling devices; but obviously, the domain E has become completely redundant.

This situation is a standard occasion to switch to the algebraic approach. The set 2^E is replaced by a plural lattice E, now comprising individuals and their fusions alike. The singletons become the atoms in E, i-sums are arbitrary elements $x \in E$. VP denotations are now sets $X \subseteq E$ of i-sums. The part-of relation $x \sqsubseteq y$ is defined between i-sums, and the fusion of a set of i-sums $X \subseteq E$ is its least upper $\sqsubseteq$-bound, i.e., i.e., $\sup_{\sqsubseteq} X$. As a result of this reform, determiners are again just 2-place relations in the power set of E. But the structures remain richer than those of pure GQ theory because now the elements of E are partially ordered under $\sqsubseteq$. We can exploit this fact to give representations of some plural determiners, including some critical non-monotonic determiners that have recently been the focus of discussion in semantics.[10] The latter are special in that they take scope over some material in the sentence they occur in (this material has usually to be determined from context). Therefore the definition of these determiners contains an extra clause relating to the material in their scope. This clause is a universally quantified expression, where the quantifier runs over arbitrary i-sums. Thus while is hard to find overt universal quantification over genuine i-sums in language,[11] that type of quantification is nonetheless needed as a theoretical tool.

[9]Chapter 4 of this book.

[10]See van der Does (1992), and Chapter 7.

[11]For a discussion, see Link (1987b)

Except for Equation 20 the function $|\cdot|$ gives the cardinality of the atoms of the i-sums it takes as arguments.

Pluralic determiners.

(12) $[\![\text{some}_{pl}]\!] = \lambda X \lambda Y. \exists x\,(\,|x| \geq 2 \,\wedge\, x \in X \cap Y\,)$

(13) $[\![n]\!] = \lambda X \lambda Y. \exists x\,(\,|x| = n \,\wedge\, x \in X \cap Y\,)$

(14) $[\![\text{at.least } n]\!] = \lambda X \lambda Y. \exists x\,(\,|x| \geq n \,\wedge\, x \in X \cap Y\,)$

(15) $[\![\text{more.than } n]\!] = \lambda X \lambda Y. \exists x\,(\,|x| > n \,\wedge\, x \in X \cap Y\,)$

(16) $[\![\text{not.more.than } n]\!] = [\![\text{at.most } n]\!] =$
$\lambda X \lambda Y. \neg\exists x\,(\,|x| > n \,\wedge\, x \in X \cap Y\,) =$
$\lambda X \lambda Y. \forall x\,(\,x \in X \cap Y \Rightarrow |x| \leq n\,)$

(17) $[\![\text{(some.but) at.most } n]\!] =$
$\lambda X \lambda Y. \exists x\,(\,|x| \leq n \,\wedge\, x \in X \cap Y \,\wedge\, \forall y\,(\,y \in X \cap Y \Rightarrow y \sqsubseteq x))$

(18) $[\![\text{exactly } n]\!] =$
$\lambda X \lambda Y. \exists x\,(\,|x| = n \,\wedge\, x \in X \cap Y \,\wedge\, \forall y\,(\,y \in X \cap Y \Rightarrow y \sqsubseteq x))$

As Equations 16 and 17 show there are two representations for the non-upward-monotonic determiner *at most*. The first is the purely logical one that does not carry an existential entailment. It is equivalent to Equation 19 which, in turn, corresponds to the classical GQ definition 20 (in this equation only, X and Y run over subsets of a flat domain, and $|\cdot|$ is ordinary cardinality).

(19) $[\![\text{more.than } n]\!] = \lambda X \lambda Y.\,|\,\sup(\,X \cap Y\,)\,| \leq n$

(20) $[\![\text{more.than } n]\!] = \lambda X \lambda Y.\,|\,X \cap Y\,| \leq n$

Equation 17, by contrast, is designed to deal with the fact that quite often phrases like *at most three men* do carry an existential commitment and are thus suited to introduce a discourse referent into the representation which can later serve as antecedent for anaphora. An example is the following sentence.

(21) At most four squatters were left in the building. They were dragged out by the police.

Here we see very clearly that it is not enough to uncover the "logic" of structural phrases like NL determiners. Rather, proper attention has to be paid also to their role in discourse. This is a point that has been forcefully argued for some time now by Hans Kamp and his discourse representation theory.

8.5 Ontology of Plurals

In the final section I'd like to apply the general methodological remarks in the introduction to the ontology of plurals. Suppose your favorite plural theory has *sets* as the denotations of plural terms. This idea might have occurred to you naturally, given that you were brought up with set-theoretic modeling; and anyway, it has always been around since Russell (1903). Now does that mean that ontologically, you committed yourself to the existence of sets or classes? To be sure, you have "singularized" the plural talk by trading it for the set talk; instead of letting the plural term 'those animals in the backyard' stand for *your neighbor's cats* you take its denotation to be *the set of your neighbor's cats*. And when you use the pluralic quantificational locution "there are some cats in the backyard" you presumably mean to employ a singular quantification over sets. And this is what many linguists thought they have been doing. Take, for instance, Lauri Carlson, here quoted from (Boolos, 1984b, p. 446, fn), who writes: "I take such observations as a sufficient motivation for construing *all* plural quantifier phrases as quantifiers over arbitrary *sets* of those objects which form the range of the corresponding singular quantifier phrases." David Lewis expresses this view by saying: "Plurals, so it is said, are the means whereby ordinary language talks about classes. According to this dogma he who says that there *are* the cats can only mean ... that there *is* a class of cats" (Lewis, 1991, p. 65). Also Boolos [1984b, p. 448f.] spies this attitude among "some who theorize about the semantics of plurals", viz. that they conclude "just from the fact that there are some Cheerios in the bowl that ... there is also a set of them all." But he protests: "It is haywire to think that when you have some Cheerios you are eating a *set* — what you're doing is: eating THE CHEERIOS." I obviously couldn't agree more with Boolos here;[12] but if the plurality *the Cheerios* is not a set, then some alternative account of its nature has to be given. The account that Lewis gives is the mereological one:

[12] I used this kind of argument *ad hominem* in my Hydras paper Link (1984) but people commenting on it, e.g., Cresswell (1985) or Landman (1989), weren't particularly impressed by it.

> ...if we accept mereology we are committed to the existence of all manner of mereological fusions. But given a prior commitment to cats, say, a commitment to cat-fusions is not a *further* commitment. The fusion is nothing over and above the cats that compose it. It just *is* them. They just *are* it. ... If you draw up an inventory of Reality according to your scheme of things, it would be double counting to list the cats and then also list their fusion. (Lewis, 1991, p. 81)

I for one have been trying to advocate the algebraic approach for precisely this reason: pluralities do not have a different ontological status from their singular counterparts; they are mereological fusions of the latter, not classes of them Link (1983a). Some circles in linguistics in particular have thought the algebraic theory of plurals to be an unnecessary gimmick, but it's not. Mereology is a sound philosophical attitude; in fact, applied to plurals it can avoid serious problems in the philosophy of logic, as Lewis (1991) argues. But more importantly for our present purposes, the usefulness of the algebraic approach has been demonstrated through genuine linguistic work, as the references in Sections 8.3 and 8.4 show.

It is instructive to read what Lewis has to say about the principle of *Unrestricted (mereological) Composition* against which many people expressed their reservations when they were invited to consider the individual sum of a certain collection of objects.

> Most of all, it is the axiom of Unrestricted Composition that arouses suspicion. I say that whenever there are some things, they have a fusion. *Whenever*! It doesn't matter whether they are all and only the satisfiers of some description. It doesn't matter whether there is any set, or even any class, of them. ... There is still a fusion. So I am committed to all manner of unheard-of things: trout-turkeys, fusions of individuals and classes, all the worlds styrofoam, and many, many more. We are not accustomed to speak or think about such things. How is it done? Do we really have to?
>
> It is done with the greatest of ease. It is no problem to describe an unheard-of fusion. It is nothing over and above its parts, so to describe it you need only describe its parts. Describe the character of the parts, describe their interrelation, and you have *ipso facto* described their fusion. The trout-turkey in no way defies description. It is neither fish nor fowl, but it is nothing else: it is part fish and part fowl. It is neither here nor there, so where is it? — Partly here, partly there. That much we can say, and that's enough. Its character is exhausted by the character and relations of its parts. (Lewis, 1991, p. 79)

This is not the place to enter the discussion pertaining to the use of pluralic locutions in the philosophical interpretation of second-order logic and the conception of classes.[13] Suffice it to say that Boolos, and with him Lewis, takes the pluralic idiom as it is found in natural language to be an ontologically irreducible tool of theoretical philosophy. Here is a typical passage from Lewis (1991) concerning the principle of *plural comprehension*[14] that illustrates this view: "If there is at least one cat, then there are some things that are all and only the cats. (Regimented: ... then there are some things such that, for all x, x is one of them iff x is a cat.)" [p. 63] The locutions *there are some things, x is one of them* etc. are not analyzed any further but are considered as understood by any competent speaker of the language. Now it is here that the algebraic plural logic could step in and, by providing a theory of the pluralic part of our linguistic competence, could vindicate the coherence of that philosophical position.

Let me hint at a possible misunderstanding that might be engendered by the way Lewis speaks about the "singularist position" which according to him turns pluralic quantification into singular quantification about sets or classes. It seems to me that there are two claims in this: one is the assertion that we can replace the pluralic idiom by the singular one; and the second says that by embracing plural quantification you commit yourself ontologically to universal entities like sets. Now you might be a *pluralist* and still use second order logic (PL2) as your representation language *without* committing yourself to classes. This is Boolos's view; the standard semantics of PL2, however, which he works with, uses "singular" quantification over second-order objects. But that is just one way to *model* the pluralic locution "is one of them", viz. by set-theoretic membership (LP is another way to model this locution, by using i-sums). In order to avoid confusion about his position Boolos prefers to use the informal pluralic talk. But I think that is unnecessary: as long as it is made clear that the sets in PL2 are only modeling devices no extra commitment is produced. Again, the i-sums of LP fare much better here: they are just fusions and ontologically nothing beyond first-order individuals; so no confusion arises in the first place. I think that the emphasis on the singular mode of speech as diagnostic for ontological implications is a red herring; it is just mixing up grammatical and philosophical categories. The very term *the fusion* is singular in grammatical number and could not be otherwise in a language admitting the number distinction; but of course, no universalist commitment comes along with using that phrase.

Another source of error is Boolos's vague use of the word *collection*. If collections are classes then it is true that pluralities are not collections; they

[13] Important references are Boolos (1984b,a, 1985a); Lewis (1991); Shapiro (1991). My own views on these matters are layed out in greater detail in Chapters 13 and 14.

[14] Its formal analogue is the completeness property of the plural lattice; cf. Lønning (1989) and Chapter 6.

are just fusions. So in this sense there is no quantification over collections, which Boolos rightly rejects. But plural theory does admit ontologically innocent quantification over fusions; thus, if the term *collection* is just another word for *fusion* then of course there *is* quantification over collections.

8.6 Conclusion

I hope to have shown that the algebraic perspective is a fruitful one in the semantic analysis of natural language. In taking up the mereological tradition in philosophy it also goes along well with formal philosophy. Finally, it fits both the spirit and practice of much work that has been done in the AI field of knowledge representation.

Chapter 9

Quantity and Number

1. In this chapter I am going to focus on the relation between the *syntactic* notion of number and the *semantic* notion of quantity from the perspective of logical semantics. The chapter consists of two parts. In the first I discuss the conceptual issue of *individuation* as a prerequisite to counting. In this context classifier constructions will serve as illustration for some remarks on the philosophical problem of *ontological relativity*. The second part of the chapter is concerned with a certain kind of *structural relativity* in quantifying phrases. Following Keenan a distinction is made between *autonomously referring expressions* and those with dependent reference. It is only the former that indicate the true amount of individuals or material in the world that they refer to; the amounts given by the dependent expressions have to be derived from the denotations of noun phrases taking scope over them, and/or often from an iterative interpretation of the given sentence. Each part ends with a semantic model that is meant to capture the basic logic of the linguistic phenomena discussed.

9.1 Ontological Relativity

2. It is of course no accident that the English word 'number' (like its Latin origin 'numerus') stands both for the mathematical concept of number and the grammatical distinction between singular and plural. The paradigmatic function of the singular of a given count noun N like *apple* is to pick out a single object of the kind N; the corresponding plural stands for some collection of objects of the same sort which can be *counted*, that is, numerically specified. For the process of counting to make sense at all there has to be some appropriate *discrete unit* available. In the case of count nouns this unit is naturally given: it is the objects themselves that such a

noun is true of. Mass nouns like *gold* or *water*, on the other hand, need some special lexical item expressing the unit in terms of which the given quantity is measured (e.g., *three ounces of gold*, *two cups of water*). So the process of counting is but a special case of the process of *measuring quantities*: in general, we have to fix a particular unit first, and only then we can proceed to take multiples or proportions to determine the measure of the quantity at hand. This is the procedure in science (for an account of basic measurement theory, see Krantz *et al.* (1971)), but (experimental) science is no more than theoretically guided and sophisticated cultural practice. This is why we find the practice of measurement extensively encoded in language. Relative to a given unit the absolute amount is represented by the class of *numerals* (NL; let us ignore proportion expressions like *three quarters of*); with its strict mathematical meaning their structure shows little cross-linguistic variation. What counts as unit, however, is far less predetermined conceptually, and outside the class of count nouns the *measure words* for mass terms or *transnumeral nouns* (for the latter term see Biermann 1981) abound accordingly.

An object may be measured along different dimensions (e.g. length, width, volume, weight); within the same dimension there may be several lexicalized units that are multiples of each other (*inch*, *foot*; *ounce*, *pound*; discrete objects,too, can be taken in *pairs*, *dozens*, etc.). In less standardized contexts we have *cups*, *mugs* and *bottles*.

Those measure words or *numeratives* (NM; Krifka 1989b) are obligatory with mass terms. The reason is, of course, that it is the measure word that divides up the continuous stuff denoted by the mass noun into discrete, countable units. The situation is slightly different with *collective terms* like *cattle* and *furniture*. If somebody speaks of "the cattle over there" we are able to count the animals referred to, and yet, *cattle* has to be preceded by the numerative *head of* before we get a well-formed measure phrase like *six head of cattle*. One might wonder now what the reference (the "extension") of a term like *cattle* is; it cannot be the class of bovine animals, or the measure word *head of* would have no semantic function. While this is possible in principle it is not very likely; rather, the grammaticality of the definite NP *the cattle over there* gives a clear indication as to what the proper extension is: the term *cattle* refers not to single animals but to collections of those. It is only when we want to single out individual animals that we have to use the measure word *head of*. The difference, then, between collective terms and mass terms is the feature ±*discrete*: the former stand for discrete, the latter for continuous totalities.

Usually, *three head of cattle* and the like is considered as one of the rare examples of a *classifier construction* in English, a pervasive phenomenon in the so-called classifier languages (for a survey see Allan 1977). The nouns in those languages, it is said, refer to unstructures wholes, and it is the *classifier* (CL) that fabricates the units for counting. Thus, even nouns standing for objects that are perceptually clear-cut enough to be referred to by regular count nouns in Western languages like English have to be preceded by such a classifier.

3. It is interesting to note that W. v. O. Quine used the phenomenon of classifier languages as an example to illustrate his famous philosophical thesis of the inscrutability of reference. In his *Ontological Relativity* (Quine 1969: 35–38) he writes:

> In Japanese there are certain particles, called "classifiers," which may be explained in either of two ways. Commonly they are explained as attaching to numerals, to form compound numerals of distinctive styles. Thus take the numeral for 5. If you attach one classifier to it you get a style of "5" suitable for counting animals; if you attach a different classifier, you get a style of "5" suitable for counting slim things like pencils and chopsticks; and so on. But another way of classifiers is to view them not as constituting part of the numeral, but as constituting part of the term - the term for "chopsticks" or "oxen" or whatever. On this view, the classifier does the individuating job that is done in English by "sticks of" as applied to the mass term "wood," or "head of" as applied to the mass term "cattle."

Quine says here that there are two possibilities of parsing a classifier phrase NL + CL + N, either (i) [NL + CL] + N or (ii) NL + [CL + N]. Version (i) would mean conceptually to "parametrize" the notion of number; thus we would get, for instance, "animal-five" or "slim-thing-five" as different counterparts of the number Five depending on what is there to be counted. Version (ii), on the other hand, is modeled after the English 'head-of' paradigm: here the classifier in an operator on nouns which does the "individuating job" of forming countable units.

Now Quine thinks that there is no fact of the matter as to whether we say that the Japanese word for 'ox' refers to the class of individual bovines, and the parametrized numerals are just an extra, albeit obligatory, linguistic device to specify them numerically—this is version (i); or else that that word refers to, as he puts it, "the unindividuated totality of beef on the hoof." We could only say that the English NP *five oxen* and its Japanese translation correspond to each other as wholes; on the level of single constituents reference may fail to be invariant.

Could there possibly be linguistic evidence to decide Quine's philosophical claim one way or the other? The answer is obviously negative if we talk about the inscrutability thesis in its whole breadth; seen in this context the

Japanese case is only an example. But the universal linguist might indeed try to give arguments for or against a particular way of parsing the classifier construction and then see if Quine is justified in deriving support from it for his philosophical point.

Now we could try some familiar tests for constituency in order to settle the issue of the proper parse for NL + CL + N. In the case of German, for instance, if we are prepared to argue from analogy and take the German NM for CL, the tests all seem rather inconclusive, as Krifka (1989b) notes. Both parses (i) and (ii) are possible as the examples (1) and (2) show; in (2), the a) sentence corresponding to the parse [NL + CL] + N is better than the b) sentence, but they are both grammatical.

(1) a. [[*drei* $_{\text{NL}}$ *Barren* $_{\text{NM}}$] *und* [*zwei* $_{\text{NL}}$ *Säckchen* $_{\text{NM}}$]] *Gold* $_{\text{N}}$
[[three $_{\text{NL}}$ ingots $_{\text{NM}}$] and [two $_{\text{NL}}$ sacks $_{\text{NM}}$]] (of) gold $_{\text{N}}$

b. *drei* $_{\text{NL}}$ [[*Barren* $_{\text{NM}}$ *Gold* $_{\text{N}}$] *und* [*Säckchen* $_{\text{NM}}$ *Silber* $_{\text{N}}$]]
three $_{\text{NL}}$ [[ingots $_{\text{NM}}$ (of) gold $_{\text{N}}$] and [bags $_{\text{NM}}$ (of) silver $_{\text{N}}$]]

(2) a. *An Atommüll* $_{\text{N}}$ *wurden* [*250* $_{\text{NL}}$ *Fässer* $_{\text{NM}}$] *gezählt.*
of nuclear waste $_{\text{N}}$ were [250 $_{\text{NL}}$ barrels $_{\text{NM}}$] counted

b. *An* [*Fässern* $_{\text{NM}}$ *Atommüll* $_{\text{N}}$] *wurden* *250* $_{\text{NL}}$ *gezählt.*
of [barrels $_{\text{NM}}$ (of) nuclear waste $_{\text{N}}$] were 250 $_{\text{NL}}$ counted

In Chinese, like Japanese one of the typical classifier languages, the regular order of constituents in the classifier construction is NL + CL + N. Although there are cases where the classifier appears together with the noun while the numeral is missing (see sentence (3)) the numeral and the classifier form a standard unit (see sentences under (4)).

(3) a. *qianbian shi tiao* $_{\text{CL}}$ *he* $_{\text{N}}$.
over there is [CL] river $_{\text{N}}$

	b.	*qing*	*nin*	*he*	*bei*$_{CL}$	*kafei*$_{N}$	*ba.*		
		please	you	drink	[CL]	coffee$_{N}$	[particle]		
(4)	a.	*Ni*	*you*	*haizi*$_{N}$	*ma?*	*You,*	*wo*	*you*	*liang*$_{NL}$
		you	have	children	[QPt]	Yes,	I	have	two
		ge$_{CL}$.							
		[CL]							
	b.	*Ni*	*mai*	*guo*	*shu*$_{N}$	*le*	*ma?*	*Dui,*	*wo*	*mai*
		You	buy	[past]	book(s)	[Perf]	[QPt]	Yes,	I	buy
		le	*san*$_{NL}$	*ben*$_{CL}$.					
		[past]	three	[CL]					

Now curiously, Japanese happens to be not as good a candidate as Chinese, for instance, to illustrate Quine's point. In this language the usual order of constituents in classifier constructions is not NL + CL + N, but N + NL + CL, so that Quine's example *five oxen* translates into

(5)	*o-ushi-*	*go-*	*to*
	oxen$_{N}$	5$_{NL}$	CL

Thus the noun and the classifier are separated by the numeral. That makes the notion of the classifier operating on the noun not a very natural one. Moreover, it is apparently a universal fact that out of the six possible combinations of ordering the constituents N, CL, NL, only those four occur in the world's languages in which the classifier and the numeral are contiguous elements. The most common arrangements are the ones exemplified by Chinese (NL + CL + N) and Japanese (N + NL + CL); some Pacific languages realize CL + NL + N and N + CL + NL, while NL + N + CL and CL + N + NL do not occur anywhere (Allan 1977: 288).

The conclusion to be drawn from this evidence seems to be that at least in Japanese version (ii) above is not realized, viz., the classifier does not have the individuating function after all. This is somewhat at variance with the picture that a semanticist would like to draw: he would prefer to be able to identify an item in the lexicon that has precisely that individuating function. But there is an assumption operative in the conclusion, namely that the parse of the classifier phrase is in fact a reliable indication of the classifier's conceptual role. This assumption can hardly be proved, however, and thus need not actually be embraced; there may simply not be a close enough fit between syntax and semantics that the evidence from parsing could decide the conceptual issue. (But then, Quine would have to give up his Japanese example altogether, which would in fact not be a loss after all since the linguistic facts concerning Japanese are, as we saw, not the way Quine represented them in the first place.)

A more cautious formulation of the above conclusion would be to say only that linguistic evidence makes it *more likely* that classifiers have a parametrizing than a individuating function. But it might also be that they combine those two functions, so that the reference of the pure noun is still "holistic" or "mass-like." Now there is a distinction to be made here within the notion of mass term; the fact that a noun does not have the feature *+count* can have at least three reasons: it refers either to (i) some kind of *homogeneous stuff* like *water* or *mud*; or to (ii) some *collection* of discrete objects, but in a wholesale manner (paradigm example: *cattle*); or finally to (iii) some *abstract concept* like *solidarity*. There is no reason to suppose that in classifier languages these conceptual distinctions cannot be drawn; the only thing that can be legitimately claimed is that apparently there is no reference to *single countable entities* by the noun itself in those languages. Together with (i) - (iii), let us call this the fourth *mode of reference*.

4. The semanticist who wants to model those four modes of reference in a close-to-language fashion has to provide a set of formal characteristics distinguishing four classes of objects, one for each mode. Now the traditional Tarskian style of formal semantics recognizes only one kind of entities in its domain of discourse, the only postulate being that this domain be non-empty. There are, however, systems of *multi-sorted logic* which do distinguish different sets of basic individuals (e.g., usual individuals and spatio-temporal locations); moreover, there are sometimes various cross-connections between the different basic domains in those systems which can serve as a paradigm for the modeling task under discussion here (e.g., the postulate that every concrete object occupies some location in time and space). However, apart from such *functional* connections linking two otherwise unrelated subdomains there is the need to establish certain *algebraic* connections structuring one and the same subdomain. Consider the second and the fourth mode of reference; one picks out collections, the other single objects. When we consider collections and single objects as belonging to two altogether different subdomains we obviously miss a fundamental intuitive relationship between the two: collections are *made out of* single objects. Thus, if we consider a certain group of cattle it is the individual animals that make up this collection. I have proposed elsewhere (Chapter 1) to introduce one single domain for collections and individual objects and endow this domain with the algebraic structure of a *lattice* to represent the basic *part-of* relationship between those kinds of objects. The individual objects are then called *atoms* in this domain and are but special

collections of "size" one. A general collection is also called *individual sum* or *i-sum*, for short. These theoretical devices basically take care of modes (ii) and (iv).

The first mode of reference concerns homogeneous stuff. Its manifestations are all the particular quantities of matter of the given kind; they, too, are structured by a fundamental part-whole relation, with the important difference that this ordering does not recognize smallest parts (Chapter 1, Krifka 1989b). Finally, there are the abstract concepts or properties for the fourth mode of reference. Logical semantics used to model these entities at first as sets of concrete individuals (extensional semantics), then as intension functions, i.e., functions from possible worlds into sets of individuals (intensional semantics). But it is in the spirit of the algebraic approach to go even more "intensional" and to introduce an additional subdomain of basic entities standing for concepts; these are structured with the familiar concept hierachy.

Our current ontology, then, consists of the following four subdomains (the operation '$\sum$' forms all possible i-sums over a given set):

(6)

A set of atomic individuals	[referential mode (iv)]
E set of individual sums over A, $E = \sum A$	[referential mode (ii)]
Q set of homogeneous quantities, $Q \subseteq A$	[referential mode (i)]
C set of concepts	[referential mode (iii)]

Next, we introduce some definitions. Let α be a noun, $\|\alpha\|$ its denotation, n a natural number, and c a classifier; furthermore, let the down arrow '$\downarrow$' stand for the operation that forms from a given element x in E or Q the set of all elements "below" x (in the sense of the relevant ordering relation); finally, '$\oplus$' is the summing operation on E and Q. Then we define:

(7)

$$A_c := \{x \in A \mid x \text{ classifies as } c\text{-object}\}$$

$$E_{n,c} := \{x \in E \mid \exists y_1, \ldots, y_n \in A_c : x = y_1 \oplus \ldots \oplus y_n\}$$

$$CLP_{n,c}(\|\alpha\|) := \|\alpha\|^{\downarrow} \cap E_{n,c} \qquad [\alpha \text{ in referential mode (ii)}]$$

$$Q_c := \{x \in Q \mid x \text{ classifies as } c\text{-object}\}$$

$$\textstyle\sum(Q_c) := \text{the set of i-sums of elements of } Q_c$$

$$Q_{n,c} := \{x \in \textstyle\sum(Q_c) \mid \exists y_1, \ldots, y_n \in Q_c : x = y_1 \oplus \ldots \oplus y_n\}$$

$$CLP'_{n,c}(\|\alpha\|) := \textstyle\sum(\|\alpha\|^{\downarrow}) \cap Q_{n,c} \qquad [\alpha \text{ in referential mode (i)}]$$

$$SG(\|\alpha\|) := Ext(\|\alpha\|) \qquad [\alpha \text{ in referential mode (iii)}]$$

Thus, when c is a classifier for slim things, A_c is the set of all atomic individuals that are classified as a slim object in a given language, and

$E_{n,c}$ is the set of i-sums that are built up from n atoms classified as c-objects. Let α be a noun in a classifier language in referential mode (ii); then it is reasonable to assume that its denotation $\|\alpha\|$ is the maximal collection x in E such that x qualifies as a collection of kind α. This captures the intuition of "whole-sale" reference that is connected with the bare nouns in classifier languages: the bare noun by itself does not single out any particular collection. So if α = *o-ushi* ('*ox*') then $\|\alpha\|$ is the collection or i-sum made up of all oxen. Let CLP be the classifier phrase [**n**+c] with the numeral **n** standing for the natural number n; then the semantic effect of this CLP is as described by the operator $CLP_{n,c}$: it forms all the subcollections of $\|\alpha\|$ and intersects this set with all collections that consist of exactly n atomic individuals classifying as c; thus for the Japanese CLP *o-ushi-go-to* ('*five oxen*') we get the set of all i-sums consisting of five oxen (here we assume that the numeral is not quantifying but behaves semantically like an adjective; for a discussion of this issue see Chapter 4). If a is a noun in referential mode (i), its denotation $\|\alpha\|$ is the maximal quantity of matter of kind α; again, the reference of α does not single out any particular quantity. Then the analoguous CLP operator $CLP'_{n,c}$ forms the set of i-sums of elements below $\|\alpha\|$ that are at the same time collections of n units of portions of matter classifying as c. Thus if α is *wine* and c is *glass-of* then *three glasses of wine* denotes the set of collections of quantities of wine that make up three glasses. Finally, if α has referential mode (iii) its denotation is a concept in C. This mode is perhaps not that much realized in classifier languages but fits languages better that have a *singulative* form for their otherwise transnumeral nouns. In Arabic, for instance, the transnumeral form for the concept *fly* is *dabban*; from this the singulative *dabbane* ('*a fly*') can be formed, which is the countable unit (Cowell 1964:369; cited after Biermann 1981: 234). So the singulative marker seems to play a genuinely individuating role here. If α = *dabban* its denotation in our model is an element of C, viz., the concept *fly*; then the semantic effect of the singulative is very straightforward: it is the familiar extension function *Ext* forming the set of all objects that the given concept applies to.

Now let us come back the the familiar count nouns in Western languages; their mode of reference is individuating right away, without the detour over unstructured totalities or concepts. It is interesting to note, however, that the *bare plural* in those languages tends to fill the role of the transnumeral in other languages. Thus, bare plurals can refer to collections and also to *kinds*, which are not to be identified with concrete particular individuals and are best treated in semantics as still another class of entities (for arguments see Carlson and Pelletier (1995)). Some examples are given in (8). The bare plural *tigers* refers to the natural kind TIGER, and *Otto motors* to the kind OTTO MOTOR.

(8) a. Tigers are a subspecies of wild cats.

b. Otto motors were invented by the German engineer Nikolaus August Otto.

Since it is obvious that no single tiger can be a (sub)species, and Otto motors cannot have been invented over and over again, we see that the number distinction *singular vs. plural* is neutralized in these contexts. Bare plurals have the *form* of a plural, but their *reference* is transnumeral.

The above discussion shows that universal semantics has to apply richer models to account for the varieties of referential modes that can reasonably assumed across languages. To repeat, we are not putting forward a philosophical thesis here as to what the "real" reference of a term is. Our model is based on the assumption that universal grammar gives reasonable clues for the choice of the proper ontology. But this is an assumption that is not uncontested in philosophical quarters favoring metaphysical regimentation in ontological matters. This is not a place to argue a case which could help to resolve that tension.

9.2 Structural Relativity in the Notion of Number

5. In the first part of the chapter we addressed the basic problem of individuation of nouns in language and tried to model their various modes of reference in terms of a multi-sorted, algebraically structured ontology. We now extend our discussion from the nominal to the verbal domain, and from the phrasal to the sentential level. This step opens up the possibility to discuss matters of quantity and number *relationally*, i.e., consider possible scope relations between measure phrases. Moreover, a fifth mode of reference will come into play, that to events.

To start with the latter, a very interesting linguistic phenomenon was recently discussed in the literature (Krifka 1991a), which in view of examples like (9a) I like to call the "transportation facts."

(9) a. Lufthansa carried five million passengers last year.

b. Four thousand ships passed through the lock last year.

c. The library lent out 23,000 books in 1987.

The sentence (9a) is in fact not about five million passengers, nor does (9b) speak about four thousand ships, nor (9c) about 23,000 books. Rather, some *collection of related events* is counted in each case: in (9a), *five million* counts the number of tickets sold by Lufthansa last year; (9b) claims that four thousand passages by ships took place at the lock in question; and (9c) speaks of 23,000 events of lending out a book. In each of those situations, there might well have been much less individuals involved than the figures indicate, and that is indeed the typical case.

Now counting presupposes individuation, as we said above. Thus linguistic evidence leads us the conclusion that the semantics has to provide for still another kind of discrete, countable entities - this time not individuals of some sort or other, but whole *events*. There are more classes of examples that show the need for such a move. One case in point in *Georgian reduplication*; the following data are taken from an intriguing paper by David Gil (1988).

(10) a. *Orma k'acma sam-sami canta c'aigo.*
two-erg man-erg three-dist-abs suitcase-abs carried-3sg
'Two men carried three suitcases [each / each time].'

b. *Or-orma k'acma sami canta*
two-erg man-erg three-dist-abs suitcase-abs
c'aigo.
carried-3sg
'Three suitcases were carried by two men [each/ each time].'

In Georgian, numerals can be reduplicated, which leads semantically to the distribution of the NP containing the reduplication over another NP in the sentence. That means for (10a) that each of the two men carried three suitcases (for more on the phenomenon of distributivity see below). But this is only one reading; reduplication can also have the semantic effect of introducing some reference to events. That is, (10a) can also mean that the same group of men carried the three suitcases several times or on several occasions. Note that this effect seems to be symmetric in Georgian: if the numeral of the ergative NP is reduplicated as in (10b), the interpretation of the sentence is exactly the opposite. Now this NP (*or-orma k'acma*) is distributed over the suitcase NP (rendered in English by passivization), with the same ambiguity between the internal distributive reading and the event reading. Call the NP that is distributed over the *domain of distribution*. Then we can formulate a generalization subsuming the two readings: if the reduplicated NP_2 can find another NP_1 in the sentence denoting a pluralic entity, then NP_1 serves as domain of distribution for NP2 (call it the *distributional antecedent* for NP_2). In this case NP_1 is an explicit *internal* antecedent for NP_2. But NP_2 can also refer back to an *implicit*, *external* antecedent, a pluralic event-like entity, e.g., some contextually given set of occasions at which the action expressed in the sentence takes place. So the reduplicated NP_2 looks for a suitable antecedent as its domain of distribution, which may be either internal (usual distributive interpretation over plural individuals) or else external (event interpretation).

The same ambiguity can be observed in Korean, where distribution is *asymmetric*, however. Consider the sentences under (11) – (13); the data are due to Choe (1987a).

(11) a. *[ai-tul]-i* *[phwungsen-hana]-rul* *sa-ess-ta.*
child-pl-nom balloon-one-acc bought
'The children bought a balloon.' (collective)

b. *[ku-tul]-i* *[phwungsen-hana]-rul* *sa-ess-ta.*
he-pl-nom balloon-one-acc bought
'They bought a balloon.' (collective)

(12) a. *[ai-tul]-i* *[phwungsen-hana-ssik]-ul* *sa-ess-ta.*
child-pl-nom balloon-one-AQ-acc bought
'The children bought a balloon each.' (distributive)

b. *[ku-tul]-i* *[phwungsen-hana-ssik]-ul* *sa-ess-ta.*
child-pl-nom balloon-one-AQ-acc bought
'They bought a balloon each.' (distributive)

(13) a. *na-nun* *[phwungsen-hana]-rul* *sa-ess-ta.*
I-top balloon-one-acc bought
'I bought a balloon.'

b. *na-nun* *[phwungsen-hana-ssik]-ul* *sa-ess-ta.*
I-top balloon-one-AQ-acc bought
'I bought a balloon [at several occasions].'

Sentences (11a,b) have a collective reading: the content is that there is one balloon which was bought by the children (by them). Now if the particle *-ssik* is added to the object phrase the sentences get a distributive reading according to which each of the children (each of them) bought a balloon. As in the Georgian example, the NP carrying the distribution marker *-ssik* (called *anti-quantifier* by Choe) looks for another plural NP that it can take as domain of distribution. Now interestingly, if no such NP is available, as in (13b) the sentence can still be interpreted: some contextually understood plurality of events steps in as external antecedent, so that the sentence means that I bought a balloon each day or at each store. But note that the distributive particle does not work in the same symmetric way as Georgean reduplication: in the above sentences the subject NP could not carry the particle *-ssik* with the distributional effect reversed.

It is a curious fact that a particle very much like the Korean *-ssik* exists in German (see Chapter 5). Like *-ssik*, the German particle *je* not only triggers distribution in the same way as in Korean, but can also induce an event interpretation if no NP is present in the sentence that qualifies as an internal distributional antecedent. Thus, while sentence (14a) is ambiguous between the collective and the distributive reading, *je* in (14b) forces the distributive reading much like the English *each*; but unlike *each*, *je* is much more flexible: thus in (15a) it is not necessary that the subject NP be plural for the sentence to make sense; (15b), which is (15a) except for the singular subject NP, still has the event reading, which is already present in (15a) and makes that sentence ambiguous between the internal individual and the external event interpretation.

(14) a. Die Touristen kauften eine Eintrittskarte. (collective/distributive)
'The tourists bought an admission ticket.'

b. Die Touristen kauften je eine Eintrittskarte. (distributive)
'The tourists bought an admission ticket each.'

(15) a. Die Zollbeamten prüften je einen Koffer. (distributive)
'The customs officers checked a suitcase (each/each time).'

b. Der Zollbeamte prüfte je einen Koffer. (event reading)
'The customs officer checked a suitcase (each time).'

c. Die Zollbeamten (/der Zollbeamte) prüfte(n) jeweils einen Koffer.
'The customs officer(s) checked one suitcase each time.'

The event reading in (15a) becomes stronger when we replace the particle *je* by the adverbial *jeweils*; see (15c). Accordingly, *jeweils* interchanges with *je* in sentences where an internal distributional antecedent is altogether missing, like in (16a). As for the question of symmetry, it seems to be the case that in certain contexts *je* behaves kataphorically; thus, in (16b) the NP *den Körben* serves as antecedent, and the sentence means that we have one situation with several baskets in each of which there is an apple. But then, world knowledge excludes the external event reading according to which at each occasion, one apple was "collectively" in the baskets. (16c) remedies this flaw, and the sentence is again ambiguous between the external and the internal reading (preference is for the internal reading on pragmatic grounds).

(16) a. Je ein Apfel war faul. (event reading)
'An apple each was rotten.'

b. Je ein Apfel lag in den Körben. (internal distribution)
'An apple each was in the baskets.'

c. Je drei Äpfel lagen in den Körben. (internal > external)
'Three apples each were in the baskets.'

In (16b,c) we have a definite plural NP as domain of distribution. Internal kataphoric distribution becomes worse if not outright impossible, however, when the NP is indefinite plural with a numeral in it. Thus I think that (17a) has only the event interpretation meaning that for each "case" of two diplomats in need of an interpreter one such person was actually dispatched. In (17b) the internal reading is there again; but here the regular order of constituents is reversed, the subject NP and the object NP are switched.

(17) a. Je ein Dolmetscher begleitete zwei Diplomaten. (external)
'An interpreter each accompanied two diplomats.'

b. Je einen Dolmetscher erhielten zwei Diplomaten. (external, internal)
'[An interpreter each]$_{\text{OBJ}}$ got [two diplomats]$_{\text{SUBJ}}$'

For more discussion on *je* the reader is referred to Chapter 5. Here I would like focus now on the question of autonomous vs. dependent reference of numerative phrases in relational plural sentences. Consider (18) and (19).

(18) a. Two interpreters accompanied four diplomats.

b. 'Four diplomats were accompanied by two interpreters.'

(19) a. Three ensembles played two symphonies five times.

b. 'In five performances, three ensembles played two symphonies.'

Sentence (18a) speaks about two interpreters whose job it was to deal with a certain number of diplomats. Without further information the sentence does not say precisely how many diplomats were involved; if the intended reading is collective then the interpreters accompanied a group of four diplomats, but otherwise, in its distributive reading there may have been up to $24 = 8$ diplomats around. The NP *four diplomats* has *dependent reference*: its numeral *four* does not give the true amount of the individuals involved. This number can be anything between 4 and 8. Thus in general, let NP_2 be the dependent NP in a relational plural sentence and NP_1 a noun phrase taking scope over it; let **k** be the numeral in NP_1, **n** the numeral in NP_2, and k, n the respective natural numbers expressed by those numerals. Then the exact number i of individuals in the denotation of NP_2 cannot be completely determined since it depends on pragmatic circumstances; but on semantic grounds alone the interval $n \leq i \leq k \times n$ can be specified. (We ignore here the further complication that arises in connection with the transportation facts: thus in (20), not even the lower bound n for the interval can be taken for granted.)

(20) Two lock-keepers counted 3,000 ships.

On the other hand, the numeral **k** in NP_1 can be taken absolute: the denotation of NP_1 does indeed involve precisely k individuals. NP_1 has *autonomous reference.*

Now the question arises what it is that determines autonomous reference for a noun phrase in such relational plural sentences. (18) and (19) show that it is the *leftmost* NP that refers autonomously; autonomous reference does not seem to be connected with any particular thematic role. However, it seems intuitively plausible to assume that autonomous reference has something to do with *topicality*: the NP which is the topic of the sentence refers autonomously, and the leftmost position in a sentence is only an indication for the sentence topic. Before such a hypothesis can be seriously entertained as a possible universal (as suggested by Dietmar Zaefferer), many more examples from various languages have to be discussed, but this is something that cannot be done here.

6. I conclude this section with the rough sketch of a second semantic model which incorporates the fifth mode of reference, that to events. So far we have a multi-sorted domain of individuals modeling nominal reference, comprising atomic and proper sum individuals, quantities of matter, and concepts. If we want to add a subdomain for events we have to say what the relations are between this new subdomain and those we already have. Now it is hard to define explicitly what an event is; the best we can do is to give some *axioms* that mirror the structure inherent in our intuitive conception of events. For a formulation of those axioms see Chapter 11; for a related one, which antedates mine, see Krifka (1989a). Both approaches are algebraic in spirit, i.e., they consider events as inherently structured objects. Thus, the summing device for individual objects is carried over to the subdomain of events. In this way the notion of a *pluralic event* becomes available, the need for which is demonstrated by linguistic data like the above transportation facts and the iterative interpretation of relational sentences. But pluralic events do work in much more central areas of grammar. I would like to illustrate this claim by modeling, within

the event theory of Chapter 11, the *collective*, the *distributive*, and the so-called *cumulative* reading of simple relational plural sentences. Consider the sentences in (21).

(21) a. Denys and Tania shot a lion. (collective/distributive)

b. Denys and Tania shot two lions. (cumulative)

Then the basic semantic intuition to be represented here is this. Imagine Denys Finch-Hatton and Baroness Tania Blixen are out in the wilderness where they meet a lion and collectively shoot it (e.g. with Denys aiming the gun, Tania pulling the trigger; call this *scenario 1*). Then (21a) in its collective reading is true. We have *one single event* of shooting a lion but a *"collective" agent*, which is modeled by the individual sum DENYS $\oplus$ TANIA consisting of the atomic individuals Denys and Tania. Now let us change the scenario and assume that both Denys and Tania have a gun and shoot their respective lion (*scenario 2*). Then, according to the above rule, we have $2 \times 1 = 2$ lions shot, which makes the distributive reading of (21a) true. There are now *two events* with different agents (Denys, Tania) on the atomic level. But then, again, those two events are one in some sense, for instance, when this double achievement filled Denys and Tania with pride (assume that neither subevent alone would have done so). So this is a sum event with singular agents. The need for sum events seems particularly clear with examples like (21b). That sentence is also true under scenario 2, but in a different reading, the cumulative one. The only thing that matters now is the total number of lions shot, which is two. As in the distributive case, there is no single event having Denys and Tania as agent, nor one having the two lions as theme. What is characteristic of the cumulative reading, rather, is a situation in which the number of events involved is completely unspecified; the only constraint is that the total number of objects filling the agent role in those events is as indicated in the subject (agent) NP of the sentence, and similarly with the other roles that might occur.

The representations in event theory of the examples in (21) are as follows. The material in brackets stands for the *event type* that is described by the appropriate reading of the sentence. Such an event type is characterized by a certain number of *roles* ρ_i (ρ_1 for 'agent' , ρ_2 for 'theme', τ for temporal location); these are filled by the various objects involved in the event. The boldface letters are parameters in a quantifier-free set-up of the system, where the upper indices carry the dependency information. '$\oplus$' is the i-sum operator of the plural logic in Chapter 1, '$\sum$' is the sum operator for event types; '*' is the plural operator on predicates, again taken from Chapter 1, and '^{D}' is the distributivity operator that can be found, e.g., in Link (1987b) (Chapter 4). 'Π' means 'part of', '$\cdot\Pi$' 'atomic part of'. '$\mathbf{t}_0$'

is the (free) parameter for present time, and '$\triangle$' is the conjunction operator for event types. With this notation, the event types $\|S_1\|$, for instance, can be read thus: "the type of event $\mathbf{e}$ such that (i) the agent of $\mathbf{e}$ is an object $\mathbf{a}$ which is the i-sum of two objects $\mathbf{a}_1$, $\mathbf{a}_2$ with $\mathbf{a}_1$ being Denys and $\mathbf{a}_2$ being Tania; (ii) the theme of $\mathbf{e}$ is an object $\mathbf{b}$ which is a lion; (iii) the temporal location of $\mathbf{e}$ is a $\mathbf{t}$ which is earlier than the present time; and (iv) $\mathbf{e}$ is a shooting." The representation for the distributive reading of (21a), the event type $\|S_2\|$, basically says that there is an i-sum $\mathbf{a}$ consisting of Denys and Tania such that every atomic part of $\mathbf{a}$ has the property of being in the role of the agent in an event that is of type 'shot a lion'. Finally, the cumulative reading $\|S_3\|$ of (21b) describes the type of an event $\mathbf{e}$ which is the sum of two shooting events $\mathbf{e}_1$, $\mathbf{e}_2$ in the past such that Denys and Tania are their respective agents, and their themes are i-parts of a certain i-sum $\mathbf{b}$ of lions that consists of exactly two atomic individuals (this is the cumulative information provided by that kind of sentences).

(21a) $S_1 = {}^0[$ **Denys and Tania shot a lion** $]$; (collective)

$$\begin{aligned}\|S_1\| = [\,&\mathbf{e}\,;\rho_1(\mathbf{e}) = \mathbf{a}^0 \;\&\; \textsc{denys}(\mathbf{a}_1^0) \;\&\; \textsc{tania}(\mathbf{a}_2^0) \;\&\; \mathbf{a} = \mathbf{a}_1 \oplus \mathbf{a}_2 \;\&\\ &\rho_2(\mathbf{e}) = \mathbf{b}^0 \;\&\; \textsc{lion}(\mathbf{b}) \;\&\\ &\tau(\mathbf{e}) = \mathbf{t}^0 \;\&\; \mathbf{t} < \mathbf{t}_0 \;\&\\ &\textsc{shooting}(\mathbf{e})\,]\end{aligned}$$

(21a) $S_2 = {}^0[$ **Denys and Tania** ^{D}**(shot a lion)** $]$; (distributive)

$$\begin{aligned}\|S_2\| = [\,&\textsc{denys}(\mathbf{a}_1^0) \;\&\; \textsc{tania}(\mathbf{a}_2^0) \;\&\; \mathbf{a}^0 = \mathbf{a}_1 \oplus \mathbf{a}_2\,] \;\triangle\\ &\textstyle\sum(\,[\,u^1 \cdot \Pi\,\mathbf{a}\,] \mid [\,\mathbf{e}\,;\rho_1(\mathbf{e}) = u \;\&\\ &\qquad\qquad\qquad \rho_2(\mathbf{e}) = \mathbf{b}^1 \;\&\; \textsc{lion}(\mathbf{b}) \;\&\\ &\qquad\qquad\qquad \tau(\mathbf{e}) = \mathbf{t}^1 \;\&\; \mathbf{t} < \mathbf{t}_0 \;\&\\ &\qquad\qquad\qquad \textsc{shooting}(\mathbf{e})\,]\,)\end{aligned}$$

(21b) $S_3 = {}^0[$ **Denys and Tania shot two lions** $]$; (cumulative)

$$\begin{aligned}\|S_3\| = [\,&\textsc{denys}(\mathbf{a}_1^0) \;\&\; \textsc{tania}(\mathbf{a}_2^0) \;\&\; \mathbf{a}^0 = \mathbf{a}_1 \oplus \mathbf{a}_2\,] \;\triangle\\ [\,&(\mathbf{2}^*\textsc{lion})(\mathbf{b}^0)\,] \;\triangle\\ [\,&\mathbf{e}\,;\mathbf{e} = \mathbf{e}_1 \oplus \mathbf{e}_2 \;\&\; \rho_1(\mathbf{e}_i) = \mathbf{a}_i \;\&\\ &\rho_2(\mathbf{e}_i)\,\Pi\,\mathbf{b} \;\&\\ &\tau(\mathbf{e}_i) = \mathbf{t}_i^0 \;\&\; \mathbf{t}_i < \mathbf{t}_0 \;\&\\ &\textsc{shooting}(\mathbf{e}_i)\,]\end{aligned}$$

The discussion has shown that the problem of properly locating quantitative information in natural language is quite a complex matter. The above examples and the sketch of their formal representation in a multi-sorted algebraic model give an indication of how a promising system of universal semantics might look like.

Acknowledgment. I would like to thank the members of the workshop on universals in Wuppertal for informative and stimulating comments. Among them, I mention in particular Dietmar Zaefferer and Manfred Krifka with whom I have had intriguing discussions about semantic matters for years.

Chapter 10

The French Revolution — a Philosophical Event?

10.1 Introduction

In 1989 the French Revolution still cast its long shadow onto the present. If you believed newspaper reports you could get the impression that historians, in particular in France, of course, but also in other countries, were impatiently awaiting the 200th anniversary of the storming of the Bastille in order to commemorate the revolution, to interpret it anew—even to continue the fight. In France that amounted to the culmination of a phenomenon that the historian François Furet, in a brilliant analysis some years ago, described as the recurring application of the old social battle lines to their putative counterparts in the political present. Almost imploringly, Furet gave his analysis the heading: "The French Revolution is over!" — a reassuring reminder, which will give us the right to speak of that event as a closed episode of history.

Furet's analysis is contained in his book "Penser la Révolution Française," which carries the English title: "Interpreting the French Revolution" (Furet (1981)). The historian takes the event for granted and investigates its historical meaning. It is for the philosopher to inquire into the concept of event in general. The example of the French Revolution will be used in this chapter as a reservoir of vivid examples to illustrate the terminology which will be introduced along the way. It was my intention to counteract a tendency which is widespread among philosophers, especially working in the analytic tradition: that of using the same relatively simple and time-worn examples over and over again. But a philosophy of language in particular that deserves its name has to cope with the full richness of language. Its descriptive power should go beyond the mere analysis of "linear" facts, as it were; suitable concepts should be introduced for dealing with multi-layered contexts and states of affairs, which in the case of historical events overlap or follow each other in space and time and can relate to various levels of historical analysis. It seems worthwhile to confront the abstract, philosophical conception of event with that field of inquiry that most naturally employs the language of events: the field of history.

Making frequent use of historical dates from the French Revolution, then, I shall start by shedding some light on the various components connected with the notion of event: its location in time and in space, persons and things that trigger an event to happen, or are affected by it, and finally the formation of complex events and their possible structure. I shall then go on to talk about the distinction between particular events that are historically unique, and those that seem to be repeatable, and I shall show that the impression of recurrence is rather due to the perception of a pattern, a uniformity. This observation will lead to the notion of *event type* whose relationship to individual events will then be discussed. Here I shall sketch the elements of a definition of truth that is suited to a theory of events.

The transition is thereby made from the descriptive to the theoretical levei of analysis. I shall map out the project of Algebraic Semantics, which, roughly speaking, tries to answer ontological questions in *structural* and not metaphysical terms. A mathematical model for event structures, the so-called "Aether model," will conclude this chapter. This model will be further pursued in the following chapter.

10.2 The Structure of Events

10.2.1 Events in Time and Space

Let me start with the easiest question, which is about the temporal course of events, in our case: When did the French Revolution take place? Well, at least its beginning seems to be clearly fixed: the revolution began with the storming of the Bastille by the people of Paris on July 14, 1789. That was no doubt a revolutionary act. But weren't the following already such: the calling of the Estates General the year before, with the doubling of the representation of the Third Estate, the Third Estate assembly proclaiming itself a National Assembly on June 17, or the Tennis Court Oath three days later, "never to separate until the constitution has been firmly established, and to leave only at the point of a bayonet?" The calling of the Estates

General, for instance, is considered by the historian Albert Soboul as part of a *judicial revolution.*

Here already we see a phenomenon which is all-pervasive in the characterization of processes of all kinds, not only of those that are declared historical events by historiographers: that of various levels of description. A process can be broken up into subprocesses, and what goes for a fairly precise dating of the process itself has to be specified more exactly when we focus on the subprocesses. For Soboul, for instance, the 14th of July is really but the beginning of a subprocess of the revolution, which he calls the *popular revolution*, and the French Revolution as a whole is brought on its way only through the combination of those "sub-revolutions." Thus one could as well speak of 1788 as marking the beginning of the revolution. The issue of the levels of description I shall call the *granularity problem*, to which I return below.

The question as to when the revolution ended is more problematic (let us ignore Furet's polemical jest that apparently its end still escapes some of his contemporaries): it could have been September 22, 1792, when the French Republic was proclaimed, or 9 Thermidor with the fall of Robespierre, or else—another famous date of that time— 18 Brumaire of Year VIII in the revolutionary calendar, when Napoleon Bonaparte seized power through a coup d'état. The answer to these questions depends, of course, on the historical interpretation that is attached to the event of the French Revolution: it can be the process of establishing the republic, or the revolutionary chaos and its termination, or again the period of the democratic constitution up to the establishment of the Napoleonic military dictatorship. It all depends on the common denominator we assign to the events. That will decide which events will be counted among that collection of events that make up the historical episode at hand.

Having seen the difficulties in dating that well-studied historical example, we should expect even more trouble with episodes in the history of ideas. Let us consider the Scientific Revolution (now that we are at revolutions): when did it begin, and when did it end? Usually one speaks about the 16th and 17th centuries, say from the time of Copernicus via Kepler and Galileo up to Newton. However, I will mention two extreme opinions just to elucidate the problems involved: Pierre Duhem claims that the scientific revolution began in 1277 when important tenets of the aristotelic-averroist tradition were condemned in Paris which, among other things, put natural reason on a par with divine revelation; as a side-effect, that freed the sciences from the useless ballast of a realist metaphysics which, in Duhem's eyes, was a precondition for scientific progress (here the instrumentalism of the 19th century shows up). And its termination? Inveterate adherents of Popper's falsificationalism could entertain the idea that it was the year

1838 when the astronomer Bessel discovered the first fixed-star parallax and thereby really falsified the geocentric world-view for the first time.

Is it only historical complexes of events or socio-cultural processes whose dating is that notoriously vague? For instance, the eruption of a volcano is a natural process and should be easy to date since it barely lends itself to different "interpretations:" the eruption of Vesuvius in the year 79 or the eruption in 1980 of Mount St. Helens in the state of Washington. But what about less violent eruptions like those of the Hawaiian shield volcanos that throw out lava for months and are active more or less continuously? Even here criteria have to be developed according to which such eruptions are declared to be over (for instance on legal grounds) even if the geological process continues.

Regarding the *localization in space*, it often seems even more difficult to say exactly *where* an event took place. Did the French Revolution take place in Paris, or in France? Didn't the revolutionary war spread it over the whole of Europe? After all, the European powers tried hard to prevent the revolutionary spark from flashing over, and the development of the war had a crucial impact on the course of events inside France. Thus all those places could be taken as parts of the spatial location of the revolution. Again, where did the plot take place (let us assume the facts were the way the Gironde temporarily claimed them to be) involving Danton and the army general Dumouriez, who in March 1793 lost Belgium for the Republic and who, even on the very day he defected to the Austrian lines, met with some friends of Danton's? If Danton took part in it and stayed in Paris, didn't the plot occur in Paris, too, and not only at the front in Dumouriez's quarters? Incidentally, modern times have aggravated the problem of spatial location of events; imagine a business man in Tokyo placing an order at the New York Stock Exchange for some Union Carbide stocks which happens to belong to the Deutsche Bank—where this deal takes place is not easy to say.

What adds to the problems of spatial locations is the fact that the expression *event e occurs at place s* is used with systematic ambiguity. Let event e be the storming of the Bastille; e took place at the Bastille, in Paris, but of course also in France and in Europe. Here we have to bar the imminent conclusion that all events occur "everywhere." Let me explain that. First of all, it won't do if we take places to be geometrical points in space; usually, processes or events are extended in space (and in time, for that matter). Thus we let the parameter s range over whole regions of space. Now the area of 18th century Paris is a region (call it s) that includes the area around the Bastille where the action was, which is why we can truthfully say that e occurred at s. But then, e occurred at s' where s' stands for the whole universe. It is in this sense that every event can be said to occur everywhere. However, the formal analysis provides the simple means to avoid that consequence: the place of an event e is defined as the set-theoretic *intersection* of all spatial regions s of which it can be truthfully said that e occurs at s.

Now let us count among our events also facts, which describe states of affairs rather than changes (so, for instance, the fact that Danton and Robespierre had a strained relationship); then it doesn't make much sense anymore to ask where this relationship took place. Presumably their rivalry first showed up in the political clubs and only later in the Convention, but even such a statement speaks more about the *temporal* phases of their relationship than its spatial location. Therefore, no location in space will be assigned to pure states in what follows; but they can arise in time and also come to an end.

In spite of all the difficulties pertaining to their precise location it should be clear now that events manifest themselves in space and time. One could take care of the inevitable problem of vagueness in a more systematic fashion by applying a different "grid" to the various events which assigns a unit of spatio-temporal precision according to how fine-grained the event happens to be. Thus the storming of the Bastille is measured in hours or days, and its spatial location in square miles perhaps; the financial crisis of the Ancien Régime is measured in months or years, and the Scientific Revolution in decades or even centuries. Those units of precision I call the *granularity parameter* of events.

Now a first attempt at systematizing the notion of event has been to *identify an event with the stretch of time during which it takes place*. These are the kinds of argument that were advanced to support such a representation: (i) the notion of time stretch is well understood and free from the vagueness and obscurity surrounding events; (ii) for many purposes the only relations on events that matter are *temporal precedence* and *temporal overlap*, and for them time stretches are all that is needed. However, we saw above that even the temporal extension of events can be vague, and when we have complete overlap the distinction between two simultaneous, but different events is obliterated. The next move, then, is to include spatial regions in the representation. An event is now *the (minimal) spatio-temporal region in which it occurs*. Incidentally this is basically the proposal that Quine makes. But there are problems with this proposal, too, and apparent counterexamples come to mind easily. It seems that many events that we clearly distinguish intuitively are coextensive in space and time. The lowering of the drawbridge was at the same time the surrender of the fortress's governor. Again, on July 21, 1791, Louis XVI traveled to

Varennes. This was the king's attempt to flee, but it might as well have been a kidnapping (as some royalists later tried to make people believe). Or take Robespierre's speech in the Convention on April 3, 1793, denouncing Dumouriez's conspiracy, which launched the attack on the Gironde. The pattern is always the same: when a person does A he or she thereby does B; when you stretch out your arm on a bike you thereby indicate an intended change of direction.

A few years ago, during the high tide of Richard Montague's paradigm of formal philosophy, the next move was almost predictable at this point. If spatio-temporal regions are too coarse-grained to discriminate between events then one resorts to *intensionalization*: events become so-called intension functions from possible worlds into regions of space-time. The identity conditions for events are thereby made much stricter. Thus, for example, we can now say that the trip to Varennes was an attempt to escape only in the real world; in a different world it could have been a kidnapping or the king's traveling innocuously to his hunting-ground. Therefore, the traveling event and the flight cannot be equated since they don't co-occur in all possible worlds. As a matter of fact, Montague himself made a proposal along these lines. In Montague (1969) he starts out from tensed predicates like "star x rises at time t." Now suppose that the fixed star Arcturus rises above the horizon at a certain time t_0; then t_0 has the property of being a moment of time at which Arcturus rises. But since the rising of Arcturus describes an event, events can be *equated with* properties of moments of time, according to Montague (he ignores the spatial component). Now moments of time are individuals, and properties of individuals are intension functions in the Montagovian framework; thus events come out as a certain class of intension functions with the set of possible worlds as their domain. However, while gaining an extra degree of freedom, such an approach is rather "formalistic" and doesn't address the structural properties of events that have to be dealt with even in one and the same world, in particular, that one event can be part of another event. So I won't pursue it here, but I shall return to a discussion of the various modeling devices below.

The conclusion I want to draw from this discussion is this: The best policy seems to be to distinguish events from the spatio-temporal regions in which they take place, but to make such regions part of the notion of event. That means that modulo a particular granularity, time stretches are assigned to events, and also certain regions in space (in case we are not dealing with pure states). I shall speak of temporal and spatial *traces* of events. The temporal trace of the king's flight to Varennes is July 21, 1791, and its spatial trace is the route taken by the carriage (passing Châlon sur Marne, for instance). The temporal trace of the French Revolution is presumably contained in those ten to twelve years at the end of the

18th century, but its exact extension depends, as we saw, on the nature of subevents that history considers part of the revolution.

10.2.2 Roles

The notion of event has been in the focus of this discussion; yet I do not intend here to embark on a reductionist project that would try to translate our talk of objects into a pure event language, thereby reducing the ontology of objects to an ontology of events. Even if modern physics should have every reason to base its discourse on an event ontology (in the sense of interactions of force fields of the various sorts, say) that would not be a sufficient philosophical argument for reduction. The ontology of things is too deeply entrenched in our language-mediated view of the world for it to be dismissed as a mere *façon de parler*. I am not saying that the notion of thing is altogether unproblematic; but neither is reductionism *per se* good philosophy.

Thus I prefer to leave the issue of a possible reduction unresolved, and rather address the question as to how things and events relate to each other. To begin with, things, e.g., people or material objects, are "involved" in events in various ways, filling different *roles*. To be sure, Louis XVI was still the agent when in 1788 he recalled the Genevese banker Jacques Necker, after the country had definitively gone bankrupt; that cannot quite be said of the circumstances of his return from Versailles to the Tuileries: it was the people of Paris that escorted the king, who himself had to accept a rather passive role in this, back to his castle in Paris. The women's march to Versailles on that 5th of October, 1789, which prompted the return of the King to Paris the following day, is also the kind of event that we will be concerned with in more detail below: the march had a "collective subject:" it is true that individual people went to Versailles that day, but that alone doesn't constitute a march or a procession. The same applies to the storming of the Bastille and the invasion of the Tuileries three years later. History is rich in collective agents of events, and they call for a theoretical account.

Thus we can distinguish the role of an *agent* and the role of an object that is acted upon, the *patient*. Furthermore, many actions are performed with an *instrument*: The king used for his flight a sturdy carriage that had been built just for this purpose (but turned out to be too heavy), and a number of fresh horses along the way. The role of instrument assumes a more bloody reality in the knife of Charlotte Corday who murdered the sans-culotte hero Jean Paul Marat, or in the cruel routine of the guillotine during the Terror Regime.

There are more roles that can be considered but these might suffice for the purpose of illustration. Thus for every event there are a number "players" or *characters*, that is, persons and objects in their various roles. In keeping with the theatrical metaphor I shall call the collection of all individuals that are involved in an event its *cast*. The roles that are actually filled in an event may of course vary from case to case. The king's soliloquies during his detention at the *Temple* have a speaker, but no hearer, and the event of receiving his death sentence has no proper agent (the king is here only the experiencer).

Let us come back to the concept of collective agency that was briefly mentioned above. The deputies of the National Assembly, with a bare majority, voted for the death penalty for "citizen Capet," the former Louis XVI. That event consists, of course, of a number of subevents in which every single deputy casts his vote; but none of these subevents can be equated with the verdict. The decision of the National Assembly is therefore an event *sui generis*, and only it, not the single votes, is *causally responsible* for the ensuing execution. I propose to deal with this situation by introducing so-called *sums of individuals* that are the collective agents of such events; as a formal explication of this notion we can make use of the theory of plurals developed in Chapter 1.

Here is a metaphysical argument that was advanced by Lawrence B. Lombard (1986) against the assumption of collective agents in events. Lombard characterizes events as *changes in objects*. But since changes according to him can take place only in individual objects there cannot be such an event as the verdict of the National Assembly. Now first of all, the idea of endogenous changes in objects will hardly be enough to account for all kinds of event: for instance, imagine a satellite that is launched into space and thereby loses its weight—it has simply escaped the gravitational field of the earth, but that is not a change that is internal to the satellite. More generally, it seems to me that Lombard's conception of event is *subject-centered* and doesn't take into account the *relational character* of events. Finally, Lombard is completely lost when it comes to bigger complexes of similar events that constitute a new kind of event on a higher level. Take the repeated famines that accompanied the various phases of the revolution. The hunger of the individual people is necessary, but not sufficient for the state of famine. Again, it is the interaction of the single states of hunger plus a common structural cause that make up the event.

Thus I believe that a sound analysis of events will hardly get off the ground without the assumption of collective "characters." Yet I would agree with Lombard on one point. It has to do with the notion of *intention* of human agents who are involved in collective actions. The question is whether we have to introduce, as John Searle once suggested, a collective intention over and above the individual intentions of the participants.

Consider one of his mundane examples in which John and Mary are making dough where she is pouring and he is stirring. This has to be a coordinated process, of course, since otherwise the mixture won't be soaked through homogeneously. Now Searle assumes that while John has the intention to stir and Mary has the intention to pour they together have to develop an additional, collective intention, which is to act in a coordinated way so as to produce the dough. One thing is uncontroversial, of course, namely that the two are able to coordinate their respective intentions in such a way as to achieve their common goal. However, there is hardly such a thing as a *collective subject* that has the collective intention even if John and Mary harmonize perfectly. So there is a collective intention, but it is shared by John and Mary as individuals. Intentions are confined to individual minds. But that doesn't mean that an analysis of events involving collective characters is unnecessary: cooperation is one of the basic social techniques, and an account of events that either overlooks this aspect or seeks to avoid it will have trouble in analyzing richer phenomena than individual actions.

There is a possible objection that can be raised against my account of collective characters, on the grounds that it is just too close to language. It grants that there might be a collective role of the Triumvirate Robespierre – Saint-Just – Couthon in the liquidation of Danton's faction in Germinal (of Year II). By the same token then, the argument could continue, there is also a collective execution of the three men in Thermidor four months later; but even if the executions take place simultaneously, obviously each of them has to die for himself. So language with its pluralic expressions leads us astray.

My reply to this argument is that the analysis given here is not forced to speak of a collective dying. The occurrence of a *plural term* in a report of the Thermidor executions does not mean that it has to refer to a collective *character* in some single event. That is only the case with genuinely *collective* processes which are typically described by collective predicates or collective readings of predicates. Now 'die' is a distributive predicate since it invariably refers to the distributive process of dying. Yet it can take a plural term, as in *"Robespierre, Saint-Just and Couthon died"*. However, in my theory of plurals the predicate *died* triggers a reduction of this sentence into three singular statements that express events with singular characters. Thus the death of Robespierre and his followers describes a situation that is logically opposite to their common role towards the Dantonists: instead of a single event with a collective character we have a *sum of similar events with individual characters*. This is the topic of the next section.

10.2.3 The Structure of Complex Events

The French Revolution can be taken as the sum of its phases: the Constituent Assembly, the Legislative Assembly, the war of the First Coalition, the times of the Convention and the Great Terror, the Directories. But not all events line up in such a clear-cut way. In 1793, as the failure of the Girondists in leading the country became apparent in Paris, the Revolutionary Army had to suffer setbacks on various fronts, and Dumouriez defected to the enemy as his attempt to march on Paris failed; at the same time there was inflation and famines, and insurrections broke out in the provinces. All these factors *together* led to the fall of the Gironde and the establishment of the Terror. Monocausal developments in history are rare. The interaction of different factors in producing an effect is the rule. Like events with collective characters, complexes or sums of events must be considered part of the causal history of other events.

The formation of sums of events mirrors on the event level the formation of collective characters on the level of individuals. The former stands to the latter in a chiasmatic relationship: a singular event with a collective character here, a collection of events with singular characters there.

Lombard, whom we already saw criticize collective characters, doesn't like sums of events either. According to him, there is neither the sum of events consisting of the latest eruption of Vesuvius, the flowing of ink from his (Lombard's) pen, and the moving of Jupiter in its orbit in 1944, nor even the sum of two similar events like any two occasions on which he feels the urge to sneeze (Lombard (1986):236f.). Since he refuses to consider such wild collections of events (which are obviously not very useful), and has no principled way to exclude them, he gives up on a theory of composition of events.

My way of going about it is more liberal. In a first step, any old combination of events is admitted, however bizarre (it doesn't do any harm, but it might not do any work, either). But not every such collection can in a significant sense be considered *a coherent part of the world*: for that to be the case it is necessary that a sum of events be closed or saturated with respect to all *lawful constraints* that organize reality and hold it together. The execution of Louis XVI is lawfully connected with the fact that his body ceases to live; a sum of events that contains the first event but not the second would not count as a coherent part of the world. By the same token, Lombard's spurious collection of events, while being an admissible sum of events in the mereological sense, isn't a coherent part of the world either since it doesn't contain the event of Jupiter's exercising a gravitational pull on its satellites.[1]

[1] Coherent parts of the world correspond to what I call *world chunks* in Chapter 11.

The notion of sum of events, which I will also call a *complex event*, brings along its order-theoretic counterpart, a part relation. So one complex event can be part of a bigger complex event. That leads again to the notion of *subevent*, of which we now have to distinguish *two senses*: a spatio-temporal sense and a complexity sense. The vote of the Montagnards in the National Assembly is part of the total ballot of the Convention in the complexity sense, whereas the first hour of a speech of Robespierre's that lasted several hours is a subevent of the complete speech in the sense of the temporal part-relation. Two events that stand in the part-relation in the complexity sense (call it *c-part*) may behave in an arbitrary way along the spatio-temporal dimension: they may be completely disjoint or they may possess identical traces. Thus it is important to keep these two concepts apart.

At this point we may formulate a structural axiom of the kind that will be part of our formal model below. Clearly, there is the following connection between the spatio-temporal trace of an event sum and the traces of its c-parts: *The trace of a complex event is equal to the sum of the traces of its c-parts* (for that to make sense we have still of course to make precise the notion of sum of spatio-temporal traces). Formally, this relationship is encoded by the mathematical notion of *homomorphism* between the algebraic structure on events and that on spatio-temporal regions.

10.3 Uniformities: Types of Events

Up until now I have only considered concrete, particular events; they are historically unique and belong to the past once they have occurred. But in everyday life, and the more so in the historical sciences, we classify the stream of events into similar ones, we recognize the same or a similar pattern. We also say that *the same event occurs repeatedly*. But taken literally, this way of speaking is certainly difficult to explicate. There is, however, a valid intuition behind it, which can be analyzed by introducing the notion of *event type*. The king calls and dismisses the banker Necker twice: the two actions are of the same type, respectively, even though they differ in time and circumstances. Every time a priest takes the oath on the Constitution (many refused, as is well-known), an event of the same type takes place. In the case of Louis and Necker the characters were the same, whereas in the case of the priests, the agent changes. But the individual priests always fill the same role. Or let us come back to 18 Brumaire: It was Karl Marx who, in *The Eighteenth Brumaire of Louis Bonaparte*, made this date the synonym for a certain *type of event*, the *coup d'état*. Using a

There is also a notion of coherence defined in that chapter, but there it is a property of the "is of type" relation.

quote from Hegel who once said that all facts of world history occur twice, as it were, he ridicules the 1848 revolution as a new edition of the first revolution which, however, turned the tragedy into a farce: "Caussidière for Danton, Louis Blanc for Robespierre, the Montagne of 1848–1851 for the Montagne of 1793–1795, the Nephew for the Uncle. And the same caricature occurs in the circumstances attending the second edition of the eighteenth Brumaire!" Here the theatrical metaphors, which stand behind our terminology, are explicit: the characters slip into their old roles only to stage the same play all over again.

Returning to the modern world, what we come across everywhere these days is *scenarios*; they can be thought of as schemata of whole courses of events and can easily be modeled in our approach by sequences of event types whose temporal traces satisfy the condition of linear precedence. In times where things change rapidly and the types of old experiences can no longer be applied to new situations, scenarios are often the only basis for decision; in such circumstances scenarios, and hence event types, seem to gain the same degree of "reality" as particular events. Thus we are led to treat event types on a par with particular events and ordinary individuals in our model by supposing that they form a special sort of basic entity in the domain of discourse.

Finally, there has always been the need for something like event types in scientific theorizing, of course. For instance, probabilities in mathematics are real-valued functions that are defined on an algebra of "events" (that's what the textbooks say). But such events are certain sets that rather model *types* of events, e.g., the generic event of a coin coming up heads. Probability theory is not concerned with particular events.

Now what is the relationship between particular events and their types? What is at issue here is the relation of *subsumption*. We classify events as belonging to one type or other. This relation closely resembles the more common relation of predication in which an individual falls under a property. It is more general, however, in that not only individuals but possibly rather complex parts of the world can be subsumed under an event type. Using this idea a generalized truth scheme can be developed which goes back to John Austinand taken up again in Situation Semantics. In his paper *Truth* (1950) Austin distinguishes two kinds of linguistic conventions: *demonstrative conventions*, which we use to refer, in our case, to particular, "historical" events, and *descriptive conventions*, which serve us to characterize a certain type of event. Then the truth scheme for a statement S is roughly this: when S is uttered at a certain occasion then the demonstrative conventions help us to retrieve from the circumstances of the given utterance the particular event e that the statement is about. The truth claim is that e is subsumed under, or an instance of, a certain event type

θ; θ is in turn read off the specific form of S by means of the descriptive conventions. We will see in the next section how this idea can be made formally precise.

Here and in the next chapter I am going to explore a close-to-language conception of the relationship between events and their types, which was inspired by Situation Theory and which, by traditional standards, would no doubt be called highly intensional. In Chapter 12 I'll try to develop a more balanced view; but the main perspective is, as always, algebraic.

Events, according to the way they are understood here, are rather like *aspects of the world.* When the editor of the *Vieux Cordelier*, Camille Desmoulins, is executed together with Danton and the others, then this event makes his wife Lucile a widow. Yet the execution and Lucile's becoming a widow are two different aspects of the world and hence two different events in our sense. Recall the trip to Varennes: it had been carefully prepared for months, and was planned as a flight, and as such it was certainly an action that was taken deliberately by Louis XVI; let us further suppose for the sake of example that the story of his abduction were true, and Louis had unintentionally stepped into a trap. Then one and the same action (the trip) was performed both intentionally and unintentionally. This dilemma follows a pattern that is much discussed in the literature: Coming home I intentionally flip the light switch, but unintentionally thereby alarm the burglar who searches through my belongings; or I intentionally lift my arm as part of my physical exercise, thereby unintentionally giving a signal. I analyze the situation like this. There are a number of constraints that connect two aspects of the world like the lifting of the arm and the signaling that comes with it under certain circumstances. Now a given part of the world that contains a particular event of lifting in those circumstances must also contain the signaling since world parts are closed under constraints, as I mentioned earlier. Given this it is possible for me to direct my intentions towards a certain event e that I bring about without paying attention to all the aspects of my action that are connected with e via the obtaining constraints. This is the "logic of excuses" as Donald Davidson once called it: you admit to have brought about e intentionally but claim you were unaware of the constraints that link e to a certain e' which was realized together with e but which entails sanctions of one sort or other.

10.4 Algebraic Semantics for Events

10.4.1 Modeling

In this section I proceed to sketch a model structure for events that summarizes in a more formal way the features of the notion of event discussed so far. But before I do this I want to say a few words about the choices that are available in giving a representation for events. This is the *modeling problem* that is a continuous theme in this book. The usual modeling tools in analytic philosophy and linguistic semantics are the concepts of set theory, usually supplemented by a collection of *urelements* or individuals, which are not sets and are thought of as simple objects of one kind or another. This is the conception of Tarskian semantics. Properties are represented by sets of individuals, and the predication relation is modeled by the epsilon relation. Now in the case of events there are two main branching points in the representation process. The first concerns the question whether to take events as urelements or to construct them set-theoretically from other objects in the universe of discourse. For instance, I mentioned above that Montague defined events as properties of moments of time. Properties of objects of a certain kind are intension functions from possible worlds into sets of those objects; thus the Montagovian notion of event presupposes moments of time and possible worlds as urelements in the ontology. Events are thereby *reduced to* those other entities. The problem with such a reductive approach is always whether the the typical properties of the entities being reduced are faithfully represented by the construction given for them. For a discussion of the merits or drawbacks of the reductive approach the reader is referred to the following two chapters. Here I simply opt for taking events as urelements. But then there is the second branching point coming up: how are we going to model event types? When event types are likened to properties of events then again we have to decide how to treat properties of individuals in general. The *extensional* approach is to take them as sets of individuals. This choice is the usual one, in particular in theoretical contexts that are not specifically philosophical. In the case of events, for instance, event (types) in probability theory are sets of possible outcomes. Also in linguistic semantics many event-based approaches take types as sets of events. Possible-world semantics of the Montagovian variety, while ceasing to be extensional, still models the event types set-theoretically, for instance as the kind of intension functions mentioned above. The third option is again the urelement option, which is chosen here.

The important point now is that it will not do to simply declare that events and event types are urelements. That would amount to giving no theory at all. Rather we have to say something about the way events and their types behave. That is, we have to set up rules or postulates that capture their behavior. Here the algebraic perspective comes in again. Structure is not internal, but external to the urelements; thus the various subdomains in our ontology carry a structure. For instance, regular individuals consist of individual sums that carry the kind of (semi)lattice structure that was described in the earlier chapters. Similar algebraic structures for space-time regions, events and their types will then express the information about the network of relations both among the various kinds of entities and across them. Thus the slogan is: *A detailed structuring of urelements replaces set-theoretic modeling.*

10.4.2 The Aether Model

By way of summarizing the discussion of this chapter I am now going to describe the various components of the model structure. A model structure is the sort of thing in which the formal representations of pieces of discourse are interpreted. It does not contain the interpretation function that maps the representations into semantic entities built up from the model structure. So it is just an elaboration of the domain of discourse D in the usual Tarskian semantics. To begin with, we have a multi-sorted domain instead of Tarski's universal set D; we distinguish regular individuals, events, stretches of time, regions of space, and event types. Accordingly, there are five different sorts of entities that are collected in pairwise disjoint, non-empty sets: the set A of individuals, the set E of particular events, the set T of time stretches, the set H of spatial regions, and the set $\mathcal{E}$ of event types. Recall that I counted plural individuals among the individuals. That means that the set A carries the kind of algebraic structure that was developed in the theory of plurals in previous chapters of this book. For definiteness, let us assume that it is an *complete atomic Boolean algebra without zero element*. Since Boolean algebras are particular lattices I call this structure a *CAB^- lattice* for short.[2]

I skip the set E for the moment and discuss the set T of time stretches first. Typically, we think of time stretches as *intervals* in a linearly ordered set, i.e., sets which contain with any two points any other point lying between them. Since time stretches are to serve as temporal traces of events we will have reason to consider the case where a trace is not an interval, for instance, the trace of a complex sum event. That is, we would like to be able to form sums or unions of time stretches as well. But if T is a set of intervalls then it will not be closed under the formation of unions since the union of two intervalls is not, in general, an intervall. By contrast, the

[2]See the Appendix for the relevant technical concepts of lattice theory. In Chapter 6 CAB^- lattices are simply called *plural lattices*.

union of two unions of intervalls is a union of intervalls again, so we do have a simple (join) semi-lattice structure on unions of time intervalls. Thus a natural candidate for the elements of the set T would be such unions of intervalls in a linearly ordered set, say the set $\Re$ of real numbers. Now the algebraic approach is not to presuppose a certain ready-made structure like $\Re$ since it may contain information that is not needed for the modeling purpose at hand (for instance, the completeness of $\Re$ or even its denseness). Rather, we start out with the simple semi-lattice property and gradually bring up stronger axioms which have to be tested in turn for qualifying as necessary conditions for the notion of temporal trace of an event. A further axiom that certainly does quality is the closure under meets since we need the notion of temporal overlap; so in addition to its being a join semi-lattice T should be also be a meet semi-lattice, which makes it into a lattice *simpliciter*. If we add the natural requirement of closure under *arbitrary* joins and meets we arrive at a C *lattice*, i.e., a complete[3] lattice of objects we call "time stretches." Thus T provides sums, but also meets of time stretches. If such a meet is non-zero we speak of *temporal overlap*. Is T also a plural lattice, then? Not quite. To begin with, every complete lattice has a zero element, so there is the "empty" time stretch. That is perhaps not a proper trace, so we could throw it out and go over to a C^- lattice. However, we still have to decide the questions of *atomicity* of T and its *Boolean* character before we have a regular CAB^- lattice. As to the first, one might think that atomicity is not called for since for every stretch of time which is non-zero there is always a non-zero time stretch contained in it that is strictly smaller (this property is sometimes called *divisibility*). If we ignore the granularity aspect then divisibility is certainly a property to opt for, but otherwise we might think of certain minimal stretches of time of varying size according to the granularity parameter that comes with a given event. This idea needs to be elaborated on, of course, but I won't do it here.[4] Next we have to discuss whether the lattice T has also the Boolean property, that is, whether it is distributive and complemented. While distributivity is certainly a desirable property,[5] it is not clear whether we need to postulate the existence of a complement for every time stretch. The reason is that the existence of complements for events, to which I turn presently, is in doubt: if events are not closed under complements their temporal traces don't have to be either. So let us settle

[3]Complete now in the sense of the lattice structure; this has to be distinguished from the completeness of the real numbers.

[4]For some indication of how one could proceed, using an inverse system of lattices, see Chapter 11; it is partly for this reason that the Aether model in that chapter assumes atomicity for the both the lattice of temporal traces and the lattice of spatial traces.

[5]It excludes, for instance, the existence of "diamonds" in the sense of Grätzer (1978); see the Appendix.

here with T as a CD^- lattice ('D' for "distributive"). Now similar considerations should apply, *mutatis mutandis*, to the set H of spatial regions, so we simply postulate the same CD^- lattice structure for H. Of course, we still have to impose conditions that enable us to tell temporal stretches from spatial regions, so T and H should not just be isomorphic copies of one another.

Next comes the set E of events. I tried above to motivate the introduction of complex events as mereological sums of smaller events. So we are dealing again at least with join semi-lattices, with the part relation taken in the complexity sense (I called it c-part). But we might also want to refer to a common c-part of two complex events, for instance, the sum of events that is common to the French Revolution and Napoleon's career. Then we have a full lattice again, and by a reasoning similar to the case of the other sorts of entities we might suppose that E is a CD^- lattice. The question of atomicity is a hard one. I assume it here but will defer a discussion of it to Chapter 12. Complementedness, i.e., the full Boolean character of events, is equally problematic since it is far from clear what, say, the complement of Robespierre's dying is: do we have to close this event under the obtaining constraints before we go over to the complement, and if so, is that complement "the rest of the world?" Then it is certainly different from the local event (or state) of Robespierre's not dying, which, however, would be a more accessible opposite event. Thus, while elementary events come in pairs of opposite events that are accessible by negating the verb phrase in their description, those pairs seem to be "local" and not global opposites. Furthermore, there are no obvious opposites of larger event sums. That suggests to leave out the postulate of complementedness.

Here a significant formal difference emerges between events and propositions. Propositions are Boolean, i.e., they are closed under complements. They are typically modeled by some kind of Boolean algebra, for instance, the power set algebra over a certain basic set X or more generally, a field of sets over X. Thus the standard model for the set of propositions is the power set $\mathcal{P}W$ over a set W of possible worlds. Then the proposition that Robespierre died on 9 Thermidor is the set of all worlds at which this is true. Here there is of course a well-defined complement to this: the set of all worlds at which Robespierre didn't die on 9 Thermidor (the real world being among them, incidentally: he was executed a day later). But notice that events are part of the real world and not some possible world; there has to be a real process going on somewhere, and it is precisely the question what a real complement process might consist of.

The final set of entities is the set $\mathcal{E}$ of event types. I said that event types as conceived here are highly intensional entities, that is, they are closely correlated with their descriptions. So I speak of elementary event types like

WALKING and READING. More complex ones are then built up from these by certain structural rules, which I won't specify here; they are given in the next chapter. A typical example would be the event type of delivering a speech to the National Convention during the French Revolution, which was instantiated a great number of times by various particular events, with different individuals filling the role of the speaker.

Now that I have described the kinds of entity that are part of the model structure I have to specify formally the various relations across these sets of entities. I just mentioned the role of speaker, which is the agentive role; it takes certain events as arguments and returns an individual as value. Let us call the agentive role ρ_1, and let e be the event of Robespierre delivering a certain speech; then $\rho_1(e) =$ Robespierre. So in general ρ_1 is a function from the set E to the set A. But since not every event has the agentive role defined, ρ_1 is only a partial function; similarly for the other roles that were mentioned above. Thus I add to the model structure a finite set R of *partial role functions* ρ_i from E to A. Next there are the two trace functions, τ for the *temporal trace function* and σ for the *spatial trace function*. τ is a function from E to T, and σ is a function from E to H. I indicated above that σ is only a partial function on E because E is taken to contain states of affairs that obtain or do not obtain, but don't obtain in any particular region of space; so σ isn't defined on states. One might also ponder whether to count, say, mathematical facts among the states, and those shouldn't even be assigned a temporal trace. If we agree to include them then the temporal trace function τ, too, is only a partial function. It seems reasonable to assume, however, that wherever σ is defined, so is τ, that is, there are no events with spatial, but no temporal traces.

Above I hinted at a certain homomorphic relation between events and their traces. Consider two events, e and e', and their temporal traces, $\tau(e)$ and $\tau(e')$. Denote the sums of these pairs of entities by $e \sqcup e'$ and $\tau(e) \sqcup \tau(e')$, respectively. Now if τ is a join homomorphism with respect to the summing operation $\sqcup$ then it "commutes with" $\sqcup$, that is, we have the following equation:

(1) $$\tau(e \sqcup e') = \tau(e) \sqcup \tau(e')$$

It says that the temporal trace of the sum event $e \sqcup e'$ is the sum of the temporal traces of its parts e and e'.[6] τ can also be seen to commute with the meet operation on events, so τ is a *lattice homomorphism*. Since I assumed lattice completeness for both the domain and the codomain of τ it is natural to extend completeness to the notion of homomorphism, that is, I assume that τ not only respects finite sums and meets but also

[6] For the notion of homomorphism, see the Appendix, and also Chapter 12.

arbitrary ones. That turns τ into a *complete lattice homomorphism*. The same condition is imposed on the spatial trace function σ.

There is one final element missing to our model structure. Recall that since we decided to take event types as urelements we have detached their representation from the kind of things that are subsumed by them, viz. the events. In an extensional set-up, event types would simply be sets of events, and being of type θ would mean being an element of the set representing θ. That option is not available now, so we have to posit a special *subsumption relation*, called π, between events and event types, which replaces the $\in$-relation. Actually, I shall only need to posit the relation between events and *elementary* event types; it can then be extended to arbitrary event types by a recursive definition like the one given in Chapter 11. But notice that a rather complex event might be taken by the model structure to be of some elementary event type. This feature is intended to reflect the fact that an act of walking, for instance, is actually a rather complex process, with lots of events going on that are part of the walking but that are not "mentioned" in a description of it. One of the advantages of the algebraic approach is that we don't have to specify the precise makeup of such an event, and I even doubt that we ever can; all we have is that the event classifies as being of a certain event type.

We can now put all the pieces together and arrive at the general form of the Aether model structure:

(2) $$\Im = \langle A, E, T, H, \mathcal{E}, R, \pi, \sigma, \tau \rangle$$

An *Aether model structure*, then, is a 9-tuple $\Im$ of this kind that satisfies all the conditions discussed above. These conditions are part of the formal development of Chapter 9 where a full semantics based on Aether models is given for some sample fragment of English. Chapter 10 contains a philosophical discussion of the issues involved in an event ontology.

Chapter 11

Algebraic Semantics of Event Structures

11.1 Introduction: The Project of Algebraic Semantics

A semantic theory is, like other theories, a means of representing a certain domain of phenomena in such a way as to bring out and characterize the network of structural relationships that govern their behavior. There are two prominent ways to go about such a representation: the first is that of giving a set of *axioms* describing those relationships, e.g., Peano's axioms for the natural numbers; the trouble is that in general the axiom system is not able to exclude unintended models (e.g., non-standard models). The other way directly provides a concrete *model* for the phenomena to be represented. A typical example here is von Neumann's set-theoretic representation of the natural numbers where the '$<$' relation is modeled by the membership relation. There are several problems with this approach, too. The objects that model the phenomena must not be mistaken for the phenomena themselves; and even if the model is adequate it imports a number of *artifacts* of the representation that have little to do with the phenomena described: we have, for instance, the spurious fact $2 \in 5$ in von Neumann's model of the natural numbers.

In model-theoretic semantics the usual picture is that as modeling tools at our disposal we have set theory with urelements (the intended domain of individuals). All the structure to be described is then modeled by suitable set-theoretic constructs over those urelements. Even Montague, when he emphatically proclaimed the new philosophical age of intensional logic (Montague 1974: 154f.), didn't really depart from this methodological track; events, for instance, he construed as intension functions of moments of time, thus merely extending the urelements by adding times and worlds. The structure of events, then, is to be found in those set-theoretic functions. In spite of its formal beauty, the inherent weakness of this approach was soon uncovered: the vast variety of structures of "real world" objects couldn't be captured by the uniform shape of functions from possible worlds to individuals.

To get at what I call the *Project of Algebraic Semantics*, consider plural individuals as a very simple example of objects that have to be represented in natural language semantics. In my own work on plurals (Link 1983; Link 1986 and references therein) I did not give a set-theoretic model for those individuals; rather, I extended the domain of urelements so as to comprise usual individuals along with plural objects. The structure otherwise inherent in a set-theoretic model is here expressed in purely *algebraic* relations among the urelements. As a result, the familiar domain of individuals in Tarskian semantics, which is simply a non-void set of objects, is replaced by a *relational system* in the sense of Universal Algebra: in addition to the basic set of individuals, such a system may contain operations on and relations between the elements of this set. In the theory of plurals, for instance, the relational system is basically a semilattice structure.

Algebraic semantics, then, stays away from extensive set-theoretic modeling in favor of the axiomatic approach. Writing axioms for the objects to be represented comes down to a characterization of these objects in algebraic terms. The purpose of the present chapter is to extend this approach to the theory of events.

The admission of a category of events to the domain of discourse is presumably not something these days that has to be particularly argued for. So let me merely recall some of the advantages of having events at our disposal in a semantic theory. Events can be named, counted, quantified over, and anaphorically referred to. The literature on verbal aspect even suggests that not only should events be counted among the objects of a natural language ontology, but semantic theory should be *based on events*. A familiar argument for that decision is the fact that the aktionsart of a sentence is not fully determined by the denotation of its verb phrase: while *arrive* is an achievement verb, which is incompatible with a durative adverbial phrase like *for three hours*, a plural subject transforms the achievement into a complex activity, witness the sentence *buses have been arriving at the airport for three hours*. What expresses a certain type of event, then, is nothing less than whole sentences.

The idea of a semantics that is based on events, however, requires a rethinking of the familiar model-theoretic framework. Should everything under the sun be construed as an event? Authors like Richard M. Martin (1978) seem to think so. His theory, however, is developed in a reductionist spirit deriving from modern nominalism. This is not the avenue that should be followed here I think. The elaborate system of ways in which language is able to refer to the familiar individuals strongly suggests that these individuals should be kept in our domain of discourse. Rather, "event-based" is to be understood as an alternative to "property-based": individuals are not primarily thought of as having properties or entering relations, but *playing roles in certain events*. Thus, an event-based semantics as it is understood here describes the interplay between individuals and the events that they are part of.[1] Recent work in linguistics has shown that within such a framework important generalizations can be captured that elude formulation in a more traditional setting (see, e.g., Krifka 1986, 1987). As a case in point, the well-known dichotomy between cumulative and discrete reference in the nominal domain finds its close parallel in the distinction between cumulative and "quantized" events. In fact, homomorphic relations can be established between individuals and events that show how the "referential type" (die *Referenzweise* in Krifka's wording) is transferred from one domain to the other.

An immediate consequence of a set-up along these lines is the need of a reformulation of the elementary rule of truth. What is now the condition under which, say, the sentence *John is running* is true? The work in situation theory (Barwise and Perry 1983; Barwise and Etchemendy 1986) has reminded us of an important truth scheme introduced by Austin (1961) that can provide a suitable answer to this question. Austin says that a declarative sentence S, when used assertively, contributes two things: the descriptive conventions of language yield a certain *type of event* θ that is expressed by S, whereas the demonstrative conventions refer to an actual, *"historic" event* e. The rule of truth, then, is simply that e *is of type* θ. Thus, the sentence *John is running* describes that type of event θ which is a running and has John as agent; this is a schematic event or event type since, for instance, the time of running is not specified. When a speaker truthfully utters that sentence he refers to an actual happening, one in which John is running, and asserts that this happening is of type θ, i.e., can be classified as an instance of a running of John.

If we adopt the Austinian scheme, how are we then to model types of events? Parsons (1980) and Krifka (1986,1987) choose sets of actual events, which looks like the natural option. Notorious problems with the so-called imperfective paradox, however, have convinced me that we have to go more intensional here. Following the spirit of Barwise and Etchemendy

[1]Obviously, events have to be conceived broadly enough so as to cover states like *John is intelligent*; see below.

(1986) I therefore further extend the domain of urelements by adding event types apart from events. Truth is then to be explicated in terms of an inductively defined relation between events and event types. Here is a plausibility argument in favor of admitting event types as further objects in our ontology: sequences of event types are what I like to call *scenarios*; in a world where the options of actions are tested by computer simulation it should be obvious how "real" those scenarios are: they figure prominently in economic as well as political decisions.

There are a number of problems that any theory of events has to deal with; let me mention what seem to me the most important ones: (i) individuation, (ii) negation, (iii) quantification, (iv) minimal parts, (v) the imperfective paradox. With respect to (i), some questions at least can be answered within the algebraic approach. Consider, for instance, the event of John and Bill hitting each other; is that one event or two? It looks like one single happening, and yet, there seem to be two different acts of hitting involved, with the same participants, but in reciprocal roles. Now the lattice structure on the set of events that will be introduced below allows us to say that the event is *both* one and two: it is the unique sum event consisting of those two hitting events. Two seemingly conflicting intuitions about individuation can be reconciled this way. It is assumed here, of course, that we have to start with some class of "simple" events to which we have easy access. Basically, these are the simple facts of the form *John hits Bill*. Such an event has defined roles and roughly corresponds to what John Perry calls an *aspect* (Perry 1986). I shall adopt his term together with the term *chunk* he contrasts it with, but give them a technical meaning in the present framework; while a chunk is going to be *any* sum of events in the lattice, an aspect is either a simple fact (an atomic event) like *Bill left* or a sum of atomic events that are of the *same type* like *every student left*.

Negation poses another conceptual problem for event theory. When events replace properties in our ontology then it is hard to say what kind of event a negative sentence like *John didn't come* is about: nothing happened, and in particular, there was no role for John to play. The first solution that comes to mind is to treat negation as *denial*: then the type of event that is expressed by *John didn't come* applies to all events that do not overlap with any event of type *John came*. The problem with this solution is that it presupposes the possibility of percolating every negation, regardless how deeply embedded, up to the highest sentence level. But this trick is doomed to fail: consider the sentence *every student didn't work for three hours*; this is simply not the same as *not every student worked for three hours*. Thus, the existence of durative adverbials blocks the denial interpretation of negative sentences. In order to conceptually avail ourselves of a recursive negation I propose to consider *John didn't come* as expressing a negative type of *state*, the state characterized by John's not coming. Linguistic evidence supports this move: while an achievement verb like *come* cannot be combined with a durative adverbial phrase like *for three hours*, the negative sentence *John didn't come for three hours* is fine. Thus it is only when conceived in a narrow sense (as happenings, say) that events are not closed under negation. In what follows events are taken to comprise telic and atelic processes as well as states; the latter are needed anyway for generally representing all non-dynamic states of affairs.[2]

Next I like to take up the *minimal parts problem*, which is familiar from the study of mass terms (see Bach 1986); it reappears in the present context of event structures. Consider the event of Mozart's death, referred to by the sentence *Mozart died on the 5th of December, 1791*. In the formal treatment this event is going to be represented as an atomic element e in the lattice of events, since *die* is an achievement verb denoting a durationless change; no non-zero event can be a proper part of e. There is, however, a level of description on which Mozart's death is composed of a more or less complex series of events, e.g., the physical processes involved in his dying. The formal problem here is how those intuitive subevents are to be located in the lattice structure of events with respect to the (already atomic) event e. Or consider two other, less macabre examples.

(1) a. Immediately after (landing/his return from the summit meeting) the president conferred with his advisers/?opened his seat belt.

b. Immediately after touch-down the president opened his seat belt/?prepared a speech.

(2) Soon the polarity of the earth's magnetic field will be reversed (i.e., in 2000 years).

The point here is this. Conjunctions like *immediately after* should connect two atomic events where there cannot be a third one in between. But the president certainly must have opened his seat belt before conferring with his advisers (who are not on the plane, say). In the second example, the pace of the kind of processes that are on the same level as the reversal of the earth's magnetic field must be such that a time lapse of 2000 years can be called "soon."

Observations like these suggest that a single lattice E cannot represent the multi-layered structure of events in a realistic way. Since the model

[2]In Situation Semantics, the very existence of simple facts with a negative polarity (viz., $\langle\langle \text{COME}, \text{JOHN}; \text{no}\rangle\rangle$) shows that a similar decision regarding negative events has been tacitly made.

presented below is already quite complex I shall only indicate here how it could be extended in such a way as to resolve the minimal parts problem. A whole system of lattices $(E_i)_{i \in J}$ replacing E can be introduced, which is indexed by a partially ordered set J. Each element of J represents a certain *granularity* of the events in the corresponding lattice: the events in E_i are more fine-grained than those in E_j for $i \leq j$. A family of mappings between the E_i is added to model the fact that the E_i are but different conceptualizations of the same realm of phenomena. Formally, we have an *inverse system* in the following sense: for all $i, j \in J$ with $i \leq j$ there is a homomorphism h_{ij} from E_j into E_i such that (i) for all $i \in J$, h_{ii} is the identity map on E_i, and (ii) for all $i, j, k \in J$ with $i \leq j \leq k$, $h_{ik} = h_{ij} \circ h_{jk}$. Thus, a certain event can be atomic in a coarser domain and still be the image under a h_{ij} of a complex sum of events in a more fine-grained E_j.

The granularity of events has to be treated as a discourse parameter. A piece of discourse might evolve at a certain granularity which can be lowered and raised by appropriate linguistic means. Familiar narrative structures with their characteristic changes of level can be captured this way.

I would like to conclude this introduction with some remarks on the imperfective paradox. How can a sentence like *John is building a house* be represented in a way which does not presuppose the existence of a house that John built? Using the notion of granularity the following suggestion could be made: on a more fine-grained level the event type of building a house really consists of a whole scenario $\langle \theta_1, \ldots, \theta_n \rangle$ of subevent-types θ_i such that θ_i has to temporally precede θ_j for $i < j$ $(1 \leq i, j \leq n)$. Then the event that John is building a house simply means that it can be matched with an initial segment $\langle \theta_1, \ldots, \theta_k \rangle$ of the above scenario with $k < n$, which could mean, for instance, that the house is only half-way completed. Since a scenario as a sequence of event types is an intensional entity that condition would not imply the existence of a house, as desired. But this suggestion has a serious drawback:[3] it hinges upon the assumption that every accomplishment type of event has some kind of standard "definition" which can be expressed by a scenario of the above kind. Now consider the sentence *Mary made John a millionaire*; there is really no standard way of making somebody a millionaire: the only thing that counts is final success! I shall not pursue the issue further here but leave it with the following rather weak suggestion due to Jon Barwise, which can in a similar form be also found in Parsons (1980). There is the clear intuition that if John has built a house there was a time when he had the property of building a house. This connection could be formulated as a *constraint* on types of event. The converse, however, does not hold: if John has the *property of building a house*, there is no state

[3]This objection, together with the example, is due to Barbara Partee.

in which John has built a house.

11.2 The Model Structure

I now proceed to present in turn the various elements of the event-based semantics announced above. In what follows I shall assume a basic model structure of the form $\Im = \langle A, E, T, H, \mathcal{E}, R, \pi, \sigma, \tau \rangle$ (the "Aether model"). Here, A is the set of (ordinary) *individuals* $a, b, \ldots$ of $\Im$ with parametric variables $\mathbf{a}, \mathbf{b}, \ldots$; E is the set of *events* $e, e', \ldots$ of $\Im$ with parametric variables $\mathbf{e}, \mathbf{e'}, \ldots$; T is the set of *time stretches* or simply *times* $t, t', \ldots$ of $\Im$ with parametric variables $\mathbf{t}, \mathbf{t'}, \ldots$; H is the set of *regions of space* $h, h', \ldots$ of $\Im$ with parametric variables $\mathbf{h}, \mathbf{h'}, \ldots$; $\mathcal{E}$ is the set of *types of event* $\theta, \theta', \ldots$ of $\Im$; R is a finite set of *roles*[4] $\rho_i : E \rightharpoonup A$ (e.g., agent, patient, etc.), which are partial functions from E into A;[5] π is the basic relation *"is of type"* of $\Im$ between certain events and event types; $\sigma : E \rightharpoonup H$ is the *spatial trace function* and $\tau : E \rightharpoonup T$ the *temporal trace function* of $\Im$, with $Dm(\sigma) \subseteq Dm(\tau)$.[6]

These elements of the model structure are further specified in the following way. The sets $X := A, E, T, H$ are complete atomic lattices, with lattice operations $\vee, \wedge$, the infinite join $\bigsqcup$, and the intrinsic ordering relation $\leq$; the set of atoms in X is denoted by X°. The lattice structure on A is the plural structure that was introduced in Link (1983). The extension of this structure to events was proposed by Bach (1986), Hinrichs (1985) and Krifka (1986); unlike in Hinrichs' work, no stages are considered here. The trace functions σ and τ are complete homomorphisms, i.e., they are compatible with the formation of arbitrary joins and meets.

There is a 2-place relation S defined on the set E° of atomic events: eSe' means "e specifies e'." An example is *(John reads 'Gravity's Rainbow') specifies (John reads).* S is a partial order such that if e_1, e_2 specify e they are both specified by a common e'. The idea here is that it is not pre-

[4]Jon Barwise pointed out to me that these roles can be viewed as uniformities across the roles used in situation semantics. He also argued for preferring the latter notion to the one introduced here, basically on the grounds that the specific character of the roles is quite idiosyncratic of the relation at hand. Granted this, there still remain important uniformities that justify the more general use of the notion; see, e.g., the axioms which (partly) characterize *effected* and *affected patient objects* in Krifka (1986,1987).

[5]Here and in what follows, '$\rightharpoonup$' is to indicate a partial function.

[6]This qualification is to reflect the fact that whenever an event has a spatial trace, i.e., a region of space where it occurs, it also has a temporal trace; on the other hand, stative events like that of John loving Mary, while certainly equipped with a temporal trace, do not seem to be located in space. Finally, also τ is considered to be a partial function since there are facts (e.g., those of mathematics) that are neither in space nor in time.

supposed that an event is always and inherently specified in every respect; but if *John reads 'Gravity's Rainbow'* specifies a particular event of John's reading, and *John reads at midnight* specifies the same event, then there is an event, e.g., *John reads 'Gravity's Rainbow' at midnight*, specifying both *John reads 'Gravity's Rainbow'* and *John reads at midnight.*

Furthermore, it is natural to assume a linear order $\leq_{\circ}$ on the set T° of basic times, with its strict version $<_{\circ}$. This induces a strict order $<$ on T:

(Def) $t < t' \quad :\Leftrightarrow \quad \forall s, s' \, [\, s, s' \in T^{\circ} \,\&\, s \leq t \,\&\, s' \leq t' \Rightarrow s <_{\circ} s' \,]$

That relation in turn gives rise to the *precedence relation* $\ll$ between events:

(Def) $e \ll e' \quad :\Leftrightarrow \quad \tau(e) < \tau(e')$ ("e temporally precedes e'")

Temporal overlap between events is also defined in terms of the temporal trace function.

(Def) $e \circ e' \quad :\Leftrightarrow \quad \tau(e) \wedge \tau(e') \neq \mathbf{0}$

A similar relation can be defined for spatial overlap. Note that we have $e \wedge e' \neq \mathbf{0} \Rightarrow e \circ e'$, but not the other way around. Finally, I assume the existence of *additive measure functions* $|\cdot|_{\alpha} : T \rightharpoonup \Re$ whose axioms I shall not stop to specify (α stands for **min** [minutes], **h** [hours], **d** [days], **a** [years], etc.; $\Re$ is the set of real numbers).

The roles ρ_i are necessarily partial functions since obviously, not every event has a patient, for instance. But there is another reason for partiality. A sum event or chunk which is composed of heterogeneous aspects with lots of patients also has no defined patient role since uniqueness fails. If a sum event, however, consists of atomic events which have all a certain role defined, and, in addition, this role is filled by the same object, then the role is also defined for the sum event, with the same value. That is the content of the following axiom.

(Ax) Let $\rho_i \in R$, $e \in E$, $a \in A$; then:
$e \in Dm(\rho_i) \,\&\, \rho_i(e) = a$ iff
$\forall e' \, [\, e' \in E^{\circ} \,\&\, e' \leq e \Rightarrow e' \in Dm(\rho_i) \,\&\, \rho_i(e') = a \,]$

For further axioms that specify the roles I refer to Krifka (1986,1987). The axioms are needed there to determine the referential type of an event as a function of the referential type of objects playing roles in that event. With their help it can be proved, for instance, that the cumulative reference of *wine* leads to the cumulative reference of *drinking wine*, whereas the quantized reference of *a glass of wine* makes *drinking a glass of wine* quantized, too.[7]

We will need to have a name for the set of all objects playing a role in an event. It will be called a *cast.* Let $r(e)$ be the set of all roles that are defined for $e \in E$. Then the *cast of* e, $CAST(e)$, is defined as follows.

(Def)
$$CAST(e) := \{ \rho(e) \mid \rho \in r(e) \} \qquad \text{for } e \in E^\circ$$
$$CAST(e) := \bigcup \{ CAST(e') \mid e' \in E^\circ \ \& \ e' \leq e \}$$

I proceed to characterize the important notion of an event type; this will be done inductively.

(Def) A *type of event* $\theta \in \mathcal{E}$ is of one of the following forms.

1. θ is an element of a given poset[8] $\langle \mathcal{E}^\circ, \leq^\circ \rangle$ of *atomic event types* and their opposing *negatives* (denoted by $-\theta$); examples are READING, WRITING, −READING, etc.; $-(-\theta)$ is the same as θ;

2. θ is an *elementary event type*, i.e.,

[7] A comment is in order on the relation between the lattice structure for events in Krifka (1986) and the one intended and described here. Let w and w' be two disjoint portions of wine making up the contents of a glass of wine, and consider the events e of John's drinking w, e' of John's drinking w', and e^* of John's drinking the sum $w \vee w'$ of the portions w and w' (objects like portions of wine are supposed to form a semi-lattice, which is not at issue here). Now Krifka seems to think that the sum event $e \vee e'$ equals e^*, but according to the present framework e^* is just another single atomic event, whereas $e \vee e'$ is the sum of two atoms which cannot itself be an atom again. Is there a redundancy here? I think not; the sum $e \vee e'$ models the collection of *two* acts of drinking, whereas e^* is just one, albeit protracted, act of drinking. The question cannot be whether $e \vee e'$ and e^* "really" are identical happenings—it all rather depends on our capacity to pick out different aspects of the processes around us: so in one case we might say "John drank twice," referring to the sum event $e \vee e'$, and in another "John drank the whole glass", stressing that he consumed the sum portion $w \vee w'$. The connection between the two events can be expressed in terms of the patient role function (call it ρ_2): $\rho_2(e^*) = w \vee w' = \rho_2(e) \vee \rho_2(e')$. By the same token, two consecutive (atomic) runnings e, e' might either be viewed as another atomic running e^*, where $\tau(e^*) = \tau(e) \vee \tau(e')$, or as the sum $e \vee e'$ of the two atomic runnings. It follows from this observation that (i) the sense in which activities like *running* or *drinking wine* are cumulative is *not* represented by the lattice structure on events that is assumed here, and (ii) the homomorphic property of ρ_2 and τ does not express the "referential transfer" between the activities and the domain of affected objects towards which those activities are directed; thus the structure on E is different from Krifka's event lattice. However, within the set E° of atoms of E a whole array of other lattice structures may be defined (one for each atomic activity) which does capture the aspectual phenomenon of referential transfer that Krifka is concerned with. Now although an explicit account of the aspectual classes and their interrelations is a natural desideratum for any theory of events, it will not be given here, but has to be deferred to another occasion.

[8] Poset = partially ordered set.

2.1 θ is an atomic event type; or

2.2 $\theta = [\,\mathbf{e}; Sp(\theta)\ \&\ \theta_1(\mathbf{e})\,]$, where
$Sp(\theta) := RC(\theta)\ \&\ Re(\theta)$ is the *specification* of θ, with
$RC(\theta)$ denoting the *role conditions of* θ (e.g., $\rho_i(\mathbf{e}) = \mathbf{a}$), and
$Re(\theta)$ denoting the *restrictions of* θ (e.g., BOOK($\mathbf{e}$)); and
$TC(\theta) := \theta_1(\mathbf{e})$ is the *type condition of* θ, with θ_1 elementary;

2.3 $\theta = \sum(\theta_1 \mid \theta_2)$, with θ_i elementary ($i = 1, 2$);

2.4 $\theta = \theta_1^*$, with θ_1 elementary (the *dual* of θ_1); $(\theta^*)^*$ equals θ;

3. θ is a *conditional (event type)* $\theta_1 \supset \theta_2$, with θ_i elementary ($i = 1, 2$);

4. $\theta = \triangle\{\,\theta' \mid \theta' \in \Theta\,\}$ or $\theta = \bigtriangledown\{\,\theta' \mid \theta' \in \Theta\,\}$, where Θ is a set of elementary event types or conditionals ($\triangle$ stands for the *conjunction*, $\bigtriangledown$ for the *disjunction* of types).

The semantic rules for the satisfaction relation to be defined below will make it clear why the event types are categorized just the way they are. Basically, the relation of an event being of type θ will only be defined for elementary θ, while truth is defined for all event types. Conditional event types (which formalize donkey sentences, for instance, or more generally, constraints) are not considered as elementary, since otherwise we would have to be able to say what it means for an event to be of type $\theta_1 \supset \theta_2$, or, worse, of type $[\theta_1 \supset \theta_2]^*$. Also, an event cannot be of type θ, where θ is a disjunction of types, $\bigtriangledown\{\,\theta_1, \theta_2\,\}$. Thus, rather than following the typical recursion on the usual logical symbols, the structure of event types is closely modeled after natural language. Elementary event types of form 2.2 correspond to simple (positive) predication in the highest node involving only singular NPs (like names or indefinite NPs), whereas sum event types of form 2.3 model (positive) sentences with general NPs, like *every student reads a book*. Negation is represented by the dual star, '*'.

The atomic event types under 1 are the ones which all the others are built up from. They come in positive and in negative form; this reflects our above decision to treat negation recursively. While, say, ARRIVING is an achievement event type, $-$ARRIVING is still an admissible event type, though not an achievement anymore. An event being of a negative (atomic) type, should not be located in space (it seems meaningless to ask in which region of space John didn't arrive at Munich airport), so the spacial trace function σ is not defined for it. That distinguishes such an event from an activity like RUNNING; together with the observation that negative types are compatible with durative adverbials this shows that we were right in regarding those types as states. — The partial order $\leq^\circ$ on $\mathcal{E}^\circ$, finally, is meant to represent existing *constraints* on the atomic event types, e.g., KISSING $\leq^\circ$ TOUCHING.

There is one piece left in the model structure that has to be characterized by additional axioms. The relation "is of type", π, is defined on pairs of events and *atomic* event types only. Furthermore, call two atomic events e, e' *role-trace-identical* (in symbols $e \approx_{rt} e'$) iff $r(e) = r(e')$, $\rho(e) = \rho(e')$ for all $\rho \in r(e)$, and for $\xi = \sigma, \tau$, $e \in Dm(\xi) \Leftrightarrow e' \in Dm(\xi)$ and $\xi(e) = \xi(e')$ if the traces are defined. Then π has to be coherent in the following sense.

(Def) A relation π in $\mathcal{E}^\circ \times E^\circ$ is *coherent* iff

1. $\forall e \in E^\circ\ \exists! \theta \in \mathcal{E}^\circ : \langle \theta, e \rangle \in \pi$;
2. $\langle \theta, e \rangle \in \pi \Rightarrow \langle -\theta, e \rangle \notin \pi$ for role-trace-identical e, e';
3. $eSe' \Rightarrow [\langle \theta, e \rangle \in \pi \Leftrightarrow \langle \theta, e' \rangle \in \pi]$.
4. $\theta \leq^\circ \theta' \Rightarrow [\langle \theta, e \rangle \in \pi \Rightarrow \exists e' [e' \approx_{rt} e \;\&\; \langle \theta', e' \rangle \in \pi]]$.

Since π is coherent, every atomic event is of exactly one atomic event type; in other words, every basic fact must be of a certain type, and this type is uniquely determined. Secondly, the same objects cannot play roles in trace-identical events of opposing event types. Also, if event e specifies event e', e and e' have to be of the same type. The fourth condition means, for instance, that if an event is a kissing there has to be a role-trace-identical touching.

Two final definitions. An event $e \in E$ is an *aspect (of the world)* iff e is a sum of atomic events of the same (atomic) event type. An event $e_0 \in E$ is called a *(world) chunk* if it is $\leq^\circ$-closed, i.e., if $e \leq e_0$ for an atomic event e, $\langle \theta, e \rangle \in \pi$, and $\theta \leq^\circ \theta'$, then $e' \leq e_0$ for any role-trace-identical e' with $\langle \theta', e' \rangle \in \pi$.

11.3 Translation and Truth

In the Austinian picture the descriptive conventions of language yield an event type $\theta = \|S\|$ for every natural language sentence S. This event type is what is usually called the logical form of S. There have to be rules, then, which produce $\|S\|$ from a given S. I am not going to specify those rules here. Let us simply assume that they can be specified, and that they are given in such a way as to yield event types where indefinite NPs are represented in parametrized form along the lines of Kamp (1981) and Heim (1982). Also, instead of building up trees or DRS's to encode dependence of parameters, I assume an indexing of the parameters according to some given hierarchy of *dependence levels* (for this idea see, e.g., Fine 1985). If PAR is the set of parameters $\mathbf{a}, \mathbf{b}, \ldots, \mathbf{t}, \mathbf{t'}, \ldots$, let $FR(\theta)$ be the set of "free" parameters, i.e., either of dependence level 0 or in case θ is a constituent of a larger type, of the lowest level not occurring outside of θ.[9] Then the notion of an anchor for an event type θ is defined in the following way.

(Def) Let e_0 be a chunk and θ an event type. f is an *anchor for* θ *and* e_0 ($A[f, \theta, e_0]$ for short) iff f is a partial function from PAR into $CAST(e_0)$ such that $FR(\theta) \subseteq Dm(f)$.

I need a few abbreviations to prepare the central definition of satisfaction.

(Def) Let $e \leq e_0$; then $RC(\theta) \rhd_f e$ iff $A[f, \theta, e_0]$ and the role conditions of θ are fulfilled for e with respect to f, i.e., we have, for instance, $\rho_i(e) = f(\mathbf{a})$ for $\rho_i(\mathbf{e}) = \mathbf{a}$ in $RC(\theta)$.

(Def) $e_0, f \models Re(\theta)$ iff e_0 and f realize all restrictions of θ (see below)[10]

(Def) $Sp(\theta)[e_0, f, e] \;:\Leftrightarrow\; RC(\theta) \rhd_f e$ and $e_0, f \models Re(\theta)$

(Def) Let e_0 be a chunk, e an event with $e \leq e_0$, θ an elementary event type and f a partial function from PAR into $CAST(e_0)$. Then e_0, f *satisfy the relation "e is of type θ"* (in symbols $e_0, f \models \theta(e)$) according to the following recursive clauses:

1. $\theta = \theta_0$ atomic; then $\quad e_0, f \models \theta(e) \quad \Leftrightarrow \quad \langle \theta, e \rangle \in \pi$;
2. $\theta = [\mathbf{e}; Sp(\theta) \;\&\; \theta_1(\mathbf{e})]$ with elementary θ_1; then $e_0, f \models \theta(e) \quad \Leftrightarrow \quad Sp(\theta)[e_0, f, e] \;\&\; e_0, f \models \theta_1(e)$;
3. $\theta = \sum(\theta_1 \mid \theta_2)$ with θ_i elementary $(i = 1, 2)$; then $e_0, f \models \theta(e) \quad \Leftrightarrow \quad e = \bigsqcup\{ e_g \mid g \geq f \;\&\; e_0, g \models \theta_1 \}$ such that $\forall g \geq f\, [e_0, g \models \theta_1 \;\Rightarrow\; \exists g' \geq g : e_0, g' \models \theta_2(e_g)]$
4. θ atomic; then $\quad e_0, f \models \theta^*(e) \quad \Leftrightarrow \quad e_0, f \models -\theta(e)$;

[9] For an indication of how the algorithm for assigning dependence levels works, see the examples below.

[10] The air of circularity in using this relation in the clauses of the satisfaction relation is spurious since $Re(\theta)$ is shorter than θ and "realize" could always be replaced by its definiens.

5. $\theta = [\mathbf{e}; Sp(\theta) \ \& \ \theta_1(\mathbf{e})]$ with elementary th_1; then
$e_0, f \models \theta^*(e) \quad \Leftrightarrow \quad e = \bigsqcup\{ e_g \mid g \geq f \ \& \ Sp(\theta)[e_0, g, e_g] \}$ such that
$\forall g \geq f \, [Sp(\theta)[e_0, g, e_g] \ \Rightarrow \ \exists g' \geq g : e_0, g' \models \theta_1^*(e_g)]$;

6. $\theta = \sum(\theta_1 \mid \theta_2)$ with θ_i elementary ($i = 1, 2$); then
$e_0, f \models \theta^*(e) \quad \Leftrightarrow \quad \exists g \geq f \, [e_0, g \models \theta_1 \ \& \ e_0, g \models \theta_2^*(e_g)]$.

(Def) Let e_0 be a chunk, θ an event type, and f a partial function from *PAR* into $CAST(e_0)$. Then e_0, f *realize* θ (in symbols $e_0, f \models \theta$) iff the following conditions hold:

1. θ elementary; then $e_0, f \models \theta \Leftrightarrow A[f, \theta, e_0] \ \& \ \exists e \leq e_0 : e_0, f \models \theta(e)$;

2. $\theta = \theta_1 \supset \theta_2$; then $e_0, f \models \theta \quad \Leftrightarrow$
$\forall$ chunks $e \geq e_0 \, [e_0, f \models \theta_1 \ \Rightarrow \ \exists g' \geq g : e_0, f \models \theta_2]$;

3. $\theta = \triangle\{ \theta' \mid \theta' \in \Theta \}$; then $e_0, f \models \theta \quad \Leftrightarrow \quad e_0, f \models \theta'$ for all $\theta' \in \Theta$;

4. $\theta = \triangledown\{ \theta' \mid \theta' \in \Theta \}$; then $e_0, f \models \theta \Leftrightarrow e_0, f \models \theta'$ for some $\theta' \in \Theta$.

(Def) Let θ be an event type, e_0 be a chunk, and let $Prsp(\theta)$ be the conjunction of the types that describe the presuppositions of θ. Then

$$e_0 \ admits \ \theta \quad :\Leftrightarrow \quad \exists f_0 \, [e_0, f_0 \models Prsp(\theta)].$$

(Def) Let e_0 be a chunk and θ an event type. Then θ *is true in* e_0 iff the following conditions hold:

1. θ elementary or a conditional; then θ is true in $e_0 \quad \Leftrightarrow$
e_0 admits θ & $\forall f_0 \, [e_0, f_0 \models Prsp(\theta) \ \Rightarrow \ \exists f \geq f_0 : e_0, f \models \theta]$;

2. $\theta = \triangle\{ \theta' \mid \theta' \in \Theta \}$, where Θ is a set of elementary event types or conditionals; then
θ is true in $e_0 \quad \Leftrightarrow \quad \forall \theta' \in \Theta : \theta'$ is true in e_0;

3. $\theta = \triangledown\{ \theta' \mid \theta' \in \Theta \}$, with Θ as above; then
θ is true in $e_0 \quad \Leftrightarrow \quad \exists \theta' \in \Theta : \theta'$ is true in e_0;

Remark. Truth is defined only locally. The chunks can therefore be made to play the role of the *context* in presupposition theory, guaranteeing the uniqueness of names and definite descriptions as well as the proper reference for context parameters like *I* and *now*. In fact, it can be seen from the definitions that truth is made dependent here on the condition that the context admits the event type, i.e., realizes all its presuppositions (Heim 1983, Link (1987c)). It follows that we do not have, in general, persistence of truth on chunks; i.e., if θ is true in e_0 and $e \geq e_0$, θ might cease to be true in e . There might be only one John around in chunk e_0, but a second John might enter the scene when we extend e_0 to e so that the uniqueness presupposition, and thereby truth, fails in e.

11.4 Examples

Rather than giving an explicit set of rules I shall illustrate the idea behind the mechanism of assigning dependence levels to the parameters in an event type by means of some characteristic examples. Basically, general NPs in the sense of Barwise (1985) (viz., *every student*, *no student*) give rise to pushing the current dependence one level up (example (2)), whereas with indefinite NPs the given level remains unaltered (example (1)). Proper names and definite descriptions always stay on level 0, which reflects their scopelessness (example (5)). In case an indefinite NP is intended to have scope over an incoming general NP it may stay on the current level (example (11ii)).

It is convenient to provide a natural language sentence with the relevant dependence information before it is assigned its event type, as can be seen from the following examples.

(1) John reads a book.
(^{0}[John], ^{0}[reads a book]) = ^{0}[John reads a book]

(2) Every student reads a book.
(^{1}[every student], ^{0} [reads a book]) = ^{1}[every student reads a book]

(3) John read every book.
(^{0}[John], ^{1}[read every book]) = ^{1}[^{0}[John] read every book]

(4) Not every student read a book.
(^{1}[not every student], ^{0}[read a book]) =
^{1}[not every student read a book]

(5) No student wrote a letter to the dean.
(^{1}[no student], ^{0}[wrote a letter to ^{0}[the dean]]) =
^{1}[no student wrote a letter to ^{0}[the dean]]

(6) No student read every book.

(^{1}[no student], ^{1}[read every book]) = ^{1}[no student ^{2}[read every book]]

(7) Every student read no book for three hours.
(^{1}[every student], ^{0}[^{1}[read no book] for three hours]) =
^{1}[every student ^{2}[read no book] for three hours]

(8) No student listened to a tape for three hours.
(^{1}[no student], ^{0}[^{0}[listened to a tape] for three hours]) =
(^{1}[no student], ^{0}[listened to a tape for three hours]) =
^{1}[no student listened to a tape for three hours]

(9) No student wrote more than three papers.
(^{1}[no student], ^{0}[wrote more than three papers]) =
^{1}[no student wrote more than three papers]

(10) Every German with a car polishes it.
(^{1}[every German with a car], ^{0}[polishes it]) =
^{1}[every German with a car polishes it]

(11) Every country has a government spokesman telling all journalists a lie
(viz., that the country's nuclear power plants are the safest in the world).

(i) (^{1}[every country], ^{0}[has a g.sp. ^{1}[telling all journalists a lie]])
=
^{1}[every country has a g.sp. ^{2}[telling all journalists a lie]]

(ii) (^{1}[every country], ^{0}[has a g.sp. ^{1}[telling all journalists] a lie])
=
^{1}[every country has a g. sp. ^{2}[telling all journalists] a lie]

The next step is the assignment of event types to the indexed sentences. To demonstrate it, some typical examples from the above list will be selected. The dependence level is copied to every parameter that is introduced inside its scope (indicated by the brackets) and has not yet received a level. Sentence (1) is given an elementary event type of the form $\theta = [\mathbf{e}; Sp(\theta) \ \& \ \theta_1(\mathbf{e})]$, where θ_1 is the atomic event type READING, the role conditions $RC(\theta)$ are $\rho_1(\mathbf{e}) = \mathbf{a}^0 \ \ \& \ \ \rho_2(\mathbf{e}) = \mathbf{b}^0 \ \ \& \ \ \tau(\mathbf{e}) = \mathbf{t}^0$, and the restrictions are JOHN($\mathbf{a}$) & BOOK($\mathbf{b}$) & $\mathbf{t} \geq \mathbf{t}_0$ ($\mathbf{t}_0$ is to represent the context parameter *now*). The quantifier *every* gives rise to a sum event type of the form $\theta = \sum(\theta_1 \mid \theta_2)$; whereas in principle both θ_1 and θ_2 are to be regular event types, the type θ_1 is written below in the familiar function-argument form since in all sample cases here it happens to be a state type (e.g., STUDENT($\mathbf{a}$)). Proper names and definite descriptions are treated as restrictions; as was noted above, the uniqueness condition has to be part of the context (chunk) in which the event type is interpreted. Example (9) contains the plural phrase *more than three papers*, whose parameter may be anchored to a sum individual containing more than three atoms. The star in front of the noun PAPER in (9'), which is borrowed from the plural logic in Chapter 1, expresses the transition from individual papers to arbitrary sums of papers; these are then restricted by the numeral '> 3'. Finally, the space parameter is omitted from all event types since I don't have anything specific to say about the function σ at this point; but see ter Meulen's (1987) on the topic of locating events.

(1') $S = {}^{0}[$ **John reads a book** $]$;

$$\|S\| = [\,\mathbf{e}\,;\rho_1(\mathbf{e}) = \mathbf{a}^0 \;\&\; \text{JOHN}(\mathbf{a}) \;\&\; \rho_2(\mathbf{e}) = \mathbf{b}^0 \;\&\; \text{BOOK}(\mathbf{b}) \;\&\; \tau(\mathbf{e}) = \mathbf{t}^0 \;\&\; \mathbf{t} \geq \mathbf{t}_0 \;\&\; \text{READING}(\mathbf{e})\,]$$

(2') $S = {}^{1}[$ **every student reads a book** $]$;

$$\|S\| = \textstyle\sum(\text{STUDENT}(\mathbf{a}^1) \mid \|{}^1[\,\mathbf{a}\text{ reads a book}\,]\|) = \textstyle\sum(\text{STUDENT}(\mathbf{a}^1) \mid [\,\mathbf{e}\,;\rho_1(\mathbf{e}) = \mathbf{a}^1 \;\&\; \rho_2(\mathbf{e}) = \mathbf{b}^1 \;\&\; \text{BOOK}(\mathbf{b}) \;\&\; \tau(\mathbf{e}) = \mathbf{t}^1 \;\&\; \mathbf{t} \geq \mathbf{t}_0 \;\&\; \text{READING}(\mathbf{e})\,])$$

(3') $S = {}^{1}[\ {}^{0}[$**John**$]$ **reads every book** $]$;

$$\|S\| = [\,\mathbf{e}\,;\rho_1(\mathbf{e}) = \mathbf{a}^0 \;\&\; \text{JOHN}(\mathbf{a}) \;\&\; \|{}^1[\,\mathbf{a}\text{ reads every book}\,]\|(\mathbf{e})\,] = [\,\mathbf{e}\,;\rho_1(\mathbf{e}) = \mathbf{a}^0 \;\&\; \text{JOHN}(\mathbf{a}) \;\&\; \textstyle\sum(\,\text{BOOK}(\mathbf{b}^1) \mid [\,\mathbf{e}'\,;\rho_2(\mathbf{e}') = \mathbf{b} \;\&\; \tau(\mathbf{e}') = \mathbf{t}^1 \;\&\; \mathbf{t} < \mathbf{t}_0 \;\&\; \text{READING}(\mathbf{e}')\,]\,)(\mathbf{e})\,]$$

(5') $S = {}^{1}[$ **no student wrote a letter to** ${}^{0}[$**the dean**$]$ $]$;

$$\|S\| = [\,\mathbf{e}\,;\rho_1(\mathbf{e}) = \mathbf{a}^1 \;\&\; \text{STUDENT}(\mathbf{a}) \;\&\; \rho_2(\mathbf{e}) = \mathbf{b}^1 \;\&\; \text{LETTER}(\mathbf{b}) \;\&\; \tau(\mathbf{e}) = \mathbf{t}^1 \;\&\; \mathbf{t} < \mathbf{t}_0 \;\&\; \rho_3(\mathbf{e}) = \mathbf{c}^0 \;\&\; \text{THE.DEAN}(\mathbf{c}) \;\&\; \text{WRITING}(\mathbf{e})\,]^*$$

(6') $S = {}^{1}[$ **no student** ${}^{2}[$**read every book**$]$ $]$;

$$\|S\| = [\,\mathbf{e}\,;\rho_1(\mathbf{e}) = \mathbf{a}^1 \;\&\; \text{STUDENT}(\mathbf{a}) \;\&\; \textstyle\sum(\,\text{BOOK}(\mathbf{b}^2) \mid [\,\mathbf{e}'\,;\rho_2(\mathbf{e}') = \mathbf{b} \;\&\; \tau(\mathbf{e}') = \mathbf{t}^2 \;\&\; \mathbf{t} < \mathbf{t}_0 \;\&\; \text{READING}(\mathbf{e}')\,]\,)(\mathbf{e})\,]^*$$

(7') $S = {}^{1}[$ **every student** ${}^{2}[$**read no book**$]$ **for three hours** $]$;

$$\|S\| = \sum(\text{STUDENT}(\mathbf{a}^1) \mid \|^1[\mathbf{a}\,^2[\text{read no book}] \text{ for three hours}]\|) =$$
$$\sum(\text{STUDENT}(\mathbf{a}^1) \mid [\mathbf{e}; \rho_1(\mathbf{e}) = \mathbf{a} \,\&\, \tau(\mathbf{e}) = \mathbf{t}^1 \,\&\, |\mathbf{t}|_{\mathrm{h}} \geq 3 \,\&\, \mathbf{t} < \mathbf{t}_0 \,\&\, \|^2[\text{read no book}]\|(\mathbf{e})]) =$$
$$\sum(\text{STUDENT}(\mathbf{a}^1) \mid [\mathbf{e}; \rho_1(\mathbf{e}) = \mathbf{a} \,\&\, \tau(\mathbf{e}) = \mathbf{t}^1 \,\&\, |\mathbf{t}|_{\mathrm{h}} \geq 3 \,\&\, \mathbf{t} < \mathbf{t}_0 \,\&\, [\mathbf{e}'; \rho_2(\mathbf{e}') = \mathbf{b}^2 \,\&\, \text{BOOK}(\mathbf{b}) \,\&\, \text{READING}(\mathbf{e}')]^*(\mathbf{e})])$$

(9') $S = {}^1[$ **no student wrote more than three papers** $]$;

$$\|S\| = [\mathbf{e}; \rho_1(\mathbf{e}) = \mathbf{a}^1 \,\&\, \text{STUDENT}(\mathbf{a}) \,\&\, \rho_2(\mathbf{e}) = \mathbf{b}^1 \,\&\, [(>3)^\star\text{PAPER}](\mathbf{b}) \,\&\, \tau(\mathbf{e}) = \mathbf{t}^1 \,\&\, \mathbf{t} < \mathbf{t}_0 \,\&\, \text{WRITING}(\mathbf{e})]^*$$

(10') $S = {}^1[$ **every German with a car polishes it** $]$;

$$\|S\| = \sum(\text{GERMAN}(\mathbf{a}^1) \,\&\, \text{CAR}(\mathbf{b}^1) \,\&\, \text{WITH}(\mathbf{a},\mathbf{b}) \mid [\mathbf{e}; \rho_1(\mathbf{e}) = \mathbf{a} \,\&\, \rho_2(\mathbf{e}) = \mathbf{b} \,\&\, \text{POLISHING}(\mathbf{e})])$$

I shall give now two examples for the *semantic evaluation* of event types. Let us consider first example (3'); let θ be the event type $\|^1[^0[\text{John}] \text{ read every book}]\|$ and e_0 a chunk. Then we have

(12) θ true in $e_0 \Leftrightarrow e_0$ admits $\theta \,\&\, \forall f_0\,[e_0, f_0 \models Prsp(\theta) \Rightarrow \exists f \geq f_0 : e_0, f \models \theta]$;

$$e_0, f \models \theta \quad \Leftrightarrow \quad A[f, \theta, e_0] \,\&\, \exists e \leq e_0 : e_0, f \models \theta(e);$$

$$e_0, f \models \theta(e) \quad \Leftrightarrow \quad \rho_1(e) = f(\mathbf{a}^0) \,\&\, \text{JOHN}(f(\mathbf{a})) \leq e_0 \,\&\, e_0, f \models \sum(\text{BOOK}(\mathbf{b}^1) \mid [\mathbf{e}'; \rho_2(\mathbf{e}') = \mathbf{b} \,\&\, \tau(\mathbf{e}') = \mathbf{t}^1 \,\&\, \mathbf{t} < \mathbf{t}_0 \,\&\, \text{READING}(\mathbf{e}')])(e);$$

$$e_0, f \models \sum(\text{BOOK}(\mathbf{b}^1) \mid [\mathbf{e}'; \rho_2(\mathbf{e}') = \mathbf{b} \,\&\, \tau(\mathbf{e}') = \mathbf{t}^1 \,\&\, \mathbf{t} < \mathbf{t}_0 \,\&\, \text{READING}(\mathbf{e}')])(e)$$
$$\Leftrightarrow \quad e = \bigsqcup\{e_g \mid g \geq f \,\&\, e_0, g \models \text{BOOK}(\mathbf{b}^1)\} \text{ such that } \forall g \geq f\,[e_0, g \models \text{BOOK}(\mathbf{b}^1) \Rightarrow \exists g' \geq g : e_0, g' \models \theta_1(e_g)],$$

where

$$\theta_1 = [\mathbf{e}'; \rho_2(\mathbf{e}') = \mathbf{b} \,\&\, \tau(\mathbf{e}') = \mathbf{t}^1 \,\&\, \mathbf{t} < \mathbf{t}_0 \,\&\, \text{READING}(\mathbf{e}')];$$

$$e_0, g' \models \theta_1(e_g) \Leftrightarrow \rho_2(e_g) = g'(\mathbf{b}) \,\&\, \tau(e_g) = g'(\mathbf{t}) \,\&\, g'(\mathbf{t}) \leq f_0(\mathbf{t}_0) \,\&\, \text{READING}(e_g).$$

Next let θ be the event type $\|^1[\text{every student }^2[\text{read no book}] \text{ for three hours}]\|$. This is example (7'), which contains a negation. Then we get the following truth conditions:

(13) θ true in $e_0 \Leftrightarrow e_0$ admits θ & $\forall f_0\,[\,e_0, f_0 \models \mathit{Prsp}(\theta) \Rightarrow \exists f \geq f_0 : e_0, f \models \theta\,]$;

$e_0, f \models \theta \quad \Leftrightarrow \quad A[f, \theta, e_0]$ & $\exists e \leq e_0 : e_0, f \models \theta(e)$;

$e_0, f \models \theta(e) \Leftrightarrow e = \bigsqcup\{\, e_g \mid g \geq f$ & $e_0, g \models \textsc{student}(\mathbf{a}^1)\,\}$
such that

$\forall g \geq f\,[\,e_0, g \models \textsc{student}(\mathbf{a}^1) \Rightarrow \exists g' \geq g : e_0, g' \models \theta_1(e_g)\,]$,
where

$\theta_1 = \|^1[\,\mathbf{a}^{\,2}[$read no book$]$ for three hours$\,]\|$;

$e_0, g' \models \theta_1(e_g) \Leftrightarrow \rho_1(e_g) = g'(\mathbf{a}) = g(\mathbf{a})$ & $\tau(e_g) = g'(\mathbf{t})$ & $|\,g'(\mathbf{t})\,|_{\mathrm{h}} \geq 3$ & $g'(\mathbf{t}) < f_0(\mathbf{t}_0)$ & $e_0, g' \models \theta_2^*(e_g)$,
where

$\theta_2 = [\,\mathbf{e}'\,; \rho_2(\mathbf{e}') = \mathbf{b}^2$ & $\textsc{student}(\mathbf{b})$ & $\textsc{reading}(\mathbf{e}')\,]$;

$e_0, g' \models \theta_2^*(e_g) \Leftrightarrow e = \bigsqcup\{\, e_{gg''} \mid g'' \geq g'$ & $\mathit{Sp}(\theta_2)[e_0, g'', e_{gg''}]\,\}$
such that

$\forall g'' \geq g'\,[\,\mathit{Sp}(\theta_2)[e_0, g'', e_{gg''}] \Rightarrow \exists g_1 \geq g'' : e_0, g_1 \models -\textsc{reading}(e_{gg''})\,]$,
and

$\mathit{Sp}(\theta_2)[e_0, g'', e_{gg''}] \Leftrightarrow \rho_2(\mathbf{e}_{gg''}) = g''(\mathbf{b}^2)$ & $e_0, g'' \models \textsc{book}(g''(\mathbf{b}))$.

Acknowledgment. The paper included here as the present chapter is an outgrowth of the author's research project on Algebraic Semantics, which was carried out in part during his stay at the Center for the Study of Language and Information (CSLI), Stanford University, in the academic year 1985/86. That stay was financially supported by a donation of the IBM Deutschland through the Stifterverband der Deutschen Wissenschaft, Bonn, and by the CSLI. This support is herewith gratefully acknowledged. I also wish to thank a number of people for helpful discussions and/or comments on earlier versions of that paper, in particular Jon Barwise, Hans Kamp, Manfred Krifka, Sebastian Löbner, Karl-Georg Niebergall, Barbara Partee, and Dietmar Zaefferer.

Chapter 12

The Ontology of Individuals and Events

12.1 General Remarks

This chapter constitutes a metaphysical discursion regarding the main kinds of entity that have been used in this book. I have based the discussion in this book on a "naive" ontology of things, supplemented by an ontology of pluralities and an ontology of events. Regarding the ontological status of the pluralities I will have to say more in Chapter 13. Here I will only be concerned with the nature of regular individuals (atomic ones in the sense of the theory of plurals), while taking for granted the superstructure of pluralities.

In Chapter 1 I gave an account of individuals that violates one of the paradigm examples of a traditional metaphysical principle: that no two objects can occupy the same place at the same time; call this the principle of *No Coincidence, NC*.[1] My semantics was such that the ring (one individual) and the piece of gold constituting it (another individual) coincide during the time period of the ring's existence. The reason for this was that I wanted to demonstrate how Leibniz' law could be salvaged in a close-to-language fashion while still accounting for the intimate connection between the two objects. This was done in algebraic terms by positing a homomorphic relation between the structure of the "full individuals" and that of their constitutive portions of matter. I will reconsider the structure of the constitution relation below.

[1]The term 'superposition' is used in Simons (1987) instead of the more traditional term 'coincidence'.

Later I introduced events into my ontology, on top of this elaborate thing ontology, again basically on semantic grounds. Countenancing events is more in dispute than admitting individuals; thus, for instance, there are quite a number of "No-Events" theories but only a few if any "No-Individuals" theories.[2] Among those who admit both individuals and events, primacy has usually been granted to individuals. In this chapter I will argue for an ontology of both kinds of entity that should be considered on a par. In fact I will try make it plausible that both individuals and events are special kinds of a single, albeit rather "opaque" type of object which I call *processes.* Invoking processes is far from new, too, but the way I will use this traditional term cannot really be traced back to any particular source in the literature.[3]

The kind of ontology I have in mind here as a discipline is a branch of analytic metaphysics. A further subdivision can be made by distinguishing the kind of metaphysical project that one embarks on: revisionary or reductivist on the one hand and descriptive on the other. Thus, for instance, Peter Strawson wants to do the latter; in his (1959) he says: "Descriptive metaphysics is content to describe the actual structure of our thought about the world." This says that he is concerned with the way we think about the world and not with the way we *ought* to think about it. But our thinking is of course expressed in our language, so in order to find out about the way we think about the world we should look at the way we speak about it. So we are left with a project that certainly starts out with linguistic philosophy.

Now although philosophers in the analytic tradition take language as the prime instrument of analysis they typically don't want to be mistaken for linguists. There is always the firm conviction in philosophical quarters that language cannot be taken at face value, and that some deeper level of analysis than a linguistic one is called for.[4] The structure of our thought about the world cannot be "read off" the structure of our language. I would agree with this attitude if only because it is difficult to say what "reading off" could mean. So some reform or "regimentation" of language is necessary before we are able to make meaningful metaphysical statements. But then, the enterprise isn't altogether descriptive after all.

[2]For a survey, see Thalberg (1985).

[3]Classical sources for process philosophies are Whitehead (1929) and Sellars (1981). On Whitehead, see, e.g., Christian (1959); Rorty (1963); on Sellars, McGilvray (1983); Seibt (1990). Johanna Seibt has also been developing her own process ontology; see Seibt (1997, 1998).

[4]As a case in point, the tendency to do no more than to appeal to the authority of ordinary language in an metaphysical argument is called "rather disgraceful" in Armstrong (1997), p. 100.

The other element in the above quote from Strawson is that he says that the metaphysician is not concerned with the structure of the world but rather with the structure of *our thoughts about* the world. Thus the subject matter seems to shift away from the mere concern about "what is out there" towards *ways of conceptualizing it*. It is true that concern about what is out there already presupposes a certain metaphysical position, viz. the realist position; but then the quote from Strawson brings with it a number of presuppositions, too. He wants to describe "the actual structure of our thought about the world," presupposing among other things that there *is* something like *the actual* structure of our thought about the world, and that that is a metaphysical question at all.

Regarding the latter, it is important to make clear what the source of our thinking about the world is. The main source is language, no doubt. But what language does is to convey a conceptualization that is informed by our present-day physical theories about the world. Thus it seems obvious, for instance, that Aristotle's metaphysics didn't develop from pure reasoning alone but reflects a certain view about the physical world. In particular, his notion of substance is not clearly separated from the physical notion of matter or stuff. Some have called Aristotle's physics "naive" but I prefer the term "common sense physics." (I guess that Aristotle was certainly abreast of the physics of his time, so it sounds somewhat unfair to call his views "naive.") Thus an important methodological question about metaphysics is how much physics of the day should enter a given metaphysical argument or position. To give an example, arguments for No Coincidence have been given on the grounds that physical objects "compete for space" and cannot therefore coincide completely. Presumably, that intuition has been formed with every-day observations in mind, like a bottle filled with air that cannot escape: then is is certainly hard to put water into it. But modern physics might give rise to a more subtle concept of physical object, which should in any case not be precluded by metaphysical theories. Thus while I think that it is legitimate to use examples from physics with the purpose of falsifying general metaphysical claims about the world, philosophy should be wary to rely too heavily on the content of current (or for that matter, past) physics books.

In order to steer clear of the rapid change in our physical knowledge about the world when doing metaphysics it seems mandatory to me to develop a rather *formal* conception of metaphysics; such a conception is in line with the algebraic outlook of this book.

One might think that in spite of all the changes in our physical world view we actually do have a rather robust conception of ordinary things that remains the same over time. But this is what I call the *myth of the given.*[5] What is stable with our thing concept is really rather formal; it is not the rich notion of physical object that derives from our physical theories. It is in recognition of this fact that Quine (1981):20 says: "I see all objects as theoretical. ... Even our primordial objects, bodies, are already theoretical." One could call this a modified Whorfean thesis: ontology is not really language-dependent but rather theory-dependent.

Consequently, I feel dissatisfied with philosophical views that don't distinguish the rich thing conception from the formal one, and thereby run the risk of reproducing an Aristotelian metaphysics that is actually too close to his (now out-dated) common sense physics.[6]

Once this distinction (viz. the formal vs. the rich conception of things) is drawn we see that the advantage that thing theorists seem to have over event theorists evaporates: a purely formal thing concept is on a par with a formal event concept, and it looks quite natural to have events alongside with things in one's ontology.

Thus the strategy that I want to follow here is to try to stress the common features of individuals and events while locating the apparent differences in a common context. I will do this by analyzing both kinds of entity as processes and their differences as two characteristic ways of appearance of processes. Thus on my view, individuals are *stationary processes*, whereas events are *classified processes* that undergo change and are equipped with a discernible structure of roles that are characteristic of the event in question. There have been philosophers who tried to bring together individuals and events in a similar way. Thus, Reichenbach is quoted for the slogan: "A thing is a monotonous event; an event is an unstable thing."[7] Quine's well-known four-dimensional view of physical objects likens individuals to events, too. Earlier this century a pure event ontology had been very popular, typically with authors like Whitehead and Russell who were both science-oriented and reductionist in their philosophical outlook. Later, philosophers of a nominalist persuasion also declared that individuals are just a particular kind of event.[8]

While I am willing to view individuals and events as closely related entities I have some reservations about the motives of these various authors for an event ontology. For one thing, according to my methodological point above concerning the role of science in metaphysical arguments I wouldn't want to invoke modern physics as Russell did to justify our ontology. Also,

---

[5]This phrase is, of course, originally due to W. Sellars.

[6]This is actually my main criticism of Simons' book (1987), which I otherwise consider a valuable contribution to modern formal ontology.

[7]See Bennett (1988):115.

[8]See, for instance, Richard M. Martin who speaks of the person Brutus as the "Brutus event" and accordingly says: "Events constitute all there is; there is nothing more." (Martin (1978):3)

Russell's epistemological motives for the event ontology, being phenomenalist in nature, don't carry the weight he then thought they would. Finally, those claims involved an explicit reductionist stance which I don't want to embrace either without argument. I won't enter a discussion of ontological reduction though since this issue is too involved to be properly addressed here. Thus, following Davidson, I opt for a policy of coexistence regarding these kinds of entity. Whether the claim to view both individuals and events as processes is reductionist in nature, or whether it comes down to some supervenience thesis will not be decided here. Rather I would like to give a somewhat more coherent picture of the many puzzles involving individuals and events, while stressing the algebraic point of view. Regarding the theory of events, in particular, I am going to bring together intuitions that call for a coarse-grained approach (like Quine's) and those supporting the more fine-grained approach of Kim's, by assigning each its proper place in the process-based picture. Yet my attitude will be basically Quinean in spirit since I will take Quine's view as the starting point for leveling the differences between individuals and events. But see below for differences to a number of Quinean pronouncements.

12.2 Metaphysical Methodology

Before I enter the discussion of individuals I want to make a few remarks on the way I think metaphysical investigations should be carried out when dealing with ontological questions. To begin with, these questions should be treated in a rather formal way, as I said already. That is, the focus should be on the structural properties that relate the various entities in the ontology; these properties should be described in terms of an arsenal of formal metaphysical tools. Here is a (certainly incomplete) list of what this arsenal might look like:

- identity
- predication
- abstraction
- mereology
- modality

Let me comment on these very general tools in turn.[9]

[9] Note that I am not including set theory here. Set theory is of course the underlying modeling tool for most (if not all) of our theorizing, but I don't think that set theory is a *metaphysical* tool.

Identity. For a mathematician, identity is a straightforward matter. Given that mathematics can be described in set theory, all mathematical objects are sets of some kind,[10] and the Axiom of Extensionality settles questions of identity once and for all. To the extent that concepts and objects in philosophy are modeled in set theory this clear-cut situation extends to the realm of metaphysics. However, identity is then dependent on the nature of the modeling device at hand, and in concrete cases that might prove to provide a rather skewed representation of the "real-life" objects under consideration. When we don't presuppose set theory as a metaphysical tool (and I think we shouldn't) then we should be prepared to deal with a rather loose notion of identity that takes into account the vague boundaries of the objects we deal with in the world around us.[11] Failure to consider loose identity leads to all kinds of philosophical puzzles, the Ship-of-Theseus problem and the Tibbles problem being notorious cases in point.[12] Or consider what Quine has to say about identity of events or actions: "Some things may be said of an act on the score of its being a walk, and distinctive things may be said of it on the score of it being a chewing of gum, even though it be accounted one and the same event." Quine (1981) 11f. If that view on event identity is adopted then certainly the use of Leibniz' law will be severely restricted if not made impossible. Note that the same view can be expressed regarding ordinary individuals: "Some things may be said of a man on the score of him being the composer of *Tristan and Isolde*, and distinctive things may be said of him on the score of him being an antisemite, even though he be accounted one and the same individual." What these examples suggest is that we should reconsider the role of Leibniz' law in ontology, and make it compatible with the presence of loose identity. A simple way to account for loose identity, for instance, is to regard it as similarity with respect to a given feature, X (call it *X-identity*, for short). Then X-identity is an equivalence relation, but more than that: since we will operate in richer algebraic contexts, X-identity should also be a congruence relation that is compatible with the algebraic relations at hand. With respect to some such congruence relation, '$\equiv$', Leibniz' law can

[10] Ignore the question of mathematical objects that are "too big" for fitting into the usual set-theoretical universe, like certain objects in category theory.

[11] The issue of vague identity is a thorny one. For a classical statement of how this notion should *not* be understood, see Evans (1978); Lewis (1988); a recent discussion relating to quantum indeterminacy can be found in Lowe (1994); French and Krause (1995); Noonan (1995). The conception of loose identity that will be used here is unaffected by Evans' argument, however, since it does not rely on an idea of vagueness *in re*.

[12] For a discussion of these problems see Simons (1987). As for the Tibbles story, in which a cat's tail is cut off, it is hard to imagine that if the cat is replaced by a person loosing his or her appendix that circumstance would occasion any serious doubts about personal identity.

be upheld in its weakened form,

$$a \equiv b \rightarrow (\phi[a] \leftrightarrow \phi[b])$$

According to the characteristic interplay between the notion of identity and Leibniz' law, we get back the strict form of the law when we coarsen our notion of identity by going over to the relevant congruence classes. Now what seems to me important to realize is that our identity criteria regarding objects are not fixed once and for all but vary in strictness from context to context. With such a conception in mind we will be able to handle quite a number of traditional problems about identity.

Predication. The basic tool for making judgments about the world is of course predication. Predication relates predicates and individual terms to form sentences. Semantically, that means that some individuals are said to have properties or to stand in relations to one another. In philosophical terms that corresponds to the traditional distinction between particulars and universals, brought together through the instantiation relation. I am concerned here with the "particular" side of this relation, leaving it open what the exact ontological status of the properties and relations is;[13] but I will talk freely of properties and relations, and, in particular, of event types that are instantiated by particular events or processes. Traditional philosophy speaks of substances having attributes, and I have no quarrel with this language as long as the term "substance" is understood formally and not materially.

In the writings of modern nominalists it seemed like predication could be replaced by some construction within mereology. Thus, for instance, rather than saying that the Taj Mahal instantiates the property White, nominalists would say that the Taj Mahal is part of the Earth's white stuff. Now while I have reasons to regard mereology a basic tool in metaphysics (see below) I want to keep predication around alongside with mereology. This has become clear in earlier chapters of this book, where mereology was used for building the plural lattice and predication was not only preserved but extended to cover collective properties and relations.

Abstraction. The device of abstraction is needed to form general concepts. Accordingly, I have been using the operation of λ-abstraction in the object-language to be able to form names of complex properties. λ-abstraction plus identity plus predication can be used to define quantification, as Montague showed, and also the logical connectives if there is some

[13]Thus I need not commit myself to a specific view of the nature of universals. For modern accounts on universals, see, e.g., the classical Ramsey (1925) and the more recent Armstrong (1989), as well as the introductory Grossmann (1992).

modest simple type theory available.[14] Thus in a sense those three tools are really very fundamental, making logic essentially part of metaphysics.

However, there is a second, more philosophical sense in which I'd like to use the term "abstraction" here. It has to do with the notion of "abstracting away" or "omitting details" in the description of ordinary objects. That means that things can be in space and time, thus concrete under the usual conception, and yet be "abstract" in virtue of their being only partially characterized in a given theory.[15] I think that in dealing with the objects of our world we really have no choice but to treat them as abstract objects in this sense (recall Quine's related dictum quoted above that all objects are theoretical). To say otherwise would again be subscribing to the myth of the given.

How does that notion of "leaving out" information or complexity show up in our algebraic point of view? It is involved in the relation of constitution to be discussed presently, and there the algebraic tool of homomorphism that was introduced in this context in Chapter 1 encodes the concept of loss of information.

Mereology. The theory of parts and wholes is considered common coin in ontology nowadays, and rightly so. Indeed I think that the part-whole relation is a basic and all-pervasive conceptual tool which should be considered on a par with the other metaphysical tools that I have listed here. There has been some controversy, however, about the exact nature of the wholes that are created by the mereological operation of fusion or summing. Are the sums, fusions, or aggregates (all synonyms used for those wholes) anything over and above their parts, or are they an ontological "free lunch" (Armstrong (1997)), entities simply "supervening on" their parts and adding no further ontological commitment to the one coming with the parts? The answer to this question depends on the extra burden those wholes have to carry. It often happens that when we refer to an aggregate we associate with it substantial extra structure, and it should be plain that in such a case the whole turns out to be more than its parts, and ceases to be ontologically innocuous. However, when there is a collection of things, there is always, by conceptual necessity, the "free lunch" aggregate there, which is the plurality or individual sum of these things. Formally, that corresponds to a principle of *unrestricted mereological composition*, which has been argued for in this book and is also embraced by Lewis (1991)[16] and Armstrong (1997). But it has to be borne in mind that that notion of aggregate is a completely formal one that is unsuited to ac-

[14]See Montague (1970).

[15]This second sense of the term 'abstract', which is not contrary to 'concrete', is found useful also by Bennett in his (1988).

[16]For an extensive discussion of this work see Chapters 13 and 14.

count for the "cement of the universe" (to use John L. Mackie's well-known phrase) which holds our world together. It seems to me that confusion over this fact is the main source of opposition to the principle of unrestricted composition.

Modality. This notion won't be addressed here; I mention it only to say that I consider it an important tool in metaphysics without which the above list would certainly be incomplete. It looms large in the writings of David Lewis and is also an essential ingredient in the views of other contemporary analytic metaphysicians, like Armstrong, van Inwagen, Simons, and Zalta.[17]

12.3 Individuals

Let me start with the problem of *constitution* of individuals. I already mentioned the constitution relation that I had used in Chapter 1. In that chapter I had the following simple picture: I distinguished between regular objects and the homogeneous stuff that makes them up. This picture reflects a rather traditional Aristotelian metaphysics with its distinction of form and matter: the homogeneous stuff is pure matter or substance, as it were, whereas it takes matter plus form to get a regular object. The rationale behind this was that I wanted to preserve the strict applicability of Leibniz' law. Take my paradigm example of the ring that was recently made of old Egyptian gold. By Leibniz' law, the ring and the piece of gold making it up cannot be the same: The ring is new whereas the gold is old. But these two entities are not totally unrelated; in the logic of plurals and mass terms, LPM, I accounted for their close relationship by introducing the constitution relation, '$\rhd$': the piece of gold was said to "constitute" the ring. Two individuals, a and b, were related by $\rhd$ if their material make-up coincided. I got from the individuals to the quantities of matter making them up by applying a function that behaved like a homomorphism between the part-whole structures on the domain of individuals and the domain of portions of matter. That device allowed me to account for the stability of truth value under certain predicates that were called invariant. For instance, since the place of an individual is determined by the stuff constituting it, sentences involving predicates that refer to spatial properties only didn't change truth value when the individual was replaced by its portion of matter. Similarly for predicates referring to the chemical constitution of a thing, e.g., being the element Au.

There were a number of problems with this approach. One is the question, what the nature of "pure stuff" could be. Formally, the entities in

[17]See Lewis (1986b,c), van Inwagen (1993), Simons (1987), Zalta (1983, 1997b).

the quantities-of-matter domain were (atomic) individuals; but if the piece of gold is just the ring stripped off its form, then quantities of matter are individuals without form, which is odd. Furthermore, if gold counts as pure stuff, then where do we put the elementary particles constituting a piece of gold? For instance, Emmon Bach once highlighted the difficulty with this example:[18] The snow making up a newly built snowman is pretty recent, too, but the water making up the snow is old, and the hydrogen atoms in the individual water molecules are even older. This generates a whole chain of objects standing in the constitution relation which could not be accommodated in an obvious way in the LPM framework.

Another problem that is frequently raised is the counting problem, which, however, is shared with anybody who embraces a similar bifurcation of objects here.[19] There is the ring, and there is also the piece of gold that constitutes the ring. Do I have to count those entities twice in an inventory of the world? The question, when put this way, already suggests that that would be double counting. I think we have to agree with this. One way to go here would be to note that rings and quantities of matter are not of the same kind; so we have first to decide with respect to which kind our counting goes. Thus the inventory is relative to some scheme of basic kinds according to which we cut up our phenomenal world, and in each such scheme, objects related by the constitution relation would appear only once.

In general, however, I tend to downplay objections from counting[20] and even like to speak of the *counting fallacy*. The objections fail to recognize the fact that counting is always relative to a number of parameters that have to be fixed before the problem is well-posed. To begin with, we have to fix a unit of counting; without such, for instance, it wouldn't make sense to count masses ("how many quantities of water are in my glass?"). But not only that, counting is also relative to the time considered, to the given kind or sortal (as above), to a given level of precision, to a background theory, and more. As for the ring in my example, there will be a more systematic account in the process view to be developed below: There the ring will be just a temporal part of the gold-process, with the extra stability condition regarding its shape giving rise to a new individual. Thus the piece of gold and the ring are different individuals, since their life span differs. But during the time of the ring's existence they are "loosely identical" in the sense that they stand in the constitution relation establishing some form of spatial coincidence.

---

[18] Bach (1986), p.13.

[19] See, e.g., Simons (1987), Part II.

[20] For a similar counting problem in connection with pluralities, and my reaction to it, see Chapter 13.

12.3.1 Constitution

Let me go back, then, to the issue of constitution, which I still consider an important metaphysical relation. Thus I'll stick to it while introducing amendments to my earlier account that are able to cope with the problems mentioned above. First of all, I want to give up the traditional distinction of matter and form, which in its absolute sense is just not operative under a modern world view.[21] Therefore, the homogeneous kernel of quantities of matter in the notion of model for LPM will henceforth be thought of as regions of space, which deprives it of any association with matter or stuff. The material part relation $\leq_m$ is then replaced by a the natural spatial part relation, $\leq_s$. Secondly, I will introduce whole chains of individuals that are connected by constitution. So for instance, I want to able to say that the chain created by a certain quantity of water making up, in crystalline form, a certain amount of snow forming in turn that snowman over there, links three individuals by the constitution relation. While being different individuals because of their different life spans, they are intimately related in that at any given time of their common existence, they coincide spatially. Now spatial coincidence at a time is a rather strong equivalence relation, given the way human beings interact with their environment, so much so that some philosophers have felt that not much more of identity is called for regarding physical objects. However, I would like to preserve the extra degree of freedom that is gained by a more fine-grained approach and keep the entities separate even if they are identified in many contexts.

There is one further amendment that I want to make to the notion of constitution in Chapter 1. There, objects on the left-hand side of the relation had to be atoms in the plural lattice since there were thought of as portions of matter. But what really happens is that in many cases, pluralities make up a new i-atomic individual. For instance, the members of a committee constitute the committee. Or take a rigid body, call it b, which is constituted by a finite number of molecules $b_1, \ldots, b_n$ that are subject to the constraints of equal distance described in classical mechanics. Those b_i are atoms in the metaphysical sense (i.e., in the mereological framework of the plural lattice), but no atoms in the physical sense. We could go on and take each of them as constituted in turn by a number of physical atoms or elementary particles; each move would bring us "up" again in structure of the plural lattice.

[21]This situation seems to me typical regarding Aristotelian concepts in contemporary philosophy; while it appears reassuring prima facie to link up with the philosophical tradition by making use of them, under closer scrutiny they turn out to be of no much help, to say the least, or prove to be real obstacles to deeper insights; just think of the Aristotelian pair of motion vs absolute rest, a distinction that was rightly pulverized by Galileo.

The chain of constitution thereby established should come to an end after finitely many steps when we go down along the relation $\rhd$. I am not saying that this is true of metaphysical necessity; but it is a reasonable assumption to make. Formally that means that the constitution relation, which is taken to be irreflexive, is also *well-founded*: every non-empty set of individuals has a $\rhd$-least element, or formally, with $\rhd x = \{ y \mid y \rhd x \}$:

$$\forall B\,(\,B \neq \emptyset \rightarrow \exists x(x \in B \wedge B \cap \rhd x = \emptyset)\,)$$

It is well-known that the notion of well-foundedness is not definable in first-order logic; but it can be defined in monadic second-order logic, which in turn can be embedded in the logic of plurality, LP (see Chapter 6). Thus we won't have to leave the resources of classical mereology to express this principle in our ontology. As was explained in Chapter 6, elementhood in monadic second-order logic is replaced by the i-atomic part relation of LP, $\dot{\Pi}$, and set abstraction by σ-abstraction. Thus the principle of well-foundedness of the constitution relation then reads, with '|' standing for disjointness of i-sums:[22]

$$\forall\alpha\exists x\,(\,x \,\dot{\Pi}\, \alpha \wedge \alpha \mid \sigma y.y \rhd x\,)$$

Apart from its being irreflexive and well-founded, the constitution relation should certainly be *asymmetric* (if a constitutes b then b doesn't constitutes a) and *transitive* (if a constitutes b and b constitutes c then a constitutes c). Presumably, it it also *linear*, that is, any two individuals standing in the relation are comparable (either one constitutes the other or vice versa). Moreover, the relation should be compatible with the part-whole relations of individual part (i-part; $\leq_i$) and spatial part (s-part; $\leq_s$):

$$a' \rhd a \wedge b' \rhd b \wedge a \leq_i b \rightarrow a' \leq_s b'$$

These principles will at least partially characterize the notion of constitution. With the given qualifications, the relation will also stand up to the criticisms that were mentioned above.

12.3.2 Temporal Parts

Up to now I haven't had to refer to the underlying process structure of my individuals that I announced above. But there is one controversial issue in ontology that brings us closer to it, and that is the problem of *temporal parts* of individuals. Thus, for instance, Simons (1987) says quite

[22]Note that the universal quantifier doesn't have to be restricted to non-null objects since in the free logic underlying LP, quantification runs over those objects only.

authoritatively (p.175): "A continuant [Simons' term for individual; G.L.] is an object which is in time, but of which it makes no sense to say that it has temporal parts or phases." No doubt Simons has common sense on his side, but some philosophers like Lewis and Armstrong, who developed a rather sophisticated metaphysics, remain unimpressed even if this claim is put forth "in a plonking tone of voice" (Armstrong).[23] I won't discuss the merits of those dissident views here, which cannot be addressed properly without bringing into play the metaphysical tool which I try to stay away from, viz. modality. I will rather give an example that shows why it might be rash to ban temporal parts for individuals. For instance, consider metamorphosis in biology. An insect is certainly an individual; but an insect like a butterfly, which is subject to full or holometabolous metamorphosis, goes through the well-known stages of its development, egg, larva, pupa, and adult (imago). Is it really that far away from common sense to consider each one of those stages individuals of their own? It is true that the underlying organism has a life span which is longer than any one of those stages; but anyone who already embraces the view that the piece of gold making up the ring is different from the ring (as Simons does), should have no extra problem with this case because the ring, too, is but a "stage" of the stuff it is made of.

Now foes of temporal parts do have a right intuition underlying their view: it seems obvious that at least not any old temporal part of an individual will do to form another individual. For a temporal part of an individual to count as individual itself it has to display some salient feature that remains constant throughout its own existence, like the shape of the cocoon in the case of the pupa. In the language of processes to be introduced below, it has to be a discernible stable subprocess. Moreover, the subprocess should be "maximal" with respect to that particular feature; thus I would submit that not every arbitrary temporal slice of the butterfly is another individual. Once we have processes around we will admit all kinds of subprocesses corresponding to rather arbitrary temporal slices (given some reasonable granularity parameter on the times); but those subprocesses might not display a complex of joint stable features so that their fusion can be recognized as an individual. In this respect I agree with Quinton (1979):212 who says: "The momentary phase of an object is not an object, since a persisting object ... is not a series of objects."

Once we admit temporal parts of individuals the dividing line between individuals and events is already somewhat obscured. The categorical nature of the distinction is really undermined, however, when individuals are conceived as "worms" in space-time, i.e., as four-dimensional objects of

[23]See Lewis (1986b):204; Armstrong (1997), 7.2.

some sort. Such objects have both spatial and temporal parts, as events do, and like those they cannot be said to "move around" as ordinary individuals do. Also at any given moment such objects don't seem to be wholly present any more, so it is hard to display them in their totality. Quine even goes one step further, entertaining what I like to call a *container view* of physical objects. Since he thinks that those worm-like objects can actually be identified with the space-time regions they occupy, those regions might as well go proxy for the things. What he avails himself of by that move is of course the sharp identity criterion of set theory, so that he doesn't have to deal with all the vagaries of the things in the real world. Some have felt that such regions in all their mathematical austerity don't really reflect the nature of full-fledged material objects, but then again, the objection perhaps misses the point of Quine's instrumentalist view of ontology. Leaving that view alone, I also think that our conception of individual is different from a container comprising everything that is going on at a certain region; individuals are "thinner" and hence more abstract in the sense explained above. Also, on the container view, there cannot be two different objects occupying the same regions, and we had reasons already to doubt that principle: otherwise there would be no chains of individuals related by constitution. Again, take one group of people forming two committees at exactly the same time; that creates two individuals in the same container. Moreover, we should be prepared to leave room for the possibility that among our individuals there are physical force fields that can be superimposed over one another and thus occupy the same spatiotemporal region (an example of this would be a electro-magnetic dipole planted on a gravitational mass, creating two different central force fields). That would be a case of two individuals that coincide but do not stand in the constitution relation to each other. There is another example, taken from basic quantum mechanics, that sticks even more to the thing language than the last one and yet involves coincidence. In our modern world view we regard electrons as individuals; but even though they are Fermi particles and hence by Pauli's exclusion principle, cannot occupy exactly the same state they can very well occupy the same orbital if their spins are opposite. When this condition is satisfied their respective spreads (electron clouds) can superpose, like the electrons in the helium atom do.[24] Thus they "fill

[24]This is actually a case which raises doubts not only about *NC* but also about *NC* in its weakened form due to Locke, called Locke's Principle, *LP*, in (Wiggins (1968):93) and (Simons (1987):221), viz. that no two things of the *same kind* can occupy exactly the same volume at exactly the same time: if we are prepared to include subatomic entities into our ontology (and we should) then there seems to be no way around the conclusion that two electrons are things of the same kind. — One more comment on Locke's Principle: after a lengthy and somewhat inconclusive discussion, Simons (op. cit.) remarks that there doesn't seem to be a good reason to deny it. This is due to the

the same container."[25] That means also that I would rather not subscribe to Quinton's view in (1979) which says of individuals or objects: "An object is the complete occupant of the spatio-temporal region in which it is to be found," (p. 211) stressing that this occupancy of space and time is "exclusive or proprietary." But instead of competing for their space-time location, the two electrons of the helium atom share it.[26]

In summary, then, in spite of our attempt to assimilate individuals to events, the container view is not for us, nor will it be in the context of events. The thrust of my line of reasoning is really that our regular objects are more "abstract" than it is usually assumed. While it is often granted that events are particulars in space and time and yet abstract (this is Quinton's view, for instance) regular objects are said to be concrete, not abstract particulars. What I am saying here is that if that is the only difference between objects and events then that is not much of a difference after all.

In order to get used to the idea that the there might be a difference in degree and not in principle between individuals and events regarding their abstractness, it is instructive to consider borderline cases between those kinds of entity. For instance, physicists might claim that force fields are better called events; or take a phonetic sound, which is certainly an individual, but then again also an articulatory and acoustic process or event. But the most mundane cases occur in meteorology; thus, rain is "water falling in drops condensed from vapor in the atmosphere," according to Webster's Dictionary, but also "the descent of this water;" similarly for snow, thunder and the like. My final example is from meteorology in the original sense of the word: according to Webster's, a meteor is (i) "a phenomenon or appearance in the atmosphere (as lightning, a rainbow, or a snowfall)" and (ii) "one of the small particles of matter in the solar system observable directly only when it falls into the earth's atmosphere where friction may cause its temporary incandescence." Thus a meteor is an object and then again an event, viz. the phenomenon of its incandescence, just depending on the way we look at it, either according to our present purposes, but also historically: in ancient times, a meteor was an appearance, and nowadays

fact, I think, that *LP* is more of an analytical than of a metaphysical nature: reference is made to kinds or sortals which have their origin in speaking or theorizing about the world. For obvious practical reasons they are used in a way to make the principle true.

[25]Note that in giving these examples from physics I stay in line with the methodological principle expressed above that physics should be held apart from metaphysics: I am not saying here that our ontological concept of individual should be as our current physical theories say, but rather, I give counterexamples to a notion of individual that is linked to common sense physics.

[26]Even "ordinary" logic violates Quinton's principle at times; recall Hamlet's argument prompted by the inquiry into Polonius' whereabouts: "... nothing but to show you how a king may go a progress through the guts of a beggar." (Hamlet, 4.3).

it is more of an object.

12.3.3 Processes

This is now the time to introduce *processes* as the one category that I like to think of as underlying both individuals and events.[27] Processes are complex wholes of "goings-on" that either cannot or in any case need not be fully described; descriptions typically leave something out and thus give rise to more abstract objects than the full processes at hand. So what can we say about those processes, then? On this very general level, not much; let me give an example instead that should fix our intuitions about processes and their relations to both individuals and events.

Consider two people, a and b, engaged in a conversation; one phase of it might consist of a's talking while b is listening, another of b's talking while a is listening, and yet another of a and b speaking at the same time. That conversation is a complex event e which contains those phases as subevents. Let us single out one such, say, a's speaking to b; considered as an elementary event it is "atomic," but it has a *fine-structure of underlying processes*. First of all, there are the individuals a and b that are incredibly complex systems of physiological processes making up a human organism, but playing their role in the event only as stationary processes, i.e., ones that display a certain stable set of characteristic features throughout their life-times. That is, when regarding them as individuals we abstract away from the ever-changing chemical processes going on in the persons' bodies. Now the concrete process that is classified as the speaking-and-listening event is not just the two persons taken together; it is them *plus* all the inter-active and intra-active subprocesses that constitute the talking and listening: establishing eye-contact, activating a's articulatory system for the production of speech, setting up an attentive state of mind in b towards a's words, and so forth. Thus, in addition to the stationary processes that form the stable agents in the given situation, the process and thereby the event *focuses on change*, on all the unstable subprocesses that make up this piece of verbal intercourse, but that are abstracted away from when considering individuals only. That is why the event of a speaking to b is different from the spatiotemporal container comprising a and b.

Reflecting on this example in a completely formal way, what we can say about processes, then, is that they at least have a mereology: it is

[27]The philosophical notion of process that I am discussing here is to be distinguished from the kind of processes that are found in the literature on linguistic semantics: the former are meant to express a very general ontological idea, while the latter have a rather specific meaning as atelic eventualities. Authors like Vendler (1967c); Mourelatos (1978), however, have not failed to point to the philosophical implications of these eventualities.

important to be able to consider a number of processes taken together as forming a bigger process, i.e., their fusion.[28] That mereology should not be atomistic, however (more precisely: we shouldn't take a stand on the issue of atomicity on way or the other), as we have no reason to decide this question on purely ontological grounds. Physics might give us clues to the effect that reality doesn't divide into ever smaller parts, but recall that we decided to disregard such evidence in metaphysics when making general claims about the world. A more philosophical argument for atomicity derives from Logical Atomism, but even Armstrong (1997) who wants to revive that doctrine in a modern form states explicitly that he likes to remain neutral towards Atomism (p. 1). I for one have always considered totally misguided the picture suggested by Logical Atomism that there be a perfect match between our concepts and the world and hence, between atomic terms and atomic facts or processes; thus I certainly do not want to make use of that argument to support atomicity of the process mereology, at least not in an absolute sense. But many complex processes will become atoms in special situations, typically, when bound together by a group of stable constraints that make them into the substrata of ordinary individuals; for instance, the human body is a highly complex chemical system with a vast number of subprocesses running, but in such a way that they support a human being that shows stable characteristics over some period of time.

Neutrality towards the question of atomicity reflects in a very weak sense the notion of homogeneity that has often been associated with processes.[29] One might try to formulate homogeneity in terms of a principle of divisibility, akin to the one found in the discussion of mass terms. But recall from the mass case that one will be immediately confronted with the problem of minimal parts. Even if we tried to relativize homogeneity to some notion of "granularity" there would still be not much to be said from an purely ontological point of view. It seems to me that homogeneity is an idea that should rather be addressed in epistemology (which is, of course, one of the main concerns of the classical authors), so I won't say more about it here. However, when we come to individuals there will be a principle which expresses some notion of *discreteness* (as opposed to homogeneity) that we intuitively associate with ordinary objects in the world around us.

Thus processes are mereologically structured entities that form my basic particulars. I assume that their mereology is classical, but not explicitly atomic. In particular, there is no such thing as "the null process," i.e., the

[28]This notion of "bigger" is to be taken in the complexity sense, as I called it in the previous chapters; a second type of mereologies will be induced by the spatio-temporal trace functions.

[29]Cf. Whitehead (1929); Sellars (1981).

process lattice doesn't have a bottom element.[30]

Processes occupy *space* and evolve in *time*. There have been approaches that eliminate space and time in favor of individuals and events, that is, in the current picture, in favor of processes. For instance, it is said that space is created by objects giving meaning to the notion of distance, and that instances of time can be constructed from all the events or processes occurring at them.[31] I make no attempt here to describe this additional (and downright reductionist) step; instead I will treat space and time as given, just like in the previous chapters. That is, there is a non-atomic lattice of space-time regions that splits into separate lattices by "projection," one for spatial regions (places) and one for time stretches (times).[32] Since processes are in space and in time they have temporal, spatial and spatiotemporal *traces*. Let π, π' be variables for processes, t, t' for times, s, s' for places, and r, r' for (space-time) regions. τ, σ, and ρ are the temporal, spatial, and spatiotemporal trace functions, respectively, which operate on processes to yield their corresponding containers in space and time. If r is a region, call $\pi^*(r)$ the *full process* of everything that goes on at r. The restriction of a process π to a certain time t, place s, or region r will be denoted by π_t, π_s, and π_r, respectively.

The partial ordering relations on places, times, and regions induce three natural part relations of *trace inclusion* on processes, called $\sqsubseteq_\sigma$, $\sqsubseteq_\tau$, and $\sqsubseteq_\rho$, respectively. Mutual trace inclusion yields identity of traces: two processes π and π' can be *trace-identical* again in three ways: they can have the same spatial, temporal, or spatiotemporal trace, which will be denoted by $\pi \sim_\sigma \pi'$, $\pi \sim_\tau \pi'$, $\pi \sim_\rho \pi'$, respectively.

I will also stay neutral towards the question whether space and time are continuous or discrete. But in order to be able to formulate a discreteness principle for individuals below I will need some notion of *spatiotemporal size*. Therefore I assume that the spatiotemporal lattices are equipped with a some *measure* (call it μ), which I won't specify here. The only thing that I will make use of is a qualitative relation "is at least of size," given by μ, and its corresponding equivalence relation $\sim_\mu$ reading "is of same size as."

The lattice structure on the processes is actually such that for a given pa-

[30] For a survey of various mereological systems, see Simons (1987).

[31] See Strawson (1959) for the relation of objects to space; Whitehead (1919, 1929) and Russell (1914, 1927) for the classical construction of times from events; for modern accounts of the latter, see van Benthem (1983b); Kamp and Reyle (1993). For still more recent work on relating ontology to a topological theory of regions, cf. Forrest (1996a); Roeper (1997).

[32] Recall from Chapters 10 and 11 that I think of time stretches as unions of time intervals; thus times are not instantaneous but rather extended, not necessarily connected, temporal periods.

ramcter in space and/or time there is a whole lattice of processes occurring at it. Thus we have three bundles of lattices, parameterized respectively by times, places, and regions. For instance, we might consider everything that happened on June 6, 1944 anywhere (the Operation Overlord in the French Normandy only being part of it), or everything that happened in the area of the the city of Rome (*anno urbis conditae*, before that and ever after), or else what was going on in Paris on July 14, 1789. In each case we get a whole mereology of processes going on at the given location.

In an earlier chapter I mentioned already the homomorphic relationship between events and their location in space and time. The same should hold for processes, i.e., we have the following *homomorphism principles*: Let ξ be one of ρ, σ, τ; then

$$\xi(\pi \sqcup \pi') = \xi(\pi) \sqcup \xi(\pi')$$

There is one further principle that I want to introduce in connection with processes. It is the idea that processes evolve in time according to some *causal flow of energy*. It is of course impossible to say anything specific about the notion of causal flow on the present level of generality. However, on the realist view of the world, it is those processes out there that transport energy and not, for instance, facts or other mereological aggregates; thus the causal action, i.e., the above-mentioned "cement of the universe," is located in the processes. To indicate the causal flow let us use the symbol '$\leadsto$' (to be filled with meaning in a theory of causality[33]). Then, if π is any process and $\pi_{<t}$ (π_t) is the subprocess of π up to t (at t) for any given time t, we can formulate the general principle of causal flow:

$$\forall t\,(\,t \sqsubseteq \tau(\pi) \rightarrow \pi_{<t} \leadsto \pi_t\,)$$

(*Causal Flow*)

12.3.4 The Notion of Individual

This concludes my discussion of the general concept of process. I turn now to the notion of an *individual* which, as I said, I want to conceive of as a special type of process. I am going to motivate six very general principles now that are all intended to capture some of the characteristics of an individual viewed as a process. They are: (i) *constitution*, (ii) *coherence*, (iii) *stability*, (iv) *abstractness*, (v) *local manifestation*, and (vi) *discreteness*.

If an individual is a process it has temporal parts or stages. Let $a, b, \ldots$ run over processes that are individuals. As with processes in general, let a_t denote the subprocess of an individual a occurring at t, where t is some

[33] For that see Mackie (1974); Suppes (1984); Spohn (1984).

part of the temporal trace $\tau(a)$ of a. Thus a_t is the temporal stage of a at t. Then the Principle of Constitution says: if you chop up an individual into a number of mutually exclusive and jointly exhaustive temporal stages, then those chunks, taken together in the plural sense, don't quite give you back what you started from but they do constitute the original individual. The idea behind taking constitution rather than identity is that plural sums, being ontologically neutral, cannot all by themselves "put Humpty Dumpty together again," that is, provide the causal glue that holds the individual together. To state the principle we have to use the notion of proper partition, expressed in LP, i.e., a proper i-sum ζ partitioning an individual a ('$\wr$' is disjointness, and '$\bigsqcup$' is the fusion operation in the underlying process or space/time algebra):

$$Ptn^*(\zeta, a) \;:\leftrightarrow\; \neg\mathbf{At}\,\zeta \wedge \forall x \forall y (x\,\dot{\Pi}\zeta \wedge y\,\dot{\Pi}\zeta \rightarrow x \wr y) \wedge a = \bigsqcup \zeta$$

(the ζ *properly partition* a)

Then the principle reads:

$$Ptn^*(\zeta, \tau(a)) \;\rightarrow\; \sigma\pi\, \exists t(t\,\dot{\Pi}\,\zeta \wedge \pi = a_t) \rhd a$$

(*constitution*)

In the beginning of this chapter I called an individual a stationary or stable process. There is a related idea, that of *continuity*, that is often associated with (concrete) individuals; for instance, Armstrong (1997):104 says that a physical thing "... is, roughly, a matter of spatiotemporal continuity plus the resemblance of adjoining portions of the spacetime line." We would have to introduce a lot of extra machinery to make the notion of spatiotemporal continuity precise, and at the same time fend off objections against continuity, both temporal (viz. problems of intermittent existence[34]) and spatial.[35] Thus I won't pursue the idea of continuity here. Rather, I replace it by a different, but related concept, coherence, which is to capture *causal* continuity and not just the "kinematics" of continuous orbits in space-time. I therefore repeat the causal flow principle for individuals, with the special twist that given a temporal stage a_t of an individual a, I consider not only its own history $a_{<t}$ as causally feeding into the stage, but rather some full process prior to t that contains the region of a. The reason is that individuals, being more or less "abstract" in the above sense, will not in general be physically closed systems.

$$\forall a \forall t \exists r\, (\, t \sqsubseteq \tau(a) \;\rightarrow\; \rho(a) \sqsubseteq r \;\wedge\; \pi^*_{<t}(r) \rightsquigarrow a_t\,)$$

(*Coherence*)

[34]For a discussion see Simons (1987):5.4

[35]Counterexamples from microphysics might be adduced here, e.g., quantum jumps.

Returning to the idea of resemblance of adjoining portions in the above quote, I will generalize it to the notion of *stability*: There should be a complex of properties that remain stable, i.e., apply to it all the time during the individual's existence. Typically, a stable physical appearance giving rise to resemblance will be among those properties, but that is not required (the sun will barely resemble the white dwarf it eventually turns into). Moreover, the stable properties should all be non-negative and non-disjunctive in order to avoid easy trivializations of the stability principle to be given presently. I assume a lattice structure on the properties that contains arbitrary conjunctions, so that stable property complex can be expressed by just a single property. Furthermore, since those properties are properties of processes I call them *process types*, with variables θ, θ'. The relation of a process π *being of* type θ will be expressed by '$[\pi : \theta]$'. Then the principle goes as follows:

$$\forall a \exists \theta \forall t \, (\, t \sqsubseteq \tau(a) \;\rightarrow\; [a_t : \theta] \,)$$

(*Stability*)

It is tempting to insert a modal operator here to the effect that the θ in this principle applies to all the given stages *by necessity*. However, that would open up a plethora of difficult questions about essentialism that cannot be dealt with here.

The next principle expresses the idea that individuals are more "abstract" than the full processes underlying them; a very weak formulation of this would be that at least in some region of the spatiotemporal trace of an individual its subprocess there doesn't exhaust the local full subprocess in this region:

$$\forall a \exists r \, (\, r \sqsubseteq \rho(a) \;\wedge\; a_r \sqsubset \pi^*_r(\rho(a)) \,)$$

(*Abstractness*)

Furthermore, I want to convey the common sense idea that individuals seem to be "wholly present" at any time during their existence. In the language of process-based individuals that means that every temporal stage of an individual is a complete manifestation of it . Here is a rough definition of manifestation: a process π is said to *manifest* another process π' just in case whenever a temporal stage of π' during π's existence is of some type θ then π is also of type θ at that stage; or, spelled out formally by using $\lhd$ for the relation of manifestation:[36]

[36] Just like the symbol '$\rhd$' for constitution, '$\lhd$' is intended to convey its meaning pictorially; it is not the converse relation of the former.

$$\pi \lhd \pi' \;:\Leftrightarrow\; \forall\theta\forall t\,(\,t \sqsubseteq \tau(\pi) \,\wedge\, [\pi'_t : \theta] \,\rightarrow\, [\pi_t : \theta]\,)$$

Now by the stability principle it follows from this definition that temporal stages of individuals are their "local manifestations." This is what "wholly present" means in our current framework of processes. Note also that arbitrary processes will in general not have local manifestations since they might fail to be stationary.

$$\forall t\,(\,t \sqsubseteq \tau(a) \,\rightarrow\, a_t \lhd a\,)$$

(*Local Manifestation*)

Thus the local manifestations at a time possess all the characteristic properties of the individual at that time, and hence can be said to truly represent it. For instance, when you meet Quine you will have his current local manifestation in your field of vision. The whole individual Quine cannot be completely displayed "because, as he would protest, 'of my hence and ago'," as David Kaplan once put it. However, in view of the stability and coherence principles, that stage has most anything that could be asked for of a representation of Quine's personality.

A theme that is closely related to the issue of locality is *reference*. Indeed, reference works locally, too. Take reference by ostension, for instance; we point in a certain direction and are able to pick out a much bigger object than we could possibly circumscribe by a gesture. Presumably, the way we do this is in virtue of some stable features in the object across its spatial location. By the same token, temporal stability enables us to refer to an individual that extends in time, with an ostensive gesture towards its current temporal manifestation.

The final principle tries to capture the intuition that individuals are not homogenous entities, but are processes that lump together into "discrete" units. It is not easy to capture this idea with the limited resources at hand, but one part of it is that we don't have individuals of the same kind of ever smaller size. I will express this by saying that every process type that applies to individuals comes with a spatial "gauge region" that gives the minimal size of an object falling under that type; to take care of possible growth the gauge region will be made dependent on the stages. That is, the spatial trace of such an object (stage) will contain a spatial region that is of same size as the gauge region; formally:

$$\forall\theta\forall t\exists s\forall a\,(\,[a_t : \theta] \,\rightarrow\, \exists s'\,(\,s' \sim_\mu s \,\wedge\, s' \sqsubseteq \rho(a_t)\,)\,)$$

(*Discreteness*)

There is the dual intuition about objects that they are not only not infinitesimal but also not infinite in size. Thus Zemach (1970), for instance, assumes that it is characteristic of a thing ontology that its objects are bounded in space and in time. Now it is presumably true that in an ever-changing world nothing will persist forever, and also, that manageable objects have a finite size in space. But consider electrons; fix the momentum of one such, and its position will exceed any given bounded region in space, by Heisenberg's uncertainty principle. Doubtful as the value of that kind of example may be for ontological arguments, I will nonetheless stay away from including Zemach's principle in the present list.

So much for the principles for individuals that single them out from arbitrary processes. They might have left the impression that they are lacking content; but then again, what seems to be missing here is rather the kind of rich empirical content of a world view that derives from modern science. In congruence with what I said above, however, my account tries to stay away from drawing heavily on our world knowledge in formulating ontological principles.

One final remark. All the individuals considered here are atomic in the sense of the plural lattice. Considered as processes, however, they are elements in the structure of the process lattice. But typically, individuals fail to be closed under the operations of this lattice. In particular, the Boolean operation of complementation will not in general lead from individuals to individuals, the reason being that the complement process to an individual might not satisfy the stability condition that yields another stationary process. Note that this is different from the plural lattice where every element that is not the top element has a (unique) complement, viz. the i-sum of all elements disjoint to it.

I conclude this section by addressing a number of objections that have been raised against a conception of individuals which assimilates them to occurrents like processes or events. The first one links up with the temporal part problem; it is said that individuals have properties at a time and not *simpliciter* like occurrents do. Now a property that can change its extension over time is usually represented by a relational predicate of the form $P(a,t)$, expressing a time-indexed property. I assume here that P has a "time-less" process-type correlate θ^P that applies to the t-stage of an individual a just in case P is true of a at t:

$$P(a,t) \leftrightarrow [a_t : \theta^P]$$

In this way an individual can have a property at a time and then again lack it at another, making change possible in the first place. Thus the "intrinsic properties" that Lewis (1986b):204 speaks about are really process types or, as I will call them below, event types. They apply to stages of individuals, whereas the enduring individual enters a relation to times which is derived from the intrinsic property (I consider the ontological dependence going in the opposite direction, i.e., from the event type to the temporal relation).

A related problem for the temporal-parts view of individuals is this. When Alfred Brendel plays Beethoven's sonata op. 111, we really should be saying that one stage of Brendel plays the first movement and another stage of his the second movement (cf. also Hinrichs (1985)). Such a consequence has always seemed to me implausible because it invites clutter; don't we say that it is one and the same person who plays both movements, albeit at different times? Only people can play music, but stages of people are not people (cf. the quote from Quinton above). However, in our current metaphysical framework we can rework this case in such a way as to preserve that intuition about ordinary individuals while showing how it is actually derived from the fine-structure of processes. By the locality principle, it is indeed Brendel's current stages that interact with the piano is such a way as to produce a performance of the sonata; but the principles of stability and coherence for individuals tie those stages together, yielding a continuous manifestation of the familiar kind that we call a performing pianist.

Recall that I distinguished above between the container view and the four-dimensional view of individuals. I rejected the container view as too coarse, but adopted the four-dimensional view, recast in the language of processes. Thus objections against the four-dimensional view should be addressed here, too. We have already dealt with two of them, the issues of time-dependent properties and temporal parts. There is another standard complaint concerning motion: it is said that regular objects can move around whereas processes, being four-dimensional, cannot. But of course, easy rephrasing of the term 'move' will fix that problem, too: it is the current manifestations of an individual that shift their spatial location as time goes on.[37]

Finally, let us ask what should be done with examples of the kind Quine likes to give for his four-dimensional view of physical objects. For instance, in (1981):13 he says that the president of the United States is one of those four-dimensional objects that could also be called "the US presidency" as it extends from the days of George Washington to the present term of Bill Clinton's, and on to the future as long as there will be that office in the US. In possible world semantics, the term "the US president" would rather be a paradigm case for an *individual concept*, a function from times to extensions

[37] See also the discussion in (Bennett (1988):114).

(people), returning as value at a time t the US president at t. However, in the Quinean picture these people are all "pasted together" to form one big presidential worm in space-time. That is rightly felt to be counter-intuitive, the main reason being that by using the definite description "the US president" (or "the US presidency," for that matter) we just don't succeed in referring to that worm. Either we talk about the particular president at a time, or else about the office, e.g., in statements like "The US president is the commander-in-chief of the armed forces." However, the fact that there are no ready-made terms in the language to refer to a certain particular doesn't mean that this particular doesn't exist. In fact, the stability condition above won't exclude the Quinean worm from the domain of individuals; the stationary complex of properties just wouldn't include those features that are unique to a particular human being. We happen to think that those features are rather essential, and that all those presidents were rather different from one another. But take the Dalai Lama; if reincarnation were part of our world view the succession of Dalai Lamas would be a much more homogenous individual.

Going on with one more pair of examples, how does the Ship-of-Theseus problem present itself in the light of the current analysis? Well, considered stage-wise, in the beginning there is one ship, to be sure, and in the end there are two, one still sailing the seas, with all the parts replaced, and the other being displayed in a museum, reassembled from all the original parts; the former, at the final stage, is a perfect copy of the latter. At one point in time t the temporal stage at t of the active ship will just have ceased to be a manifestation of the original ship; *when* that happens is just a matter of vagueness that has nothing to do with ontology.[38] However, considered four-dimensionally, there are two full individuals, the "form-constant" ship and the "matter-constant" ship; two different sets of stability conditions apply to them. There is overlap in the beginning, but later they split up.[39]

The last example is a variant of the Tibbles case, involving a regular person, Dion, and another object called Theon which is like Dion except for his left foot.[40] Suppose Dion actually looses his left foot; are Dion and Theon the same now? For instance, M. Burke draws the surprising conclusion from a number of supposedly common-sensical assumptions that the very moment the foot is detached Theon ceases to exist. My reaction to this example is the same as with Tibbles the cat: within the rather

[38]One could be precise about this at any time should there be need for it, e.g., by fixing some percentage of the old parts that would still have to be present for the ship to be considered the original one.

[39]The terms 'form-constant' and 'matter-constant' are due to (Simons (1987):200). Although Simons explicitly rejects the four-dimensional view, we see that his terms fit into it rather nicely.

[40]Cf. Burke (1992, 1994a); Denkel (1995).

fixed limits set by the biology of living things it is a question of vagueness how much physical mutilation an animal or a person can survive; the given quantity of matter just doesn't "define" the individual which it constitutes. Theon should be considered one such portion of matter throughout; before the separation of the foot it was a proper part of the quantity of matter constituting Dion, afterwards it happens to constitute him all by itself. Having the constitution relation available we can avoid Burke's conclusion that Theon would have to turn into the person Dion in order to survive the operation.

12.4 Events

When approaching the metaphysics of events we have to make a distinction whose importance was rightly emphasized by Jonathan Bennett in his book (1988). The distinction appears right in its title, "Events and Their Names," making it clear that on the one hand there are the objects we are concerned with in the present discussion, viz. the *events*, and on the other hand there are the linguistic terms that we use to refer to those entities, the *event names*. Now there is hardly anything more obvious and basic in semantic analysis than this distinction; so you might wonder what the point is of harping on it. In the analysis of events, however, the term and its denotation have indeed tended to be conflated to a large degree. That is not altogether surprising since there seems to be no other way of picking out an event than by giving a description of it. Thus the descriptive term is typically taken for the entity described, or at least some one-to-one correspondence has been set up between event terms and events. Such a theory basically applies the "granularity" of the descriptive terms (or of the properties coming with them) to the identity criteria of events, arriving at a rather fine-grained account of events. Since the distinctive feature of that kind of theory is the properties involved in the event name I call it the *property view*, its main adherent being Jaegwon Kim.[41] According to Kim an event is roughly a "substance" exemplifying a property at a spatiotemporal location. On the other side of the granularity spectrum there is the *container view* of events which basically equates an event with the spatiotemporal region at with it occurs; well-known proponents of that view are Lemmon and Quine.[42] Quine actually speaks of the "material content" of a region, presumably hinting at what I above called the "full process." But as Bennett rightly observes the region can stand in for its content as

[41] See Kim (1980).

[42] Lemmon (1967) and Quine (1960, 1985).

far as identity criteria are concerned.[43]

In his own careful analysis, Bennett himself doesn't want to pick just an arbitrary point in the middle between these two extremes. Rather, he argues that an event name can refer to more "content" than what is expressed by the name. He identifies this content with an instance of a complex property "that includes but may not be identical with the property expressed by the predicate in the event name" (op. cit., p. 128). Since "concrete" instances of properties are nowadays called *tropes*[44] Bennett calls this his "trope thesis." In fact, he thinks that it is more profitable to go from the complex trope to what he calls its "companion fact" because facts are more manageable than events or tropes. Thus, given an event name involving an object S, a property P and a spatiotemporal region T, that name refers to an instance of a certain complex property P^* possessed by S at T and including P; the companion fact is then the fact that S has P^* at T. Let me call this position the *inclusive trope-fact view*. Its two main points are the following: (i) An event name is able to refer to a rather complex chunk of the concrete world that has a fuller content than what is conveyed by the linguistic material in the name; and (ii) the semantics of event names should be developed in terms of *facts* rather than in terms of some other metaphysical category. Thus the upshot of that approach is that events are basically given up in favor of facts.[45]

Indeed, Bennett thinks of events as being supervenient on facts. But then it is incumbent on him to say what the ontological nature of facts is. I won't discuss what he has to say about that; instead I will dwell for a moment on one of the main applications of the event concept, i.e., *causation*. Bennett contrasts event causation with fact causation and gives a number of reasons for preferring the latter to the former (op. cit., §54). He calls the language of event causation "not very informative," "fairly indeterminate," and "poorly adapted to cope with [various] kinds of fine-graining." He considers the example of an electric motor that is connected with two power sources each of which is sufficient upon activation of a switch to start the motor. Now suppose both switches are flipped at the same time (call these events e_1 and e_2), and the motor starts (= event e); which event now actually caused e, e_1 or e_2? It can't be e_1, according to Bennett, because by the same token, we might say that e_2 caused e, and vice versa. Bennett goes on:

---

[43](Bennett (1988):104).

[44]For a primer on tropes, with useful references, see Bacon (1997).

[45]Recently, there has been another revival of facts as the basic particulars of the world, under the name "states of affairs," in Armstrong (1997). I for one prefer the more "dynamic" metaphor expressed by Sellars in his well-known phrase: "The world is an ongoing tissue of goings on" (Sellars (1981):57). Some of the systematic reasons for this preference are given below.

> Or we might say that the motor's start was caused by a spatially discontinuous event, namely the fusion of the two switch-flips. But not everyone is happy with such fusions, and in any case this account of the matter is misleading at best, because it seems to imply that the two flips collaborated in getting the motor to start, and that is not so. (Bennett (1988):140)

Bennett then explains why matters are different with fact causation; there you can say straightforwardly that is was the fact that *at least one* of the switches were flipped that caused the start of the motor (better: caused the fact that the motor started). "Fact causation," Bennett says, "gives us the luxuries of disjunction and conjunction, which are not available with event causation" (ibid.).

Now to begin with, one of the main tenets of this book has been that our ontology shouldn't be considered a loose collection of entities, but a highly structured universe. In particular, when events are viewed as a special kind of processes they come with a structure that just gives the right amount of "luxury" of algebraic operations that are needed for dealing with multiple causation and similar phenomena. It is true that on my view there are no disjunctive or negative *processes*, but on the level of events, which will be described as processes classified by event types, there will be logical operations available on the types that can emulate the tools of the fact language. Before I come to that, however, let me first finish the discussion of the example.

I think that it is a natural principle that any event which includes as part an event which is a sufficient cause for some other event e is itself a sufficient cause for e. Thus the fusion $e_1 \sqcup e_2$ of the above events is sufficient for e even if either one of the component events is already all by itself. There is nothing odd about this, and there is no implication of collaboration involved. Notice, however, that that fusion is not a disjunctive event, that is, it is not the analog of the existential fact that at least one of the switches was flipped. Thus there does seem to be some disparity there.

Here and elsewhere, I think, we have to distinguish between *causation* and *explanation*. While causation should be dealt with in the event language, the fact language is the suitable instrument for explanation. So rather than saying, like Bennett does, that the fact that at least one of the switches was closed is *the cause* of the fact that the motor started I prefer to say that that fact *explains why* the motor started. Causation is located at the level of processes, not of facts. Thus there is a characteristic *structural* difference between those two kinds of entity that has to be observed: While facts are *propositional* in character, events are not. Under the process view, events are parts of the world which can be lumped together by a fusion operation that is conjunctive in nature, but which don't lend themselves to forming disjunctions. The reason for this is that considering alternatives is really an operation on our thoughts and not one that you can perform on the world. In particular when we think of the causal flow that holds those processes together it should become clear that there are no "disjunctive" or "negative flows of energy" in the world, and thus there is no disjunctive or negative causation.[46]

But Bennett insists that having negative facts available in the analysis of causation is not only a luxury but rather an indispensable tool in many applications, for instance, moral philosophy. There you have plenty of examples in which, using the fact language, the fact that somebody *failed* to do something caused other facts to come true for which he or she is held responsible.[47] However, those examples concern cases in which typically there is a lack of causal interference (rather than causal influence) on the side of the person who is held responsible for the outcome. The fact that he or she failed to act might *raise the probability* of the outcome and thereby contribute to an explanation of it, but we have to remind ourselves that probabilistic correlation is not the same as causal influence. To give an example from recent international politics: There is certainly a positive correlation between the inactivity of the UN regarding the war in former Yugoslavia and the amount of bloodshed there. However, it was crazed people and militias there and not the UN who *caused* the bloodshed. I am not saying that it is inappropriate theoretically to find some responsibility on the side of the UN here; quite to the contrary: there is obviously responsibility for omissions. Here is how I would like to describe the general situation: people or institutions are responsible for their behavior, which, however, is always a "positive" process. An omission in action theory becomes "negative" as a process classified by a negative event type. It is in this sense that there are negative or disjunctive events. Thus the flexibility of the logical operations (including quantification) is afforded by the algebraic structure on the event types. Hence there is no reason to give up event causation on the grounds of its alleged inflexibility.

12.4.1 The Classified Process View

Let me now say a little bit more precisely, then, what events are on my present view, which I dub the *classified process view*. I want to keep the basic features of the "Aether" model structure which were described in

[46]In fact, something to this effect was observed also by Bennett who writes, speaking of a chunk of space-time (which compares to our process): "There is no chance of making that entity negative in itself: negativeness is always *de dicto*, not *de re*" (Bennett (1981):54).

[47]Bennett (1988):140f.; Bennett (1981).

the previous two chapters, but I would now like to organize the various conceptual tools there in a somewhat different way. To begin with, we still have the usual temporal, spatial and spatiotemporal lattices, with the syntactic variables for their elements that were introduced above. The lattice of event types, $\mathcal{E}$, also remains basically the same. There are four principal changes which are the following: (i) the domain E of particular events in the Aether model is replaced by the lattice of processes, called $\mathcal{P}$, which is then the domain of the spatiotemporal trace functions; (ii) the roles are now attached to the event types; (iii) a domain of new events is added which contains processes that are "classified" with respect to an event type; and finally, (iv) a constitution relation similar to the one for individuals is introduced which relates processes to events. Thus an *event* e is a process π together with an event type θ such that π is of type θ; π is then said to *constitute* e. It is from the type θ that the event e receives the structure that is typical for an event: the ensemble of individuals playing roles in the underlying process that are specific to θ.[48] In this way the *problem of atomicity* of events is solved, correcting my earlier treatment which I have been dissatisfied with for some time. When the role function operates on the particular events then such an event, if it is, say, of an atomic action type, is inherently linked to the granularity of actions, and subprocesses cannot be part of the event anymore. Under the atomistic view, for instance, if the event e is a particular walking of Mary, then e is an atomic event, and subevents as Mary's moving her right leg are precluded from being proper parts of the same complex process, which intuitively they are.

Given this set-up, let us ask ourselves exactly how fine-grained those events are now. From a purely formal point of view I defined them as entities that are basically as fine-grained as Kim-style events since my event types correspond to his properties. However, I prefer to view the matter in a different light: the "stuff" that events are made of are the processes, and in most cases when we have the strong intuition that two events are the same we are actually thinking of the underlying processes.[49] But we cannot apply Leibniz' law to those processes without qualification because that would generate all the puzzles about event identity that permeate the literature on events. Thus we will have to appeal to loose identity in the sense explained above (examples will be given presently). In some contexts two events can safely considered to be the same because what is at issue

[48] The individuals involved, considered as processes, are of course subprocesses of the process underlying e; but in the event they are singled out as its constitutive elements from the many more but mostly irrelevant subprocesses of the event.

[49] Notice that stripping types (or properties) away is a move that isn't open to Kim as that would leave him with individuals and not with occurrents, and the dynamic quality of the event conception would be lost.

there is the process underlying both of them; but then again, in another context the different types may become important, and the process view alone, or worse, the container view, just won't be able to give a reasonable account of the situation.

Another concern is how "abstract" those events are. Well, processes are concrete, to be sure, at least when they are full processes; but if our focus is on the type component in the event then it is hard to maintain the Davidsonian view that events are concrete particulars, unless types are reconstructed nominalistically as tropes. And apart from that there is almost always some degree of abstractness involved in sense of "leaving something out" (see above), that is, we rarely deal with full processes *simpliciter*. So I would say that events are, if not outright concrete, at least firmly rooted in concrete parts of the world. However, they are more abstract than individuals because there is more occasion to deal with them *qua* the types they are classified by than in the case of individuals, which are processes with prominent stability features. That is, to take up our examples from above, more confusion about identity is likely to arise when we forget to mention that a particular chewing of gum also happens to be a walk, than when we discuss Richard Wagner *qua* composer or *qua* antisemite.[50]

The looseness of event identity is a major source of problems with reference to events and their underlying processes. Usually, when we refer to regular objects we can pick out an object by using a descriptive term t, but we thereby refer to the whole individual and not just to it *qua* t. The situation is different with events: Let N be a linguistic expression that is used to refer to an event, e.g., a nominal), and let θ be the corresponding event type; then there is always the question, whether N picks out precisely the event classified by θ or the underlying process, or that process together with some other event type it falls under. Bennett (1988) has a clue from language that seems promising here although his theory is phrased in the fact language; he claims that imperfect nominals like *Quisling's betraying Norway* systematically differ in their referential behavior from perfect nominals like *Quisling's betrayal of Norway*: "An imperfect nominal names the fact that it expresses; a perfect nominal picks out a fact richer than the one it expresses." (p. 131) That richer fact includes, but is not identical with, the fact expressed by the imperfect nominal. When we translate that to our own set-up we have to make some adjustments. Bennett himself

[50]Incidentally, here is a notable difference between the business of a semanticist and a philosopher. The philosopher can always do some hand waiving regarding strict identity, and in many circumstances that is not even bad. By contrast, the semanticist has be provide a model for the language he or she gives an interpretation of, and that entails specifying a domain of possible denotations for the singular terms where no loose identity is permitted. That is why the kind of machinery like the Aether model, set up originally for semantic purposes, might look abhorrent to some philosopher's eye.

sees a very close correspondence between facts and properties construed as tropes, so he would probably be prepared to speak of tropes derived from event types. The inclusion relation on facts translates to a part relation on the tropes. That would mean, for example, that if Mary walks down University Avenue at a particular time then the imperfect nominal *Mary's walking down University Avenue* would refer to the event of Mary walking down that street at that occasion, conceived as the trope expressed by this very nominal, whereas the perfect nominal *Mary's walk on University Avenue* would refer to the richer trope that includes all the particular features which that walk happened to have, e.g., her seven stops along the way, her degree of window shopping, the conversation she had with her boyfriend escorting her, and so on.

There is one further transformation I want to make in order to get rid of the tropes. Trope theorists seem to want to have their cake and eat it, too. They want to have all the expressive power of the language of properties deriving from their linguistic descriptions and at the same time get a grip on the stuff out there in the world, rather than staying in the lofty realm of conceptualizations. While that issue leads right to the heart of metaphysics, pursuing it would lead us far astray, so, as I said in the beginning, I have to leave it undiscussed. Thus in my terms, *Mary's walking down University Avenue* refers to the "thin" event where everything except the type WALKING is abstracted away from the underlying process, and *Mary's walk* refers to that process in its totality; or still shorter: *Imperfect nominals refer to thin processes or events, perfect nominals refer to full processes.*

Suppose then that this analysis is at least qualitatively correct.[51] Then, because of the surplus referential potential of many event names exceeding their descriptive content, a lot of vagueness and indeterminacy can be predicted to impair the analysis of events, and this is exactly what happens, as the many idle disputes over the question of event identity show. In the examples below the attempt will be made to resolve at least some of the problems about identity in a way that takes account of that indeterminacy.

[51] I am somewhat skeptical about sorting things out the way Bennett does; correlating specific referential potentials with certain kinds of nominals is really a claim about language, not metaphysics; for the latter, we shouldn't say more than that the two referential modes can be expressed in language *in one way or another*. Indeed it is doubtful whether the distinction between perfect and imperfect nominals can be drawn at all in languages other than English, let alone whether their semantic function is stable across many other, possibly remote languages. The conclusion to be drawn in the present discussion, however, is unaffected by this point.

12.4.2 Formal Properties of Events

I now proceed to sketch some of the formal properties of events as envisaged here, many of which are induced from the given structure on individuals, processes, and types. The *model structure* for the event theory now looks like this: it is a quintuple $\langle A, \mathcal{P}, \mathcal{T}, E, \mathcal{R}\rangle$, where A is the set of individuals, $\mathcal{P}$ the set of processes, $\mathcal{T}$ the set of types, E the set of events, and $\mathcal{R}$ a finite set of roles, with variables R, R'. These roles are 3-place relations between types, events, and individuals. $R(\theta, e, a)$ reads: "a plays role R in e with respect to θ." For $R \in \mathcal{R}$ and $\theta \in \mathcal{T}$ I introduce a predicate that says that *R is defined for θ*:

$$Df(R, \theta) \;:\Leftrightarrow\; \exists e \exists a R(\theta, e, a)$$

A, $\mathcal{P}$ and $\mathcal{T}$ essentially inherit the structure from above or from earlier chapters. In particular, there is the *is-of-type* relation, $[:]$, between processes and types, and there are the trace functions ρ, σ, τ defined on $\mathcal{P}$,[52] together with the partial order relations on those sets. They induce the trace part relations $\sqsubseteq_\sigma$, $\sqsubseteq_\tau$, and $\sqsubseteq_\rho$ as indicated above. The relation $\sqsubseteq$ will be used both for the complexity ordering on the processes and for the partial ordering on the types. Finally, there is a two-place function symbol $(\cdot \mid \cdot)$ taking a process π and a type θ to form an event $e = (\pi \mid \theta)$, and a two-place relation of the form $\pi \rhd e$ between processes and events.

Now we come to event principles proper. All *elementary events* are of the form $(\pi \mid \theta)$ for some π, θ. An event $e = (\pi \mid \theta)$ exists just in case the process π is of type θ; that is, we have the following Kim-style *Existence Principle*:

$$\forall \pi \forall \theta \, (\, [\pi : \theta] \;\leftrightarrow\; \exists e . e = (\pi \mid \theta)\,)$$

A process-type pair gives a unique event; also, if e is an event then there is to be a unique process π and a unique type θ such that $e = (\pi \mid \theta)$. Hence we can define, for any $e \in E$, *the process underlying* e as $\pi^e := \iota\pi.\exists\theta\, e = (\pi \mid \theta)$, and *the type classifying* e as $\theta^e := \iota\theta.\exists\pi\, e = (\pi \mid \theta)$. We will say that the event e is of type θ (in symbols: $[e : \theta]$) if the underlying process is, i.e., if $[\pi^e : \theta]$. There will be other processes that constitute e, and other types θ such that $[e : \theta]$ but the terms 'underlying' and 'classifying' will be reserved for the constituents of an event. Also, a trace of an event e is defined as the trace of its underlying process, in symbols: $\xi(e) = \xi(\pi^e)$ where ξ is one of ρ, σ, τ.

[52] In what follows the symbols 'ρ', 'σ', 'τ' will also be used as function symbols in the object language; the same holds for the symbol '$[\pi : \theta]$'.

If an event e exists then it has to occur somewhere in space and in time. Again, the spatiotemporal region is given by that of its underlying process. Therefore we can write down the following simple *Occurrence Principle:*[53]

$$\mathbf{E}\, e \to \exists r\, \rho(e) \sqsubseteq r$$

There is a natural equivalence relation between two events e and e' which holds just in case e and e' share their underlying process, i.e., define:

$$e \sim e' \;:\Leftrightarrow\; \pi^{e} = \pi^{e'}$$

This relation yields a loose notion of identity between events, one that is sufficient in all contexts in which only the underlying process is at issue, for instance, at what place or what time an event occurs. That is the kind of homomorphic transition characterized by the neglect of some information: the fine-structure of events is ignored, and that is why intuitions about, say, the temporal location of events are relatively solid. However, what might be adequate for many applications of the event concept will in general not be able to provide all the necessary distinctions for a thorough philosophical analysis; for that we need a fine-structure of the kind explored here.

The first constituent π of an event $e = (\pi \mid \theta)$ is the unique process underlying e. But that process might happen to be rather "thin" in the complexity sense and might not be the full process occupying e's spatiotemporal location; so there might be many more processes that are trace-identical with, but "thicker" than, π. Any such process will be said to *constitute* e; thus the processes π related to an event e by the constitution relation are declared to be coextensive with all those processes satisfying the right-hand side of the following biconditional:

$$\pi \rhd e \leftrightarrow (\, \pi^{e} \sim_{\rho} \pi \wedge \pi^{e} \sqsubseteq \pi \,)$$

When we rewrite this slightly, we get the *Principle of Constitution for Events*:

$$\pi \sqsubseteq \pi' \wedge \pi \sim_{\rho} \pi' \to \forall\theta\,(\, [\pi : \theta] \to \pi' \rhd (\pi \mid \theta)\,)$$

Furthermore there is the following *Role Principle*:

$$\forall e \forall\theta\,(\, Df(R,\theta) \wedge [e : \theta] \to \exists a\,(\forall t\,(t \sqsubseteq \tau(e) \to a_t \sqsubseteq_{\sigma} e) \wedge R(\theta, e, a)\,)\,)$$

[53]Note that relation symbol '$\sqsubseteq$' is flanked by terms that stand for regions, and is therefore meant to refer to the ordering relation on the spatiotemporal regions; no confusion should arise with the (admittedly already ambiguous) part relation $\sqsubseteq$ on events.

It says that if a role R is defined for a type θ and event e is of type θ then there is an individual a occupying that role in e with respect to θ, and a is such that all its temporal stages during the time e goes on are to be found within the spatial location of e. That is meant to reflect the intuition that if a is the agent of an action e, say, then a's location is part of where the action takes place.

Since the relation π *being of type* θ is a non-standard predication relation we have to add rules making sure that it obeys the usual *Boolean Principles*:[54]

$$[e : \theta] \leftrightarrow \neg[e : \theta^*]$$
$$[e : \theta \sqcap \theta'] \leftrightarrow [e : \theta] \wedge [e : \theta']$$
$$[e : \theta \sqcup \theta'] \leftrightarrow [e : \theta] \vee [e : \theta']$$

Furthermore, if θ is a subtype of θ' and e is of type θ then e is of type θ'. This is the *Persistence Principle*:

$$[e : \theta] \wedge \theta \sqsubseteq \theta' \rightarrow [e : \theta']$$

On formal grounds one might wonder if there is a similar persistence condition concerning the left-hand side of the colon relation, but in general there is not, at least not for the complexity ordering on processes. Regarding the temporal ordering, there is again no general persistence: a counterexample would be that if you have a process that is a walking then the lifting of a foot is a temporal part of that process which is not a walking anymore. The reason is that a walking process needs some minimal amount of time for it to evolve in such a way that it can be counted as a walking. I call that time the (temporal) *granularity parameter* of the type.[55] But even if the temporal trace of the subprocess exceeds the length of that parameter, persistence is still not garanteed; it will be, however, if the type of the event belongs to the suitable *aspectual class*.[56] For instance, if θ is an activity like RUNNING, with granularity $\gamma(\theta)$, then we would have something like the following *principle of homogeneity*:[57]

$$[\pi : \theta] \wedge \pi' \sqsubseteq_\tau \pi \wedge \mu(\tau(\pi')) \geq \gamma(\theta) \rightarrow [\pi' : \theta]$$

(If the process π' is a temporal part of a process π which is of type θ, and the duration of π' is not less than the temporal granularity $\gamma(\theta)$ of θ,

[54] θ^* is the *dual* of θ as in Chapter 11, i.e., the (externally) negated type θ, and $\sqcap$ and $\sqcup$ are conjunction and disjunction on types, respectively. Note that it is the types that have the Boolean character, not the events themselves.

[55] In a fuller treatment we would have to add the principle that if e is of type θ then the temporal trace of e has to be at least of length $\gamma(\theta)$, the granularity of θ.

[56] For the standard classification of events into the aspectual classes *states*, *activitities*, *accomplishments* and *achievements*, see Vendler (1967c).

[57] Cf. the subintervall property for atelic events in Chapter 8, (Ax. 13).

then π' is also of type θ; here the measure μ from above gives the duration of the temporal trace of the subprocess π'.)

Many more principles could be formulated in connection with the aspect of event types, but the issue of aspectual classes is not pursued here.[58]

Let me finally consider the algebraic structure on the events themselves. As with all objects in our ontology, there is of course the mereology of pluralities of events; that is, I consider as event again the individual sum of an arbitrary number of events. We don't have to do anything more about it here except to lift the constitution relation to sums of i-atomic events. For simplicity, the definition is given for a sum of just two.

$$\pi \rhd e \sqcup e' \;:\Leftrightarrow\; \pi \sim_\rho \pi^e \sqcup \pi^{e'} \;\wedge\; \pi^e \sqcup \pi^{e'} \sqsubseteq \pi$$

(π constitutes the i-sum of two events e and e' just in case π is ρ-trace identical to and comprises in the complexity sense the fusion of the processes constituting those events.) By idempotence, this condition includes the elementary case of just one event e.

However, somewhat more interesting in the current context is the question, to which degree if any the colon and constitution relations are compatible with the lattice operations on the constituents of events. Thus suppose $[\pi_1 : \theta_1]$ and $[\pi_2 : \theta_2]$ both hold; is it true then that, say, $e_1 \sqcup e_2$ is of type $\theta_1 \sqcup \theta_2$? For instance, if π_1 is of type WALKING and π_2 is of type TALKING, is $\pi_1 \sqcup \pi_2$ then of type Walking or Talking, i.e., of type (WALKING$\sqcup$TALKING)? Let us back up and ask ourselves first if it is reasonable to say that the process $\pi_1 \sqcup \pi_2$ is of type WALKING. Well, if you have a process that passes as a walking, and add some trace-identical detail then it should still be a walking. By the same token, then, the process $\pi_1 \sqcup \pi_2$ should also be a talking. But now we can apply the conjunction principle for the colon relation to conclude that that process is also both a walking and a talking, and by persistence, we get that it is a walking or a talking. Thus we should add a principle of the following kind:

$$\forall\pi\forall\pi'\forall\theta\,(\,[\pi : \theta] \wedge \pi \sim_\rho \pi' \;\rightarrow\; [\pi \sqcup \pi' : \theta]\,)$$

(If π is a process of type θ then its fusion with a ρ-trace identical process π' is again of type θ.) Notice that trace-identity is important: we wouldn't be prepared to call the process in our example a walking if we started out with a walking process π_1 and added on to it another process

[58]See Parsons (1990) for the philosophy of these, as well as for a general study of events from a perspective which is closer to linguistic semantics than the present metaphysical point of view.

π_2 which is, however, completely unrelated spatiotemporally to the first one (in particular, π_2 could be a non-walking).

From the last principle we can now prove the (universal closure of the) following:

$$[\pi_1 : \theta_1] \wedge [\pi_2 : \theta_2] \wedge \pi_1 \sim_\rho \pi_2 \rightarrow [\pi_1 \sqcup \pi_2 : \theta_1 \sqcap \theta_2]$$

(If two ρ-trace identical processes π_1 and π_2 are of types θ_1 and θ_2, respectively, then their fusion $\pi_1 \sqcup \pi_2$ is of the conjunctive type $\theta_1 \sqcap \theta_2$.)

Finally, consider the constitution relation; how does it interact with the operations on the event constituents? It turns out that we are now in the position to prove a principle like the following:

$$\pi \rhd (\pi_1 \mid \theta_1) \wedge \pi \rhd (\pi_2 \mid \theta_2) \rightarrow \pi \rhd (\pi_1 \sqcup \pi_2 \mid \theta_1 \sqcap \theta_2)$$

(If a process π constitutes two events $e_1 = (\pi_1 \mid \theta_1)$ and $e_2 = (\pi_2 \mid \theta_2)$ then π constitutes the event e whose underlying process π^e equals the fusion of π_1 and π_2, and whose type is the conjunction of types θ_1 and θ_2.)

12.4.3 Questions about Events

With all this elaborate machinery at our disposal, let us see now what kind of questions about events can be answered that have commonly been asked. I will give a list of some of those questions and comment on each one in turn. This list draws on problems that have been raised in the literature on events, in particular, in sources like Kamp and Reyle (1993); Bennett (1988); Simons (1987).

(1) *Question*: Is an event always a change in an object? *Answer*: Changes are events, but not every event is a change in an object; for instance, an astronaut losing weight while being shot into the orbit certainly constitutes a change, but not one occurring in an object. More generally, the interaction between a number of objects is an event, but the internal changes in the objects involved may be part of, but don't exhaust the whole process underlying such an interaction. Finally fusions of changes in objects are events that are not changes in particular objects.[59]

(2) *Question*: Can events be interrupted, and then start up again, after a phase of non-existence? *Answer*: Yes, a simple example being the fusion of two processes that are temporally non-overlapping and have a temporal gap between them. But that is perhaps not the kind of example the question aims at; a more congenial example would be somebody who begins to write

[59] Cf. also the comments made on Lombard's work in Chapter 10.

a dissertation, then leaves it for two years, and takes it up again thereafter. But whether that is an interruption of an event or not is just a matter of vagueness, and nothing in formal metaphysics should try to exclude the possibility of an interruption. Similar remarks apply to the problem of intermittent existence in the case of individuals.

(3) *Question*: Do all events have a temporal as well as a spatial location? For those which do, how is it determined? *Answer*: The answer to the first question is Yes, as far as concrete events are concerned. The location is given by the temporal and spatial traces. In an earlier chapter I said that certain eventualities which are called states in the linguistic literature presumably don't have a spatial location, on the grounds that if, for example, a resents b it is hard to say where that resenting is going on. But now I am prepared to say that that is just an intra-active process going on inside a, so the location is roughly the spatial trace of a during that stage. To answer the second question we have of course to refer again to the inherent vagueness and indeterminacy of the event concept which makes it difficult to specify the boundaries of a particular event (but note that a similar vagueness, albeit not perhaps to that extent, applies to regular objects). Let us assume that locating an event in time is a relatively unproblematic matter (but see Chapter 10 for examples of vagueness, here, too), and address the problem of localization in space. In Chapter 10 I said that the place of an event e is defined as the intersection of all spatial regions s at which e occurs; but in our current theory that only means that the spatial trace of e is included in s. The problem was, however, to get at that trace in the first place. Davidson tried in (1980):176 to define the trace as "the location of the smallest part of the substance a change in which is identical with the event." I agree with Bennett here that this definition as well won't be of much use in concrete cases,[60] and what's more, it isn't for us anyway since it refers to the events-as-changes-view, which was found above to be too narrow.

Bennett (loc. cit.) has suggested a problem for anyone trying to fix the location of an event:

> The cargo in a ship has been badly stowed, and at time T one bale slides across the hold, thus changing the cargo's center of gravity. Where is *the shift of the cargo's center of gravity?*

According to Bennett, it wouldn't be a good answer to say that the shift coincides with the moving bale, since the center of gravity lies completely outside the spatial trace of that bale during T; but to say that it equals the trajectory of the center of gravity through T is bad, too, because in

[60]Bennett (1988):107.

this case the shift "may have been caused by an event that sent no energy into the zone where the shift occurred. There seems to be no third alternative." (loc. cit.) Now I think it is plain that the shift occurs along that trajectory, but I fail to see a problem here. Forces act on masses and not on geometrical abstractions like the center of gravity; so the shift is just a kind of "Cambridge change" resulting from the relational change in the relative positions of the individual bales of the cargo, and no flow of energy is required towards that abstract geometrical trajectory.

I agree, however, with Bennett's general attitude towards questions of this kind. It is just wrong to expect definite answers in an area of vagueness. In particular, the attempt is bound to fail to tell one event from another by a sufficient amount of "gerrymandering,"[61] only because one happens to subscribe to the container view of events where there is no other identity criterion available.

(4) *Question*: Can there be two different events that occupy exactly the same region in space and time? *Answer*: Yes; for instance, if a sells something to b then b buys the thing from a. These are two different events because SELLING and BUYING are different event types that switch the roles for a and b. Or consider Davidson's rotating sphere, which is spinning and heating up at the same time: the spinning is one event, the rotation another. Bennett gives the example of Leander swimming the Hellespont and at the same time catching a cold; the swimming is again different from his catching the cold. All this follows from our decision to include an event type into our notion of (elementary) event. However, in all three cases there is just one underlying process involved (or a bit more precisely: one full process constituting the different events), and it is this feature that reflects the strong intuition that there is really only one thing going on.

(5) *Question*: Are there atomic events? *Answer*: The question is multiply ambiguous because of the many part relations involved in the event concept. I modelled an elementary event e as a pair $(\pi \mid \theta)$ consisting of a process π and am event type θ. As was discussed above, there are atomic types, but no atomic processes, neither in the complexity dimension nor in the spatiotemporal dimension. However, an elementary event will be an i-atomic event in the plurality sense.

(6) *Question*: Suppose I signal by raising my hand. Is my signaling the same event as my hand raising? *Answer*: Like in the previous examples the answer depends on whether we talk about the events or just about the underlying processes. There is one underlying process to different events,

[61] This is the term Quine (1960):171 uses in his well-known definition of event.

with different event types, one being an intransitive event type (SIGNAL) and the other transitive (RAISE).

There is an extensive literature in action theory on the *by*-locution.[62] I won't be able to enter that discussion here because it involves among other things quite subtle considerations about causation; but I would like to make the general point that the distinction between processes and events gives us the necessary tools to avoid the kind of trap laid out by the idea that a person, by doing "just one thing" did a whole bunch of other things with it, whether intentionally or not. Thus let us just consider the original example which G.E.M. Anscombe gave in her (1957) where a person b kills the villagers by putting poison into their well. Anscombe says that b committed just one act (correct) *and therefore* the killing must be the same as that action, viz. putting poison in the well (Bennett calls this the *Anscombe thesis*). But that is the trap: because of missing distinctions the identification of those two actions becomes inevitable. However, we can reconstruct the case along the following lines: First, there is a certain action, e, carried out by b; e is the poisoning of the well. Some process π constituting e is such that it causally feeds into a chain of other processes leading up to the death of the villagers through their intake of poisonous water from the well.[63] Now take the fusion π' of all those processes; clearly π' is not the same as π. Thus the first observation is that the poisoning and the killing don't even have the same underlying processes, as it was the case with the events in the signaling example, and so *a fortiori* the events here cannot be the same. Yet we do classify π' as a killing (call it e'), with b in the agent role. Didn't b then do two things after all, e and e'? Well, yes, because the causal chain of processes is such that we are justified in calling it a killing. There wouldn't be anything wrong with that if it weren't for the "scholastic" puzzle about counting that we encountered before. Thus in one sense b did only one thing, e, because b invested his or her physical energy only once; but that is compatible with the fact that due to the prevailing circumstances b became *involved in* playing the agent role in another event, e'. The only adjustment that has to be made within the theory of action is defining the agent role in such a way as to comprise this type of indirect influence; but I think that the wider definition has been in use in that field all along.

(7) *Question*: On the one-process view, how can one avoid the conclusion that the process has two properties contradicting each other? So for instance, take the sphere-process again that is spinning fast and heating

[62] As with many other themes surrounding the notion of event, Bennett's excellent discussion of the matter in (1988) stands out again. See there also for further references.

[63] A suitable story has to be told about surrounding factors to the effect that none of them play a preemptive role in causing the deaths.

up slowly; doesn't that process have to be classified by both event types, FAST and SLOW? *Answer*: To begin with, let us remind ourselves of intensional predication involving regular objects. When I say, "I admire the composer of *Tristan and Isolde*" then you might reply: "Are you sure? Remember that he was an antisemite." I resolve this by the standard qualification, "Well, I admire him *qua* composer, but I don't admire him *qua* antisemite." Quite the same move can be made in our present example: the above process involving the sphere is fast *qua* spinning and slow *qua* heating up. Another way to go about this is to say that 'slow' and 'fast' are rather *relations* between processes and event types; thus the process bears the relation *Slow* to the type HEATING-UP, and the relation *Fast* to the type SPINNING. Note that this constitutes a change in Davidson's theory of adverbs: the point of that theory was that it turned adverbs into one-place predicates of events. However, a well-known stock of examples like the present one have cast doubt on the adequacy of that theory in its original form.[64] So what about the current amendment of going over to relations? That will work as long as we operate under the assumption that there is no case of a *basic* property that does and at the same time doesn't apply to a process.

Among the many examples generated in Bennett (1988) there are some that challenge even this assumption. A swim is supposed to be healthy; but Leander's swim wasn't, he caught a cold. Well, here the simple answer is that he just overdid it, and that the generic statement that a swim is healthy (under normal circumstances) just doesn't cover this particular case. A similar answer can be given to Bennett's case of a healthy quarrel "because the parties to it jogged while quarreling" whereas normally, quarreling is considered rather detrimental to one's health (op. cit., p. 126). If the healthy effects of the exercise are so intense as to supersede the negative effects like increased blood-pressure, then that case is again not covered by the generic statement.

Let us consider one more example of Bennett's, the case where two men fought on a public street while having heroin in their pockets. Suppose there is no law against fighting in the public, but against the possession of heroin. Now the question is, whether that fight was illegal (loc. cit.). In this situation fights *per se* are not illegal; but the process referred to involves an illegal activity, thus it is illegal after all. We might admit that much and yet insist that *qua* fight it was not. So it seems that the *qua*-locution is helping us out here, too.

However, it would not be good to apply the relational analysis again because that would eventually leave us with no non-relational types or

[64]See also Bennett (1985). I disagree, of course, with Bennett's overall skeptical conclusion about the feasibility of a useful theory of events.

properties. Rather, the impression that the event name 'that fight' can pick out, in a kind of surgical act, exactly that aspect of the process consisting of the fighting, and nothing else, is mistaken. That is not to say, however, that event names cannot refer to aspects of processes at all; but in this particular case reference doesn't work in the way suggested. So there is no direct link, independent of context, between the content of an event description and the event referred to. The clue is rather to be gained from the classifying predicate 'illegal': the issue of legality can only be raised in connection with a process constituting a human activity which is subject to criminal law. Thus the answer to our question is: The fight was illegal, because it was an activity which is subject to criminal law and involved the illegal carrying of heroin.

My final example takes up the notorious case of buying vs selling. Suppose a used car dealer dumps a wreck of a car on some client who just isn't smart enough to discover the lemon. Isn't the selling smart and the buying pretty stupid? And yet, the underlying process is literally the same because of the analytic relationship between buying and selling. As with many philosophical puzzles that are logically benign the problem lies in the rather superficial description of the case. A sale, i.e., a transfer of ownership for a price, is a transaction between two economic subjects x and y with incompatible utility functions f_x and f_y. That introduces a parameter which has to be fixed before the deal can be called smart or stupid. When this is done the air of contradiction is removed also in this case.

Chapter 13

The Philosophy of Plurality

13.1 Introduction

Everyone is familiar with Quine's philosophical maxim that natural language must not be taken at face value. Take the problem of ontological commitment. Our language works in such a way that it freely generates statements involving a seeming reference to all kinds of entities which a serious philosopher might rather regard with suspicion. So we might say that there is a property that is common to O. J. Simpson and judge Ito, or that there is a possibility that Newt Gingrich will become the next president of the United States. Here we seem to be talking about properties or possibilities (in the eyes of some there is nothing wrong with that). But it is hard to stop an all too free-wheeling use of language. Quine's possible fat man in the doorway, for instance, is certainly one of the more dubious creatures there are (and of course, Quine despises such entities as we know).

The remedy that Quine offers is *regimentation*. It is the well-known procedure that rephrases the sentences of language in a certain standard way such that both open and hidden references to entities become visible. He likens the parts of speech to the basic structure of a formal language, that is, first-order logic (FOL). From the FOL representation of an NL discourse you can read off its ontological commitments in a simple way: just look at the range of the bound variables. And the reader anticipates my next sentence: "To be is to be the value of a bound variable".

The obvious advantage of such a regimentation language like FOL is that there is nothing vague to it, that it has a mathematically well-defined semantics, and that it can thus be used to explain the truth conditions of NL sentences via indirect interpretation. And in this way we also "understand" our language better. Or so the story goes.

We all know that this cannot quite be true. Formalizations can be checked against an "intuitive" semantics and questions of adequacy can be raised. How would that be possible if we hadn't a certain degree of understanding of our language all along? It is true that there are always difficult cases. Who claims to immediately understand sentences like my favorite, *Many competing companies have common interests.* There are more complicated examples that are even harder to process and give an interpretation to, and that's why formalization is not at all an idle exercise. Still, in normal circumstances we know what we are talking about.

But even if we more or less understand what we say in the vast majority of cases, philosophy is made for the rarer occasions where we might be unaware of the implications of the way we speak. Specialized discourses seem to be easier to handle. Thus number theory presumably is about numbers, and set theory about sets. That much can be easily read off the range of the bound variables. The philosophical question here is whether number theory is *irreducibly* about numbers, or that set theory is *irreducibly* about sets.

But what is natural language about? Which are the entities that we commit ourselves to in using our mother tongue? This is a much more general question of which only a very special case will be addressed here. The previous chapter dealt with the ontology of singular particulars; in this chapter I will be concerned with the ontological implications of *plural phrases* and *plural quantification.*

13.2 Ontological Commitment

The first part of the book focused on the semantics of plural phrases, but the formal theory that I proposed to deal with the linguistic phenomena was inspired by a certain ontological view, viz. that pluralities do not denote sets, but rather mereological fusions of ordinary objects. Thus the pluralic particulars are considered just as concrete as the singular particulars there are built up from. A number of arguments were advanced for that view that need not be repeated here. In linguistic quarters those arguments have been less appreciated, a reaction that shouldn't surprise us since semanticists are mainly concerned with modeling and not with ontology; as a result, they typically use their set-theoretic tools to produce correct semantic rules, and in the case of pluralities we saw that these can be treated in the standard powerset model as far as truth conditions go.

But recently there has also been a discussion among philosophers of mathematics about the nature of pluralities. Thus Penelope Maddy, in her (1990), wonders how sets might be epistemically accessible in spite of their purported abstractness; she chooses the heroic course by arguing that sets can be perceived after all through what she calls a "neural 'set detector' " in the brain enabling us to "acquire perceptual beliefs about sets." In order to make her view plausible she gives examples of pluralities and calls them sets:

> Consider the following case: Steve needs two eggs for a certain recipe. ... He opens the [egg] carton and sees ... three eggs there. My claim is that Steve has perceived a set of three eggs. By the account of perception just canvassed, this requires that there be a set of three eggs in the carton... (op. cit., p.58)

Recall from Chapter 8.5 that George Boolos once called this thinking "haywire."[1] But Maddy invokes the authority of Frege who in her opinion refuted both the idea that the numerical statement is about the the stuff making up the eggs, and th e notion that it is plainly about the eggs themselves, i.e., the aggregate or i-sum of the individual eggs. Now in the passage of his *Grundlagen* which Maddy refers to[2] Frege makes the correct point that for the counting process to work there has to be a unit available, and that is given by a sortal or a concept, like 'egg in the carton'. Thus Frege suggests that a number statement involving 'three' is about a concept, *and hence*, adds Maddy, about the extension of that concept, viz. the class or set of eggs in the carton. However, both Frege's conclusion and Maddy's transition are far from obvious. What Frege is getting at in the context of the *Grundlagen* is the claim that basically numbers are second-order properties, from where he is going to develop his reduction of numbers to logic. But this line of thought, respectable as it is as part of the foundational enterprise, is already rather artificial as far as the treatment of numerals goes. What I find more natural here is the *measurement perspective*: numerals indicate the result of a measurement with respect to an underlying measure, be it discrete, as in the case of count nouns like 'egg' where the sortal gives the unit, or homogenous, as with mass nouns where the relevant unit has to be provided by a measure word or numerative.[3] Proper

[1] In fairness to Maddy, however, is has to be pointed out that she is not alone. Apart from the linguists against whom Boolos's verdict was originally directed there is also the mathematician Paul Halmos, for instance, indulging in the same kind of language: the opening sentence of his classic *Naive Set Theory* reads: "A pack of wolves, a bunch of grapes, or a flock of pigeons are all examples of sets of things." On this score, the book must be called doubly naive.

[2] Frege (1884), §§22-3.

[3] See also Chapter 9.

abstraction leads to the theory of cardinals or, more generally, of measure functions; but wouldn't it seem odd that the fact that we have *theories* about measurement makes the concrete stuff that undergoes measurement procedures abstract? Maddy's predictable move is to admit "that sets no longer count as 'abstract'. So be it; I attach no importance to the term." (op. cit., p.59) I won't comment on this piece of revisionist metaphysics but instead add two remarks here that link the present discussion to the treatment of numerals in semantics. Recall from earlier chapters that numerals are analyzed either as generalized quantifiers or as adjectives. Generalized quantifiers are usually modeled as relations between sets, but equivalently, they can be taken as relations between individual sums. An equivalence to the same effect holds for the adjectival analysis. Thus semantic theorizing doesn't entail a particular ontological claim. But the same trade-off can be observed under the more philosophical concept view. Suppose that the number property attaches to concepts after all; but now we want to be a little bit more precise grammatically and note that the numeral 'three', for instance, really operates on *starred* predicates, like '*egg', which is true of all i-sums of eggs;[4] call those properties *starred properties*. Thus the number *Three* applies to a starred property if exactly one i-sum consisting of three atoms falls under it. Thus the number property is isomorphically recast in the mereological plural language, which doesn't carry any commitment to sets. Thus it appears that Maddy has hardly succeeded in proving her point. Whatever plausibility there is to her claim, it is gained from selling pluralities as sets.[5]

Having disposed of Maddy's set view of pluralities I turn to plural quantification which has recently acquired a new status in philosophy. It was George Boolos who discovered that language could help out philosophy. For reasons that will become apparent in below Boolos sought a nominalistic interpretation of (monadic) second-order logic (ML2), that is, one in which the second-order variables are not taken to range over any set-like "collections" of objects from the first-order domain. He found that interpretation in the device of plural quantification of English. Thus a remarkable inversion of the common practice of interpretation and regimentation has taken place here: rather than explicating natural language in terms of a formal language, a formal language is interpreted in terms of certain natural language locutions. This "naturalistic" approach has been followed by David Lewis in his elegant essay *Parts of Classes.* (Lewis (1991)). So one might wonder if this is part of the world-wide ecological movement of going back

[4]Cf. Chapters 1 and 2.

[5]For an extensive critical assessment of Maddy's realist views about reference to sets, see Chihara (1990), Chapter 10.

to our natural resourses.

When I first read Boolos on this topic I felt that I completely agreed with his outlook. It is true that he explicitly rejects the idea that plural phrases like *the cheerios in the bowl* denote any singular "collection" of individuals;[6] but since he was rather vague in his use of the term "collection" I thought and I still think that there is room for interpretation here: The context shows that he does not so much argue against singular entities in general being the denotation of pluralities but rather against their class-like character. In fact I claim that *mereology* can avoid talk of collections in the sense in which Boolos rejects them, and that it can give a *theory* of the plural idiom in which non-committal ML2 is given a proper interpretation. Lewis joins Boolos' naturalistic approach to pluralities in saying that we just understand what those plural locutions mean; so he doesn't provide a rigorous formal treatment for them either. The whole issue is rather complex, so I set it aside until the next chapter, where I will provide such a treatment within LP, together with a discussion of Boolos's ideas.

There is a final topic that I want to touch upon in this section, and that is the question of the supposed "ontological innocence" of mereology. This question has a direct bearing on the nature of plurality, as I will explain.

In his most recent book (1997) David Armstrong says what he means by an *ontological free lunch*:

> It will be used as a premise in this work that whatever supervenes or, as we can also say, is entailed or necessitated, in this way, is not something ontologically additional to the subvenient, or necessitating, entity or entities. What supervenes is no addition of being. Thus, ... [m]ereological wholes are not ontologically additional to all their parts, nor are the parts ontologically additional to the whole that they compose. This has the consequence that mereological wholes are *identical* with all their parts taken together. (p. 12)

Let us use Armstrong's term *aggregate* for an arbitrary mereological whole. What Armstrong commits himself to, then, is the following: Whenever there is a collection of things then there is the aggregate of these things which is just the things "taken together." Now first of all, Armstrong seems to assume that taking things together is an act which either doesn't have the full theoretical status of an operation, or if so, is in no need of justification as other well-known comprehension principles are.[7] Secondly, it is unclear whether he wants to say that this operation *always returns a result* or that it, *if successful*, returns an entity that is of the *same ontological kind*

[6]See again Chapter 8, Section 5.

[7]Recall that *comprehendere* is just the Latin word for 'take together'.

as the entities we start out with. To give an example for the latter first, if you take the standard comprehension principle in set theory then there are collections of sets that cannot be "taken together" by that principle (like the collection of all non-self-membered sets); but when it does return a result then we get an entity (viz. a set) which is of the same ontological kind. A simple example for the first alternative is the powerset algebra of a given set, with set-theoretic union as the comprehension operation; obviously, the union set is of the same kind, too. However, in still other cases there is a result, but that is of a different kind. For instance, under the usual conception of impure set theory, we are able to form sets from non-sets, i.e., urelements, like concrete individuals. Also, a predicativist might want to say that an entity introduced by an impredicative definition from predicative objects is of a different kind (e.g., the standard definition of a least upper bound on the real line).

There is one example that Armstrong gives in the section the quote is taken from. He asks if the number 42 and the Murrumbidgee river form an aggregate. This indicates that he actually thinks of the comprehension operation as taking individual sums in the sense of plural theory. The same holds for Lewis (1991), who unlike Armstrong feels at least the need to give an explicit formulation of the operation in his *Principle of Unrestricted Composition*: "Whenever there are some things, they have a fusion."[8] Call that fusion the *pluralic aggregate*. Hence pluralic aggregates are our i-sums. What Lewis says, then, is that affirming the existence of pluralic aggregates doesn't involve any extra ontological commitment beyond the commitment to the parts. In this book I have supported the existence of arbitrary i-sums composed from a given collection of atomic individuals; thus I side with Armstrong and Lewis that mereological wholes, when they are taken in the sense of pluralic aggregates, are indeed an ontological free lunch.

However, doubts have been raised against arbitrary mereological composition from very early on,[9] and even after Lewis's vigorous intervention in favor of pluralic aggregates voices rejecting the principle have not been silenced. Thus Forrest(1996b) claims that Unrestricted Composition implies that any countably many regions in space have a fusion which is a region, which in turn is incompatible with a highly probable hypothesis about space, viz. that it is "Whiteheadian." So Forrest opts against the innocence of mereology.

Now even without going into Forrest's argument we can say a few things about this issue. First, however, let me draw a distinction. We saw above that plural structures are generated by a free lattice construction, yielding the most general lattice that can be spanned by a given set of genera-

[8] Cf. the quote from Lewis in Section 8.5.

[9] See Simons (1987), Sections 2.9 and 3.2.3.

tors. Thus the pluralic aggregates are *free-lattice fusions.* However, not every fusion or supremum in a lattice is of this sort; to have a name for those that aren't free-lattice fusions, call them *substantive fusions.* Then the following seems to be true: (i) If the innocence of mereology means that all mereologies are *complete*, i.e., closed under arbitrary fusions, then, obviously, mereology is not innocent. Example: Take a weak set theory without the Axiom of Infinity in which the natural numbers can be defined; then the union of all singletons of natural numbers cannot be shown to exist. (ii) Even if a given mereology is complete, it may happen that some of its fusions are substantive, i.e., different from the free-lattice fusion; then the question is whether Armstrong would still be prepared to call such a substantive fusion "supervenient" on its parts. The dependence will be necessary as long as we stay in the realm of pure mathematics, but even so, certain non-trivial existence axioms have to be invested to establish completeness, and assuming them might not be that innocent by some standard or other. (iii) Lewis's arguments in defense of Unrestricted Composition derive their plausibility from the fact that his examples are about free-lattice fusions. In fact, his readers are given the impression that he equates mereology with plural lattices (Forrest's interpretation of Lewis is a case in point). However, Forrest addresses mereology in general: his counterexample is about substantive fusions, not about free-lattice fusions. So Forrest and others might rightly say that if plurality is all there is to mereology then many questions about ontology (for instance, about the structure of physical space) remain unanswered; having been invited to an ontological free lunch, they might find themselves walking away hungry.

The distinction between free-lattice fusions and substantive fusions is also highlighted by a phenomenon we observed already in Chapter 1. There the model structure (the "boosk") contained as part of its i-atoms a lattice of stuff quanta or portions of matter which was assumed to be complete. Now consider a collection of such quantities; their material fusion is still an atom in the plural sense, but their individual sum is the free-lattice fusion which is not an i-atom anymore. The existence of this fusion is covered by the principle of unrestricted composition, whereas the existence of the material sum (the substantive fusion) might still be under dispute, depending on our notion of matter, or similarly, our notion of space and time.

Thus it seems that what we get for free is always the kind of fusion that "preserves" its character as a multitude, an object "as many," as Russell called it. It is only when we aim at a whole "as one" that we appear to invest an extra commitment.

13.3 The Counting Fallacy

The double perspective on mereological wholes, as one and as many, and the puzzle it creates, has a rather long history in philosophy, going all the way back to Parmenides and Plato.[10] Since the puzzle has tended to obscure our conception of pluralities I find it important enough to give it a name (the *fallacy of counting*) and to address it explicitly; the present section is devoted to it.

The fallacy consists of a one-line argument: "That which are many cannot be one."[11] As I view it, there are three sources of possible confusion, or analytic obstacles, which give rise to the fallacy: (i) the *obstacle of the unspecified unit*, (ii) the *grammatical obstacle*, and (iii) the *obstacle of double counting*. Let me deal with these obstacles in turn.

The first obstacle plays a role in the passage in Frege's *Grundlagen* that I mentioned above when discussing Maddy's views. As we saw there Frege objects to the idea that a number statement is about ordinary objects or aggregates of them because he thinks there is no characteristic way in which they can be separated into parts. Even "the number word 'one' ... in the expression 'one straw' fails to do justice to the way in which the straw is made up of cells or molecules." (op. cit., §23). What he wants to say here is that objects, either separately or collectively, just don't wear a number on their sleeves. Another example mentioned by Frege is one that was also discussed in earlier chapters, viz. the playing cards vs the deck of cards. A similar point is made by Pollard (1990):47 when he asks how many in number the Montagues and the Capulets are: is it the number of people, or the number of families? The Montagues for themselves are one as a family, and many as family members.

I think that this difficulty can be handled in the way indicated above, by adducing the notion of the unit of measurement. When the unit is given, we have a rather clear understanding of how to count objects. So the first obstacle is just a problem of underspecification and can easily be overcome.

The issue gets more confusing when it comes to singular expressions referring to or involving a plurality. This is the grammatical obstacle that led Russell in his *Principles of Mathematics*[12] to draw the distinction between *classes as one* and *classes as many*. Here is a sample of typical quotes dealing with it.

---

[10]Specifically, I have in mind Plato's dialogue *Parmenides*; for a recent interpretation of this dialogue, which uses modern logical terminology including concepts from mereology, see von Kutschera (1995).

[11]As Chisholm points out, this argument was already advance by Boethius: A man and a horse are not one thing; cf. Chisholm (1976), p.219.

[12]Russell (1903), henceforth abbreviated *PoM*.

> A class also, in one sense at least, is distinct from the whole composed of its terms, for the latter is only and essentially one, while the former, where it has many terms, is ... the very kind of object of which *many* is to be asserted. (PoM:68)
>
> Thus classes would seem to be one in one sense and many in another. There is a certain temptation to identify the class as many with the class as one, *e.g.*, *all men* and *the human race*. Nevertheless, wherever a class consists of more than one term, it can be proved that no such identification is permissible. (PoM:76)
>
> It is important to realize that a whole is a new single term, distinct from each of its parts and from all of them: it is one, not many ... (PoM:141)

Let us ignore the peculiarity here that in Russell's early work the quantifier phrase *all men* is a denoting expressing, and simply compare the examples that he gives for classes as one and classes as many. Classes as many are typically referred to by expressions like *points*, *instants*, *the soldiers*, *the sailors*, *the Cabinet Ministers* (PoM:68), and *(all) men*. Thus these are all plural phrases, either definite plural noun phrases or bare plurals. By contrast, classes as one are given by expressions like *space*, *time*, *the army*, *the navy*, *the Cabinet* (ibid.), and *the human race*. All these are singular terms now. The problem from the modern perspective is that those singular expressions don't just refer to a class or set *simpliciter*; for instance, the army is not the same as the class or set of its soldiers, even if we neglect the issue of abstractness. Thus Russell is adding content by going over from the plural to the singular expression, and so it is not surprising that he doesn't want to equate the denotations of the two. But apart from that he appears so much under that spell of grammar that he thinks that for an expression to refer to a class is has to be plural because a singular expression conveys a whole which is "essentially one."[13] For instance, at another place he says:

> In such a proposition as "*A* and *B* are two," there is no logical subject: the assertion is not about *A*, nor about *B*, nor about the whole composed of both, but strictly and only about *A* and *B*. (PoM:76f.)

What does the phrase '*A* and *B*' stand for? It is not *A*, nor *B*, nor the the whole composed of both, according to Russell; we have to conclude that it is a true plurality, i.e., an individual sum. But what is that whole supposed to be, then? Here Russell seems to be using a rather "loaded" notion of whole, for instance, the material fusion of the stuff making up

[13] A reason for this might have been Russell's notorious laxness in matters of use and mention, which was so much criticized by Quine.

A and B. But there is nothing in the logic of these concepts that forces this interpretation; the plurality could just as well be called a whole, viz. a mereological whole. If so, that would also pave the way for treating collective predication which is not available to Russell when he insists that a plurality cannot be a logical subject.

I will have to say more about the grammatical obstacle in the next chapter when I comment on Lewis and the problem of "singularism."

Finally, the third obstacle consists in the following piece of reasoning: Suppose we are looking at the night sky and see two stars in the constellation Gemini, Castor and Pollux; these are two objects; but according to plural theory, we rather see three objects, Castor, Pollux, and on top of those, their i-sum Castor $\oplus$ Pollux. But what are two cannot be three; that would be double counting. A related conception of what plural theory does leads to the following argument: Suppose the universe contains finitely many objects; if plural theory were right then the number of objects would have to be a certain power of 2, minus one; so there couldn't be exactly ten objects. But that seems counter-intuitive.

I usually reply to the argument from double counting by pointing to an analogy in the domain of stuff quantities. I have a bottle of wine in front of me; some time ago it was full, and now it is half empty. The content of the full bottle certainly constitutes a certain object, a, albeit in liquid state, of which the rest, b, is a proper part; hence a and b are different objects. Now what about the amount of wine, c, that I swallowed? It must also be a regular object since, among other things, it is apt to affect my driving fitness. Thus we end up with three respectable entities that nobody should object to; however, they are structurally analogous to the case of Castor and Pollux. That leads us to reflect on the principle which underlies our counting procedure, and we realize that the puzzle is created by two different counting rules: one that counts only pairwise disjoint objects, and one that allows one object to be part of another, or to overlap. Thus the puzzle is resolved by pointing to the ambiguity in the counting rule.

I think that an appropriate answer to the counting fallacy is to take it up head-on by asserting that some things can be *one and many at the same time*, but of course in different respects. This is not a contradiction, but rather a kind of *gestaltwechsel* which is supported by a solid mathematical correspondence. Thus, take an arbitrary element a in a plural lattice, and consider the principal ideal $a^{\downarrow}$ generated by it (recall that this is the set of all i-parts of a). If a is not an atom then this set or class "are" many, because a has more than one i-part. But note that we can go back and forth between the (non-atomic) elements and the principal ideals generated by them; they are in one-to-one correspondence, establishing an isomorphism between objects as one (the top elements of all the ideals) and objects as many (the ideals themselves). The same correspondence holds for the case when we consider only the atoms in the ideal; then we get the isomorphic correspondence $h : a \mapsto a^{\downarrow} \cap A$, where A is the set of atoms in the plural lattice.[14] Here again we can switch back and forth between the "as one" perspective and the "as many" perspective. A number of atomic individuals form a unique sum, which retains the information about these atoms, so that it can return them by the mapping h.

Thus every collection of objects brings with it their free-lattice fusion. If we choose to count with overlap then that indeed makes for $2^n - 1$ individuals if we start from n generators, where n is finite. But this is not the usual way to count: we normally apply the disjoint counting rule, which gives us just n individuals.

Let me summarize the resolution of the fallacy by way of our example. Castor and Pollux are two when we apply disjoint counting. They are one when we consider their free-lattice fusion, for instance, in the context of collective predication. The fusion is nothing over and above its constituents Castor and Pollux, taken together, because, as Lewis says, they are just them. We can go back and forth between the fusion and the atomic elements via the isomorphic correspondence, so without qualification there is no fact to the matter whether we have one or many. That means that grammatical number (singular vs plural) cannot be conceived of as an absolute distinction.

13.4 Is There a Problem with Denotational Plural Semantics?

Let us now turn to linguistic theory. As we have seen LP is committed to a denotational semantics with regard to definite plural terms. In his recent book *Plurals and Events* (1993) Barry Schein takes issue with that *objectual view*, as he calls it. He takes Boolos' reflections on plural quantification to mean that there could be no plural semantics which treats plural expressions as *denoting phrases*. This is of course directed against both of the two camps of denotational plural semantics, the set approach and the mereological approach. So I find myself put in the same corner as the proponents of the set view (Remko Scha, Fred Landman, Jaap van der Does, to name a few). I will defend mereology here, but I don't think there is reason for concern for the others either, as far as the semantic business goes. In my view, the set approach is simply a special case of the algebraic approach.

[14] Cf. the representation theorem for plural lattices in the Appendix.

As a matter of fact, Schein claims that the objectual view is self-defeating in allowing only one-membered models. In this section I will dispel any worry about that, and show that Schein's problem is actually a *Scheinproblem*.

In his book Schein says things like this: "The view that I will argue for here sticks to the naive idea about what an object is, at least, in rejecting the idea that plural terms refer to plural objects" (p. 3); "A plural term ... is a predicate. It does not refer to a plural object" (p. 4). He spies paradox for the objectual view: "The paradox of the objectual view is this: there is no way to both give a true semantics for plural terms and avoid contradiction." (p. 31) Or again, referring to mereological plural theory: "This happy theory founders on Russell's paradox." (p. 45) Isn't that a truly intimidating allegation; let us see whether Schein is able to prove his case.

Schein starts out to ask what a speaker knows about plurals. He presents intuitively valid inferences like the following ((Schein, 1993, p.24)).

(1) The elms are clustered in the forest $\vdash$ There exists an elm

(2) Every one of the elms is tall $\vdash$ Every elm is tall

(3) Every elm is tall $\vdash$ Every one of the elms is tall

All three inferences come out valid in LP, too, as can be easily shown. Schein wouldn't deny that, but he writes: "On the objectual view, to acquire the knowledge of plurals reflected in the above inferences, a speaker must have mastered a family of extra-logical axioms in addition to the semantics of predication and quantification". Yes, indeed, and those additional axioms concern the nature of part-whole relations in language. Thus it is not the case that speakers of certain languages only (viz. those which have the singular-plural distinction) have, under the objectual view, to learn "more" than speakers of other languages. Rather I claim that a grasp of the part-of idiom is part of the semantics of any language. It is true that linguistic semantics of the 1980s has developed a separate subject matter of plural studies; but I for one have always tried to stress the common conceptual frame regarding plurals and mass terms. Keeping that perspective would have saved Schein from a number of pitfalls in his argumentation.

But let us proceed to his attempt to actually derive a contradiction, and see exactly where the argument fails. From the intuitively valid inference

(4) There exists an elm $\vdash$ The elms exist

he abstracts a "comprehension principle" for plural objects:

(5) $\exists x N(x) \rightarrow \exists y \forall x((At(x) \wedge x \leq y) \leftrightarrow N(x))$

where 'N' is a predicate, '$At(x)$' stands for "x is atomic", and $\leq$ is the *part-of* relation. This schema is interpreted thus: "For any predicate 'N', [the schema] (conditionally) asserts the existence of an object y that collects together in some sense the objects that 'N' denotes." (p. 31) Now Schein reads the schema $At(x) \wedge x \leq y$ as a two-place relation $R(x, y)$ and forms, in a Russellian manner, the negated diagonal property $\neg R(x, x)$ to replace 'N' in (5). The antecedent then is equivalent to $\exists x \neg At(x)$ (because of Refl($\leq$)) which is satisfiable in all domains with more than one atom. The consequent, however, reduces to a straightforward contradiction:

(6) $\exists y(At(y) \leftrightarrow \neg At(y))$

All this couldn't be otherwise since under the above substitution the "comprehension principle" becomes

(7) $\exists x \neg R(x, x) \rightarrow \exists y \forall x(R(x, y) \leftrightarrow \neg R(x, x))$

but its negated consequent,

(8) $\neg \exists y \forall x(R(x, y) \leftrightarrow \neg R(x, x))$

is the well-known theorem of first-order logic expressing the Russell paradox (think of the barber who has to shave exactly those who don't shave themselves).

If the above "comprehension principle" were part of the theory of plurals it would confine that theory to domains with one singular object, a blatant failure indeed. But happily, Schein's principle is neither an axiom nor a theorem of LP. So what is going on, then?

The real existence axiom in LP reads as follows:[15]

(Ax. 1) $\exists x \phi[x] \rightarrow \mathbf{E}(\sigma x \phi[x])$

(If there is an object with property ϕ then the i-sum $\sigma x \phi[x]$ of all objects that are ϕ exists.)

Now the i-sum $\sigma x \phi[x]$ has the characteristic properties of a supremum (Π is the individual part relation in LP):

[15]See, for instance Theorem (T.16) in Chapter 1, although no attempt was made there to give a Hilbert style axiomatic presentation of the theory, the focus being on semantic issues.

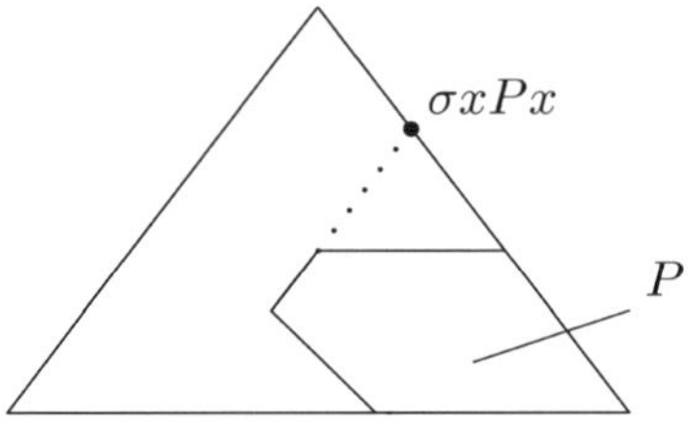

Figure 13.1: Illustration of a mixed predicate and its i-sum

(Ax. 2) $\exists x\phi[x] \rightarrow \forall y(\phi[y] \rightarrow y\Pi\sigma x\phi[x])$

(Ax. 3) $\exists x\phi[x] \rightarrow \forall y(\forall x(\phi[x] \rightarrow x\Pi y) \rightarrow \sigma x\phi[x]\,\Pi\, y)$ (σ-*properties*)

(Every object y which is ϕ is below the i-sum $\sigma x\phi[x]$; and whenever some object y is above all x that are ϕ it is also above $\sigma x\phi[x]$.) Thus $\sigma x\phi[x]$ is an upper bound for the elements having ϕ, and it is the smallest upper bound for them. Ax. 1 and 2 say, then, that $\sigma x\phi[x]$ denotes the supremum of all objects falling under ϕ whenever it exists.

Observe now that in forming the supremum some information is lost, which can be best illustrated in the mass case: pour some cups of water together, and there is no way of telling anymore which water quanta you started from. Likewise, if you form the sum of two proper sums the atoms below the supremum don't give you any information about the constituent sums. The example used in earlier chapters was those two families, the Leitches and the Latches: their sum is no different from the sum the individual members of the two families. The same holds for mixed predicates: if the sum consisting of Bill and John on the one hand and Obelix on the other are in the extension of the predicate *lift the piano* then the supremum of the elements of this extension is the sum made up of Bill, John, and Obelix, and there is no telling from this sum alone who did the lifting. There might be an empirical problem with that approach, which I am well aware of, and which has been discussed extensively under the heading of "intermediate readings"; but the issue here is consistency, and as for that, supremum formation is as innocuous as it could be.

An illustration might be of help that uses the familiar pyramid picture; it is taken from Chapter 6 and repeated here as Figure 13.1.

The base line of the pyramid consists of the atoms of the structure. The extension of the predicate P as shown in the picture contains atoms as well as non-atoms, but it does not contain all the atoms of each element. As a result there are atoms below the supremum which do not have the property P.

So much for technicalities. Let us return to the conceptual point again. Schein says that Boolos has put the current problem "sharply" in examples like these:

(9) If there is a set that does not contain itself, then there are the sets among which is every set that does not contain itself.

(10) If there is a set that does not contain itself, then there is a set a member of which is every set that does not contain itself.

Schein claims that the objectual view could not tell the two sentences apart. But this is obviously not so. The first sentence is a special case of the very general mereological principle that whenever there are some entities then there is also a fusion of them. This fusion is "just them", and it does not add to the ontological commitments of the theory. The second sentence is of course an instance of a comprehension principle in the highly specialized context of set theory, and that happens to be contradictory in its unrestricted form, as we know.

Schein also tries to "translate" the set-theoretic examples explicitly into the mereological context:

(11) If there is a plural object, then there are the plural objects.

(12) If there is a plural object, then there is the plural object that every plural object is an atomic part of.

Schein comments: "Statement (11) is obviously true, but no believer in plural objects can believe (12). Yet the objectual view cannot tell apart the plural term in (11) and the singular term in (12)." Now if there is a plural object or non-atom (i.e., there are at least two different atoms in the domain), then there is the supremum of all plural objects which equals the unit element **1** of the whole domain. If the term *plural object* is expressed in the object language by the predicate $\overline{At} = \lambda x \neg At(x)$, then the plural term $\sigma x \overline{At}(x)$ denotes **1**. Is **1** a plural object? Linguistically no, because it's a single entity; but semantically yes since its atomic parts are more than one.

Now what about *the plural object that every plural object is an atomic part of*? This phrase doesn't even come close to anything the objectual view might be committed to, and it has nothing to do with Russell's paradox anymore. The air of absurdity is engendered by the fact that an *inherently relative concept* is taken as absolute. So this is another instance of the counting fallacy. The cats and dogs and tables and stones persist across various domains of discourse, at least to a large extent. Not so for plural objects; you have to tell me first what "regular" things you are prepared to include in your domain of discourse, and then it's easy to say what the plural things are in your ontology. A plural thing here becomes a new singular thing there as it often happens; so for instance, the two hydrogen atoms and the one oxygen atom combine to form one water molecule. The subject phrase of this sentence denotes a plurality, whereas *the water molecule* stands for an atomic individual in the domain.

Schein himself sketches a number of "ways out" of his paradox, only to conclude that we have to abandon the objectual view altogether. All these options are rather artificial, and having disposed of the principal objection there is no need to consider them.

Finally, of course, another problem does remain which is of real semantic interest: even if the objectual view is coherent, can we do without? Schein's book is an attempt to show that a Davidsonian event theory in which plural terms are treated as second-order predicates is able to cope with all plural phenomena in language that have been described so far. An answer to this claim would enter a different debate, and must be deferred to a different occasion.

13.5 Cross-linguistic Evidence

I'd like to conclude this chapter with a reflection on some universal characteristics of natural language: it consists in a (rather modest) case study of the part locution in a language which is genetically remote from English, that is, the classifier language Chinese.[16] I try to give evidence for my view that conceptually, the basic element of plural quantification is not the "atomistic" notion of plurality, but the portion-taking idiom *(part) of*, and hence that plural quantification should be treated as part of mereology.

Let us begin by contrasting a singular existential statement with its "plural"[17] counterpart. The singular has the noun *māo* ('cat') modified by the numeral 'one' plus the obligatory classifier, whereas the plural version

[16] I wish to thank Wang Kanmin for helping me with the Chinese data.

[17] More precisely: the Chinese translation of the pluralic existential statement in English.

translates to the bare noun *māo*.

(13)

Yoŭ	*yì*	*zhī*	*māo.*
there is	one	CL	cat

"There is a cat."

(14)

Yoŭ	*māo.*
there is	cat

"There are cats."

The three examples concern the attributive construction, which is effected by the particle *de*. The noun phrases contain the translations of an animate and an inanimate count noun and a mass noun, respectively. Semantically, these noun phrases are definite descriptions showing two things about Chinese: (i) There is no definite article in the language, and (ii) there is no plural marking, so that the count and the mass case are indistinguishable; hence the pluralic sense has to be inferred from the context.

(15)

Zhōngguŏ	*de*	*māo*
China	ATTR	cat

"China's cats"

(16)

nǐ	*de*	*shū*
you	ATTR	book

"your books"

(17)

Chángjiāng	*de*	*shuǐ*
Yangtse	ATTR	water

"the water of the Yangtse river"

The following sentences show how the phrase 'one of them', which is the central locution in Boolos's interpretation of plural quantification, is rendered in Chinese. The plural noun phrase *sān bĕn shū* ('three books') is referred back to by the impersonal *qí*, which together with *zhōng de* means something like the English 'thereof'. Thus the i-sum consisting of three books is conceived of here as some kind of homogenous stuff of which a certain portion is taken. However, Example (24) shows that Chinese does have a plural pronoun (viz. *tāmen*) that can be used in contexts in which numerically precise anaphoric linkage is called for.

(18)

Lǐ	*xiāngsheng*	*xiĕle*		*sān*	*bĕn*	*shū.*	*Qí*	*zhōng*
Li	mister	write	PERF	three	CL	book.	That	among

de	*yì*	*bĕn*	*shì*	*yòng*	*yīngyŭ*	*xiĕ*	*de.*
ATTR	one	CL	is	language	English	written	ATTR

"Mister Li has written three books. One of them is written in English."

(19) *Nǐ de shū zài nǎr? Qí zhōng de yì*
you ATTR book at where? That among ATTR one
běn wǒ xiǎng kànkan.
CL I want look at
"Where are your books? One of them I want to look at."

The next examples try to get at the Chinese of expressing the notion of a definite pluralic existence claim concerning a countable property like 'cat' (*māo*) or 'thing' (*dōngxi*). When these sortals are considered in isolation there is again no way to express either the intended definiteness or the plurality. That seems to indicate that Chinese nouns refer to homogenous concepts that don't have a countable unit built into them.[18] However, when the "comprehension principle for pluralic aggregates," (23), is given a logically precise linguistic form as in (24) then the anaphoric pronoun chosen is the pluralic *tāmen*.

(20) *suǒ cúnzài de māo*
EMPH exist ATTR cat
"the cats there are; the cats that exist"

(21) *suǒ cúnzài de dōngxi*
EMPH exist ATTR thing
"the things there are; the things that exist"

(22) *Rǔguǒ yoǔ yì zhī māo, nàme jiù yoǔ māo.*
If there is one CL cat, then there is cat
"If there is a cat, then there are (the) cats."

(23) *Rǔguǒ yoǔ yì zhī māo, nàme jiù cúnzài zhe māo.*
If there is one CL cat, then exist DUR cat
"If there is a cat, then there are (the) cats."

(24) *Rǔguǒ yoǔ yì zhī māo, nàme jiù yoǔ yī leì*
If there is one CL cat, then there is one sort
dōngxi, tāmen zhōng de meǐ yī ge doū shì māo,
thing, they among ATTR each one CL all is cat,
ér meǐ zhī māo yoù doū shì tāmen zhōng
whereas each CL cat in turn all is they among

[18]Cf. also the relevant discussion in Chapters 11 and 2.

de *yī* *ge.*
ATTR one CL.

"If there is a cat, then there are some things such that each of them is a cat, and each cat is also one of them."

By way of conclusion, I want to say that there is evidence for the hypothesis that (i) the part-locution is universally entrenched in language, and (ii) that it is ontogenetically prior to plural reference. This comports with the general algebraic perspective according to which the common mereological lattice structure of the material ontology is emphasized, and the atomic domain of plural objects is but a special case in this structured ontology. By contrast, a set-theoretic outlook is bound to be "atomistic" and thereby tends to neglect the more basic part structure.

Chapter 14

Mereology, Second-Order Logic, and Set Theory

14.1 Introduction

This chapter is about the role of mereology in the foundations of mathematics. More specifically, it deals with a nominalistic interpretation of (monadic) second-order logic where the second-order variables run over mereological fusions of entities in the first-order domain. This interpretation should comport with what I take to be the *spirit* of George Boolos's approach to plural quantification as an alternative semantics of second-order logic insofar as it avoids the commitment to any abstract class-like entities that are built up from the first-order individuals; however, it differs from him in two respects: (i) it sticks to a denotational view of the second-order entities; and (ii) it is spelled out within a completely formal theory of plurality, called LP, thereby providing the ground for rejecting Boolos's contention that the alternative semantics has to be irreducibly based on a locution from ordinary language. In a second step, I'll develop David Lewis's nominalistic set theory ("megethology") within the theory of plurality. It is shown that megethology can be treated in a rigorous way within LP, including all the coding that is necessary to obtain the crucial concept of a singleton function.

This is, then, how I'd like to proceed. First, I want to defend what I called the "spirit" of Boolos's approach to second-order logic. That is, I want to side with Boolos against the main criticism that was advanced by Michael Resnik on this topic. Having done that I will go on to explain why I feel that Boolos hasn't gone far enough. While he thinks that plural quantification is a self-understood notion I want to argue that this idiom is both in need and capable of a theoretical explanation, which I submit is mereology. The upshot, then, is that what Stuart Shapiro calls "the logical notion of set" can be captured in mereological terms, thereby indeed breaking the cycle of presupposing set-like entities when it comes to formulating set theory within second-order logic. In his book *Parts of Classes* (1991) Lewis adopts the device of plural quantification from Boolos without any discussion, so the same reform I propose for Boolos holds for Lewis, too. I have to admit, though, that my interpretation of Lewis faces a somewhat disconcerting question: while Boolos is silent on matters of mereology, Lewis's metatheory consists of plural quantification alongside with mereology; so why didn't Lewis extend the latter tool to plural quantification? I think that one of the reasons could be that his notion of "singularism," which is a red herring in my opinion, stopped him from taking this step, but also probably the attitude of "no explication called for" towards the linguistic device of plural quantification. Now granted that plural quantification can be treated mereologically, the reconstruction of megethology leaves us with two mereologies, and the remaining problem is how to relate them. My answer will be that the situation is analogous to class theories in axiomatic set theory which also display a coexistence of two sorts of entities, classes and sets. So I should not worry more than anybody embracing such a class theory. However, a better reason for entertaining two mereologies is that, conceptually, they serve two different purposes: one is to model the logical notion of set, and the other to give a nominalistic reconstruction of the iterative conception of set along the lines of Lewis's treatment.

14.2 A mereological interpretation of monadic second-order logic

Let me begin by recapitulating Boolos's basic motivation for giving an alternative semantics to monadic second-order logic, ML2. His paper (Boolos (1985a)) starts out with Frege's definition of the ancestor relation, which reads:

(1) x is an ancestor of y just in case x is in every class that contains y's parents and also contains the parents of any member.

A straightforward representation of this relation in ML2 is (with 'Pxy' for 'x is a parent of y'):

(2) $Anc(x, y) \leftrightarrow \forall X\, [\forall w\, (Pwy \rightarrow Xw) \wedge \forall z \forall w\, (Xz \wedge Pwz \rightarrow Xw) \rightarrow Xx]$

According to the standard semantics of second-order logic, unary predicate variables range over subsets of the given first-order domain. Thus any time you *negate* the ancestor relation between two objects you make an existence claim involving a set or class. Boolos regards that as an excessive commitment to abstract objects which is not supported by the subject matter at hand, viz. the genealogy of real people. So he looked for an alternative interpretation of the second-order variable that avoids such a commitment, and came up with a paraphrase of (1) which apparently speaks only about people rather than classes:

(3) x is an ancestor of y if and only if (I) someone is a parent of y and (II) it is not the case that there are some people who are such that (a) each parent of y is one of them, (b) each parent of any one of them is also one of them, and (c) x is not one of them.

Boolos admits that this is a rather clumsy way of putting things, even if the intended interpretation is properly expressed. The reason for this detour through the existential mode is that Boolos wants to "refer plurally" to a number of people without using the term *collection of* since that might invite unwelcome commitments to classes again. Pluralic reference is supposedly achieved by means of the existential phrase *there are some A* together with the expression *is one of them*. But this locution doesn't have a universal counterpart in language, i.e., a phrase that quantifies both plurally and universally. Boolos goes on to say that universal second-order formulas like the one in (23) are mere abbreviations of their suitable existential counterparts, since it is only the latter that acquire meaning through their pluralic interpretation.

Here we have the situation that a formal language, viz. monadic second-order logic, is semantically interpreted not in a reasonably formal metalanguage but in plain ordinary language. The relevant part of language, the plural idiom, is thought of as being perfectly well understood, so that in fact it can form the basis of an explication in formal philosophy. When I first read about this I was struck by the remarkable *reverse methodology* applied here: The typical approach in analytic philosophy would be that given a pre-theoretic concept you want to become clear about you'd try out an explication of it in some regimented formal language. But here we do exactly the opposite: we rest an interpretation of a completely formal device, second-order quantification, on some linguistic locution. What is even more interesting is the fact that this "naturalistic" reversal went virtually unnoticed and was followed by quite a few philosophers in the field, most noticeably Lewis.

Still Boolos must have felt that something was missing here, because in Boolos (1985a) he does give a formal truth definition for the second-order logic. The critical context in which this is done is a second-order language of set theory, where the first-order variables run over the usual sets and the second-order variables are normally thought of as having (possibly proper) classes of sets as values. "This [truth-]theory," Boolos says, "is formulated in a second-order language of set theory together with a new predicate containing two first-order variables "s" and "F" and one second-order variable "R": R and the sequence s satisfy the formula F" (Boolos (1985a):336). Let us call this truth theory $\mathcal{T}$ and the basic relation just mentioned $Sat(R, s, F)$. The key clause of $\mathcal{T}$, then, is the condition for elementary predication Vv, where v is a first-order variable and V is a monadic second-order variable:

(4) if F is Vv, then $Sat(R, s, F)$ iff $R\langle V, s(v)\rangle$

The intended reading for $R\langle V, s(v)\rangle$ is obviously that "$s(v)$ *is one of them*$_V$," or else, as is explicitly stated on the next page, that the set $s(v)$ "is *a* [my emphasis, G.L.] value of the second-order variable V with respect to R." So R isn't a functional assignment of a (class) value to V any more but relates plurally to V several sets $s_1(v)$, $s_2(v)$, etc. Thus this theory is not committed to the view that whenever there are some sets x such that $R\langle V, x\rangle$ there is also a single object whose members are all and only the sets x such that $R\langle V, x\rangle$.

In such a way Boolos hopes to keep the power of second-order logic without committing himself to classes of entities in the first-order domain. Michael Resnik took issue with this proposal. In Resnik (1988) he argues that Boolos doesn't succeed in proving his point. Let me single out three kinds of objections which Resnik advances: they concern (i) technical drawbacks of Boolos's proposal; (ii) differing logical/linguistic intuitions; and (iii) questions about the exact shape of the above truth theory. I'll come back to the third objection below.

Let us first consider the technical drawbacks of Boolos's set-up. Resnik says three things again here: that a) "[s]o far at least, no one has been able to find a direct translation of second-order universal quantification in terms of plural quantification;" that b) even the rule of existential plural quantification is not particularly natural in that it has to make up for the fact that it carries existential import lacking in second-order logic; and finally that c) the rule works only for monadic second-order quantification. The last point isn't too serious since the theory admits of a *pairing function* (viz. ordered pairs), and n-ary relations can be played back to the monadic case. I agree with Resnik regarding the first two points, and they will be addressed when I discuss plural theory.

The second type of objection concerns the kind of examples that Boolos gives to support his view and the intuitions behind them. Resnik doesn't share those intuitions; his own rather "differ sharply from Boolos's" [p.76]. Take, for instance, the Geach-Kaplan sentence *Some critics admire only one another*, and its Boolos style translation,

(5) There are some critics such that any *one of them* admires another critic only if the latter is *one of them* distinct from the former.

Resnik thinks that this sentence makes explicit reference to collections, and adds the rhetorical question: "How else are we to understand the phrase 'one of them' other than as referring to some collection and as saying that the referent of 'one' belongs to it?" [p. 77] Boolos could say that Resnik just refuses here to leave the set-theoretic frame of mind, but I have some sympathy with Resnik's stubbornness: It perhaps doesn't have that much to do with diehard set-theoretic convictions but with some inherent drive in language towards "singularism," to which I will return presently. For the moment, try to replace the plural *some critics* by *some group of critics*, yielding the following paraphrase:

(6) There is some group of critics such that any one in the group admires another critic only if the latter is also a member of this group distinct from the former.

Would anyone who hasn't undergone a lasting *déformation professionelle* think that we talk about classes here? A group of people just isn't a class of people. The term *group*, used informally here, but with a sufficiently clear linguistic meaning, is ontologically innocuous. We could also apply it in the definition of *ancestor* and regain the original universal format:

(7) x is an ancestor of y just in case x is in every group of people that contains y's parents and also contains the parents of any member.

It is obvious where I'd like this to be going to lead us. In the final analysis, the informal word *group* will be replaced by the technical term *(mereological) fusion* or *individual sum* in plural theory.

I am aware that many people who have followed Boolos in embracing plural quantification will think that I am missing the whole point here. As Geoffrey Hellman says, "the pluralities of plural quantifiers cannot be collected for they are not objects in their own right (on pain of losing 'their' point)" (Hellman (1994):247). So let me elaborate a little here on the issue of "singularism." In his *Parts of Classes* Lewis writes:

> It is customary to take for granted that plural quantification must really be singular. Plurals, so it is said, are the means whereby ordinary language talks about classes. According to this dogma he who says that there *are* the cats can only mean ... that there *is* a class of cats. And if you say that there *are* the non-self-membered classes, you can only mean, never mind that you know better, that there *is* the class of non-self-membered classes. (Lewis (1991):65)

And a few pages later:

> Plural quantification is irreducibly plural. It is not ordinary singular quantification over special plural things — not even when there are special plural things, namely classes, to be had. (Lewis (1991):68f.)

I think there are two issues here that should be kept apart. One is the pluralic mode of speech in language and the linguistic tendency to "singularize" certain pluralities by introducing some lexical item of singular number, for instance *those people* vs. *that group of people*. However, it is quite a different matter whether we can form from any given plurality a *class* of exactly those objects that make up this plurality. Let me call the position that endorses such a principle *(set-theoretic) comprehensionalism.* For well-known reasons that position is untenable, as Lewis argues at length. So when he attacks singularism he really attacks comprehensionalism.

What about singularism, then? It is a good question whether the linguistic process of singularization creates new objects that are different from those we start out with. A group of people ("that group over there") is certainly different from each individual member of it, but it should not differ much from "those people over there." For instance, the way of referring to this group by ostention while uttering these two expressions would be the same: it would typically consist in the same sweeping gesture towards that group. In fact, mereological plural theory says that the technical notion of *individual sum* (i-sum) replaces the informal term 'group' and gives a denotation for definite plural noun phrases like *the cats*. There are many linguistic reasons for such an approach, one most perspicuous being that it is the very same definite article *the* that is used both in singular and plural terms. Thus just like singular definite descriptions denote subject to the usual existence and uniqueness conditions, plural definite descriptions also denote under the same conditions, appropriately generalized. In technical terms, LP contains a *summation operator* σ that forms singular terms[1] like $\sigma x P x$, standing for "*the Ps*". If the extension of P contains exactly one object, the σ-term $\sigma x P x$ is provably equal to the ι-term $\iota x P x$.

[1] Here, 'singular' is used in the logical sense!

So singularization indeed takes place here, but it is the mereological process of forming fusions. As Lewis (1991) eloquently shows in his discussion of mereology, the process of taking fusions doesn't carry any ontological commitment beyond the commitment to those entities we start from.

We cannot avoid a denotational (and thereby "singularizing") approach towards pluralities anyway. *Linguistic singularization is inevitable.* For instance, Boolos (1985b:343) notices that we can say,

(8) The rocks rained down

but that it doesn't make much sense to paraphrase that by saying that there are some rocks each of which rained down. The reason for this is that the predicate *rained down* is true of the rocks only *collectively*. But Boolos doesn't go beyond this mere observation. Let us suppose the rocks that rained down on the road weigh 50 tons. Is there something that weighs 50 tons, or not? Consider the following "squish" (think of the history of an avalanche that starts out with a few huge boulders and ends up as a heap of gravel):

(9) a. The boulders weigh 50 tons.
b. The rocks weigh 50 tons.
c. The stones weigh 50 tons.
d. The pebbles weigh 50 tons.
e. The pieces of gravel weigh 50 tons.
f. The gravel weighs 50 tons.

Should we make *the gravel* denote, but not *the boulders*? And where should we draw the line? The pair of shoes denotes, but not the shoes? My glasses denote, but not the lenses in my glasses? You see my general strategy here: whether we have a singular or a plural is just a matter of grammar, not of ontology. The same material object, possibly somewhat scattered, is referred to by a plurality, and then again by a singular mass expression. 'gravel' is a mass noun, and the definite article in front of it produces a singular definite description that undoubtably denotes. Thus, there is no stable correlation between the number distinction in language and matters of ontology. That is not to say, however, that there is no conceptual difference between plural forms of count nouns and singular mass nouns. The main difference lies in the property of *atomicity* that is present in the count case and absent in the mass case. Apart from that, the structural similarities are striking; they can be used to give a uniform "algebraic" semantics of both domains, as I have shown elsewhere (Link (1983a)).

By taking the mereological approach to plurality I am able to gain the expressive power of second-order statements without any commitment to classes. However, for reasons having to do with both ontological considerations and facts about natural language semantics, the mereology of pluralities is developed in a *first-order language*, LP. Let me briefly describe again what this system looks like.[2]

The basic logic is a free logic with λ-abstraction. The individual variables range over singular objects and pluralities alike; I will use the syntactical variables $\alpha, \beta, \gamma, \delta$ for them here. The objects in the range of these variables are indiscriminately called *i-sums*, and the special i-sums that consist of just one singular individual are called *atomic i-sums* or *i-atoms*. I will reserve the symbols u, v, w, x, y, z for variables ranging over i-atoms only. The basic non-logical symbols are the *individual part-of relation* (*i-part relation*, for short), denoted by 'Π', and the σ-operator mentioned above. Π carries existential import. Let me also mention a few defined symbols in LP (ζ, η are individual terms in LP, 'P' is a variable for one-place predicates; variables α are not free in clause-mate terms ζ, η):

(10) $\mathbf{E}\,\zeta \leftrightarrow \exists\alpha.\alpha\,\Pi\,\zeta$ (*existence predicate*)

(11) $\zeta = \eta \leftrightarrow \zeta\,\Pi\,\eta \wedge \eta\,\Pi\,\zeta$ (*identity*)

(12) $\mathbf{At}\,\zeta \leftrightarrow \mathbf{E}\,\zeta \wedge \forall\alpha(\alpha\,\Pi\,\zeta \rightarrow \alpha = \zeta)$ (ζ *is an i-atom*)

(13) $\zeta\,{}^{\cdot}\Pi\,\eta \leftrightarrow \mathbf{At}\,\zeta \wedge \zeta\,\Pi\,\eta$ (ζ *is an i-atomic i-part of* η)

(14) $\zeta \bullet \eta \leftrightarrow \exists\alpha\,(\alpha\,\Pi\,\zeta \wedge \alpha\,\Pi\,\eta)$ (*i-overlap*)

(15) $\zeta \mid \eta \leftrightarrow \neg\zeta \bullet \eta$ (x *is i-disjoint from* y)

(16) $\zeta \oplus \eta \;=\; \sigma\alpha\,(\alpha = \zeta \vee \alpha = \eta)$ (*the i-sum of* ζ *and* η)

(17) $\zeta \ominus \eta \;=\; \sigma\alpha\,(\alpha\,\Pi\,\zeta \wedge \alpha \mid \eta)$ (*the i-difference of* ζ *and* η)

(18) ${}^{*}P(\zeta) \leftrightarrow \zeta \;=\; \sigma\alpha\,(P(\alpha) \wedge \alpha\,\Pi\,\zeta)$ (*the (plural) star operator*)

(19) ${}^{D}P(\zeta) \leftrightarrow \forall x\,(x\,{}^{\cdot}\Pi\,\zeta \rightarrow P(x))$ (*the distributivity operator*)

(20) $\mathbf{Distr}\,(P) \leftrightarrow \forall\alpha\,(P(\alpha) \rightarrow \mathbf{At}\,\alpha)$ (P *is a distributive predicate*)

The axioms of LP are those of a classical mereology.[3] They describe a certain lattice structure that is bottomless, atomic, free, and in which

[2]See also Chapter 6. However, I changed the syntactic variables for individual sums here in order to prepare for the application of the logic to set theory that will be given below.

[3]Cf. Chapter 6. For a standard set of axioms, differing inessentially from ours, see Simons (1987).

σ has the properties of a supremum operator (in fact, the structure boils down to an atomic Boolean algebra minus its zero element). Moreover, the structure is complete in the following sense:

(21) $\exists\alpha\phi[\alpha] \rightarrow \mathbf{E}\,(\sigma\alpha\phi[\alpha])$

When the matrix $\phi[\alpha]$ is distributive, i.e., when all its satisfiers are atomic then (21) yields the following *comprehension principle*:

(22) $\exists x\phi[x] \rightarrow \exists\alpha\forall x\,(\,x\,{}^{\cdot}\Pi\,\alpha \leftrightarrow \phi[x]\,)$

It is now straightforward to give a representation of the ancestor relation in LP. Let us begin with Boolos's condition (3), which involves plural locutions like *there are some people* and *is one of them*. These are precisely the kind of phrases for which LP had originally been developed; the latter is truly "mereological" in the etymological sense of the word. Hence it is formalized by the i-atomic i-part relation in LP. Here is a translation of (3), with '*Anc*' and '*P*' as above.

(23) $Anc(x,y) \;\leftrightarrow\; \exists uP(u,y) \wedge \neg\exists\alpha[\forall u(P(u,y) \rightarrow u\,{}^{\cdot}\Pi\alpha)$
$\wedge\,\forall u\forall v(P(u,v) \wedge v\,{}^{\cdot}\Pi\alpha \rightarrow u\,{}^{\cdot}\Pi\alpha) \wedge \neg\,x\,{}^{\cdot}\Pi\alpha]$

However, since in LP we can freely quantify over sums also universally, this contrived formula is actually not needed. The following is a perfect rendering of Frege's definition:

(24) $Anc(x,y) \;\leftrightarrow\; \exists uP(u,y) \wedge \forall\alpha[\,\sigma uP(u,y)\,\Pi\,\alpha$
$\wedge\,\forall w(w\,{}^{\cdot}\Pi\alpha \rightarrow \sigma uP(u,w)\,\Pi\,\alpha) \;\rightarrow\; x\,{}^{\cdot}\Pi\alpha\,]$

In the same vein we can represent the critics sentence. Let C stand for *is a critic*; then C is a distributive predicate, i.e., it can only be true of i-atoms, and $^{*}C$ is true of i-sums of critics.

(25) a. Some critics admire only one another.

b. $\exists\alpha[{}^{*}C(\alpha) \wedge {}^{D}\lambda u\forall v[A(u,v) \rightarrow v\,{}^{\cdot}\Pi\,\alpha \ominus u]\,(\alpha)]$

$\Leftrightarrow$

$\exists\alpha\,[\forall u(u\,{}^{\cdot}\Pi\alpha \rightarrow C(u)) \wedge \forall v[A(u,v) \rightarrow v\,{}^{\cdot}\Pi\,\alpha \wedge v \neq u]\,]$

Another example is Boolos's ingenious Zev sentence (Boolos (1984b):434), which under an arithmetical interpretation can tell standard models from non-standard ones. Since it can be represented in LP this is a way to see that LP is semantically second-order and incomplete.

(26) a. There are some horses that are all faster than Zev and also faster than the sire of any horse that is slower than all of them.

b. $\exists\alpha[{}^*H(\alpha) \wedge {}^D\lambda u F(u,a)\,(\alpha)$
$\wedge\ \forall v\,({}^D\lambda w S(v,w)\,(\alpha) \rightarrow {}^D\lambda u F(u,s(v))\,(\alpha))]$

$\Leftrightarrow$

$\exists\alpha[\ \forall u(u\dot{\Pi}\alpha \rightarrow F(u,a)) \wedge$
$\forall v[\forall u(u\dot{\Pi}\alpha \rightarrow S(v,u)) \rightarrow \forall u(u\dot{\Pi}\alpha \rightarrow F(u,s(v)))]\ ]$

This sentence can be made more perspicuous by the use of a "collective operator", viz. *uniformly*, which expresses the condition that the relation it operates on is to hold of all atomic parts of the sum in question. Thus sentence (26a) reads as under (27a). It should be obvious that this is something rather different from the singular sentence given in (27b). (The arithmetical interpretation of the latter is simply a false statement of number theory, whereas the former comes out false in the standard model and true in any non-standard model, so it cannot be first-order.)

(27) a. There are some horses that are faster than Zev and also uniformly faster than the sires of all the horses uniformly slower than them.

b. There is a horse that is faster than Zev and also faster than the sire of any horse that is slower than it.

The final example is a sentence that Boolos took from Quine's *Methods of Logic* (1972).

(28) a. Some of Fiorecchio's men entered the building unaccompanied by anyone else.

b. $\exists\alpha[{}^*F(\alpha) \wedge E(\alpha) \wedge \forall\beta(A(\beta,\alpha) \rightarrow \beta\Pi\alpha))]$

After discarding Quine's first-order representation (29a), Boolos gives a ML2 translation (29b) which again cannot be "firstorderized":

(29) a. $\exists u[F(u) \wedge E(u) \wedge \forall v(A(v,u) \rightarrow F(v))]$

b. $\exists X[\exists x X(x) \wedge \forall x(X(x) \rightarrow F(x)) \wedge \forall x(X(x) \rightarrow$
$E(x)) \wedge \forall x \forall y(X(x) \wedge A(y,x) \rightarrow X(y))]$

Comparing Boolos's (29b) with the LP representation given in (28b) we see how much simpler the latter comes out, because it can take advantage of uniform sum quantification. What it actually says it that there is an i-sum α consisting exclusively of Fiorecchio's men which collectively entered the building and was such that any i-sum β that accompanied α was an i-part of α.

After the discussion of this sample of sentences from natural language, which gives some indication of the expressive power of LP, I return now to the main line of argument. Let us take up Resnik's first kind of objection, concerning the technical drawbacks of Boolos's approach; we are now in a position to answer it without giving in on Resnik's set-theoretic convictions. First, we do have a "nominalist" version of second-order universal quantification in LP, and second, we can see more clearly in LP why we have an existential restriction on the denotation of terms: in mereology, an "empty" part just doesn't make any sense.

Having dealt with Resnik's second objection earlier I come now to a logical issue that Resnik raises as his third major criticism. Boolos's truth theory seems to avoid quantification over classes; but on closer inspection, Resnik argues, such quantification is merely shifted to the metalanguage, so Boolos doesn't really get rid of it. Now first of all, I think Boolos's paper cannot be blamed to be just an idle exercise, amounting to no more than the repetition of his old point one level higher up in the hierarchy of metalanguages. The advance over his earlier papers on the topic is that he not only says that second-order quantifiers can be read plurally; he also manages to give a *truth theory* for second-order logic that avoids the assignment of classes to its second-order variables. So far, so good. But what about the interpretation now of the quantified second-order variables of this very truth theory $\mathcal{T}$, as they occur in the critical clause (Boolos (1985a):336):

(30) if F is $\exists VG$, then R and s satisfy F iff
$\exists X \exists T\, (\forall x(Xx \leftrightarrow T\langle V, x\rangle)$ &
$\forall U\, (U$ is a second-order variable &
$U \neq V \rightarrow \forall x(T\langle U, x\rangle \leftrightarrow R\langle U, x\rangle))$ & T and s satisfy $G)$

To begin with, there is a certain looseness of notation involved here. U and V should be second-order variables of the object language, suitably coded up; but X and T are variables of the second-order language of $\mathcal{T}$, and here we do have the kind of shift to the metalanguage that Resnik refers to. However, Boolos still can, and indeed does, claim that the same device of plural quantification is available on the meta-level; this is what he must have in mind when he says that "we needn't interpret either our original second-order language of set theory or the new truth theory for this language in this manner", viz. that classes are the values of the second-order variables (Boolos (1985a):337). Boolos can argue that his position is coherent in the sense that he is not forced to make ontological assumptions in the metalanguage which he repudiates for his object language. But I think that Resnik has a valid point, too: the shift to the metalanguage doesn't make plural quantification *more convincing* in any way. This is because of Boolos's brand of "naturalism", which seems to suggest that plural quantification is actually in no need of a theoretical explanation. However, there is a story missing here that has to be told.

Having worked on the linguistic semantics of plurality to a considerable extent I have come to realize that a more careful analysis of plurals is necessary that makes use of all the relevant logical tools at our disposal. Also, our usually pretty stable semantic "intuitions" tend to fade away when confronted with the vast array of pluralic phenomena in language. This is not the place to go into it, but consider for just a moment the following English sentence that seems to make perfect sense at first blush:

(31) Most competing companies have common interests.

Thus, on this score I agree with Resnik that "plural quantification [is] in need of logical analysis" (Resnik (1988):77).

There are also formal problems that this kind of naturalism has to deal with. For instance, how could Boolos possibly make the distinction between first-order and second-order logic (which he obviously wants to make) on the level of explicitness that is typically associated with natural language locutions? We might perhaps be able to pick out arbitrary pluralities in our phenomenal world, by sweeping ostensions or other means appropriate for "dividing" our reference; but they are certainly not suited to delineate infinite pluralities. But then it is hard to see how Boolos can be sure that he actually makes reference to a range of pluralities that correspond to the full model of the standard semantics for second-order logic rather than, say, to some generalized model in Henkin's sense, which amounts to a first-order structure after all. So it is not enough to give the syntactic format that we are used to, with upper case symbols for predicate variables, to arrive at second-order logic; this has been made perfectly clear for instance by Stewart Shapiro in his recent study on the topic (see Shapiro (1991)). Rather we have to give the full standard semantics along with the formalism. It should be plain that the plural idiom won't help here; we can hardly avoid "going theoretical".

But maybe the situation isn't exactly the way we have portrayed it. Quine for instance reads Boolos as saying that monadic second-order logic regiments plural talk! In his 1991 *Dialectica* paper[4] Quine writes: "[Boolos] shows that [second-order logic] can be interpreted as a mere regimentation of plurals, and these seem ontologically innocent". It looks like we have

[4]See Quine (1991).

come full circle now: Boolos invites us to understand monadic second-order logic in terms of plurals, but Quine suggests that in order to gain a grasp of plurals we'd better regiment them by second-order logic. Now, I think that Quine's suggestion, rather than Boolos's, hints toward the right direction, *if properly interpreted*: we do stick to Boolos's basic intuition to read quantification over simple collections of individuals nominalistically, i.e., in terms of the plural idiom.[5] However, the plural idiom has to undergo regimentation like any other pretheoretic locution. Quine says: take second-order logic; but that is misleading at best, and conceptually confusing. So I propose to take LP as regimenting theory and thereby to interpret plurals mereologically.

Let me add a final consideration here. In first-order logic we read instances of juxtaposition or function-argument-structure like 'Px' or '$P(x)$' as expressing the basic relation of predication, and rightly so because this syntactic form is the core of classical "predicate calculus". The difference between the lower and the higher predicate calculus only concerns the range of quantified variables; what is kept fixed is the interpretation of elementary predication. Thus it strikes me as somewhat *ad hoc* to change the meaning of this basic relation when we go over to the second-order level. Therefore I suggest that we leave second-order logic alone, or maybe hand it over to that respectable group of theorists who are prepared to quantify over properties. Our concern should not be a reinterpretation of predication. Rather we are dealing with the notion of plurality or multiplicity of objects, which is part of our natural logic and has entered language ages before the highly technical concept of iterative set was eventually developed. It is what has surfaced recently under the name of "logical set" mentioned above. I take Boolos's main point to be that these multiplicities, albeit usually *modeled* by sets or classes, are conceptually different; being typically "flat" as far as their set-theoretic rank is concerned, they don't add a level of comprehension but are of the same ontological type as the singular objects they consist of. And they can do much of the work for which in our set-theoretic bias we have been using classes.

My reinterpretation of Boolos's naturalism, then, runs as follows. The basic issue is the meaning of the second-order set variables. I replace monadic second-order logic by the logic of plurals as indicated in the above examples; thus plural talk is regimented by LP, which was designed, and hence is much better suited, for this task. There is, however, a close relationship between the two formalisms that should come as no surprise: the reason is the atomic Boolean structure that is common to the power set algebra and the mereology of plurals (always modulo the null object). This

[5] Even Quine, in the passage quoted, considers that a palatable option.

is why classes are useful modeling devices. But in the present context the algebraic structure is all that matters.

Speaking formally, monadic second-order logic can be embedded into LP (see Lønning (1989)): The atomic formula '$X(x)$' expressing a predication is translated into the LP formula '$x \cdot\Pi\, \alpha$', which is not a predication anymore but the expression of a part-whole relation between an atom x and an individual sum α. While this is a rather simple logical fact it is the philosophical "gestaltwechsel" behind it that is important. It comes down to the notion that logical sets can be dispensed with in favor of a mereology of singular and plural entities. The only thing that is lost under way is the empty logical set, but that is certainly not that much of a loss.

The next section shows how this interpretation can be put to work in David Lewis's megethology,[6] a theory of classes based on mereology and plural quantification.[7]

14.3 LP as a framework for mereological set theory

I now proceed to use Lewis's nominalistic theory of classes as a testing ground for my above interpretation. That is, I want to develop that theory within LP, thereby demonstrating how it can be rigorously regimented without appeal to natural language locutions. Our sketch basically follows Lewis's set-up, as given in Lewis (1991, 1993).

Before I go on, a remark is in order about the question of "stacking" mereologies, Lewis's mereology of classes and on top of it, the mereology of pluralities. To begin with, it is clear that in general there will be a difference between individual sums, which are free-lattice fusions, and other substantive fusions. In real number theory, for instance, the σ-term $\sigma x.x < 1$ denotes "the real numbers < 1" but their fusion (supremum) with respect to the standard linear order on the real line is the number 1. This number is a single entity again and *no* plurality any more. Or take "the even numbers" as objects in the theory of ordinals, denoted by $\sigma x.\exists y(x = 2y)$; their fusion (supremum) is the first infinite ordinal ω, which is obviously quite another thing as the plurality we started from.

Again, think of the stacking of mereologies this way: The mereology

[6]Derived from Greek *megethos* = large. Thus megethology is the theory of (large) size.

[7]A variation of megethology was recently proposed by E. Martino (1995), who replaces the (in his eyes) controversial device of plural quantification by a primitive pairing function. While this is an interesting move, it is obviously not for those who like me want to make sense of plural quantification.

of pluralities is there to begin with, with varying domains D of regular individuals as subject matter. Now D may happen to be a domain which is itself structured by a mereology, as the algebra of classes is. It is plain, then, that we can, and in fact have to, distinguish between the i-sums and the fusions in the domain D.

The singular objects of the theory, then, i.e., the i-atoms, are *individuals* and *classes*. Arbitrary i-sums represent pluralities of classes. The plural phrase "the things that ϕ" translates into '$\sigma x\phi[x]$'. We write '$\iota x\phi[x]$' if $\phi[x]$ is true of exactly one x. In what follows, usual variable conditions are assumed to be in effect. Let us further note that for distributive monadic predicates P, an i-sum ζ is a $^*P$ just in case its i-atomic i-parts are P:

(32) $\quad {}^*P(\alpha) \leftrightarrow \forall x(x\,\dot{\Pi}\,\alpha \to P(x))$

The mereology of individuals and classes is developed within the realm of i-atoms. Its primitive part-whole relation is $\sqsubseteq$. Variables for i-atoms are u, v, w, x, y, z, as above, and I will use a, b, c for terms. The most important *defined symbols* are the following, which mimic the ones for LP:

(33) $\quad a \sqsubset b \;:\Leftrightarrow\; a \sqsubseteq b \wedge a \neq b$ (*a is a proper part of b*)

(34) $\quad a \circ b \;:\Leftrightarrow\; \exists x(x \sqsubseteq a \wedge x \sqsubseteq b)$ (*overlap*)

(35) $\quad a \wr b \;:\Leftrightarrow\; \neg\, a \circ b$ (*a is disjoint from b*)

(36) $\quad \mathrm{At}(a) \;:\Leftrightarrow\; \mathbf{E}\,a \wedge \forall x(x \sqsubseteq a \to x = a)$ (*a is an atom*)

(37) $\quad a \cdot\!\sqsubseteq b \;:\Leftrightarrow\; \mathrm{At}(a) \wedge a \sqsubseteq b$ (*a is an atomic part of b*)

(38) $\quad a = \bigsqcup\zeta \;:\Leftrightarrow\; \forall x(x\,\dot{\Pi}\,\zeta \to x \sqsubseteq a) \wedge \forall x(x \sqsubseteq a \to \exists y(y\,\dot{\Pi}\,\zeta \wedge y \circ x))$ (*fusion*)

(39) $\quad a \sqcup b \;:=\; \bigsqcup a \oplus b$ (*the fusion of a and b*)

(40) $\quad Ptn(\zeta, a) \;:\leftrightarrow\; \forall x\forall y(x\,\dot{\Pi}\,\zeta \wedge y\,\dot{\Pi}\,\zeta \to x \wr y) \wedge a = \bigsqcup\zeta$ (*ζ partition a*)

The *axioms* for the mereology of individuals and classes are the following:

(41) $\quad a \sqsubseteq b \to \mathbf{E}\,a$ (Existential Import)

(42) $\quad a \sqsubseteq b \wedge b \sqsubseteq c \to a \sqsubseteq c$ (Trans($\sqsubseteq$))

(43) $\quad \forall\alpha\exists x\; x = \bigsqcup\alpha$ (UnrestrComp)

(44) $\forall\alpha\forall x\forall x'(x = \bigsqcup\alpha \wedge x' = \bigsqcup\alpha \rightarrow x = x')$ (UniqComp)

(45) $a \sqsubset b \rightarrow \exists x\,(x \sqsubseteq b \wedge x \wr a)$ (WeakSuppl)

Let $\mathcal{R}$ denote *Reality*, the fusion of all there is; formally:

(46) $\mathcal{R} \;:=\; \bigsqcup\sigma x.\, x = x$ (*Reality*)

In order to be able to put constraints on the *size of Reality* a distinction is made between "small" and "large" things (singular mode) and between "few" and "many" things (plural mode). A *large* thing has as many atoms as there are in the whole of Reality; without resorting to functions this can be expressed by saying "that something has at least as many atoms as there are in all the *rest* of Reality" ((Lewis, 1991, p. 89)). Thus, using the symbol '$\leq$1' for 'at most one':

(47) $\text{Large}(a) \;:\Leftrightarrow\; \exists\alpha[Ptn(\alpha,\mathcal{R}) \,\wedge\, \forall x(x\,\dot{}\Pi\alpha \rightarrow \exists! u(u \cdot\!\sqsubseteq x \wedge u \sqsubseteq a)\wedge$
$\leq\!1\, v(v \cdot\!\sqsubseteq x \wedge v \not\sqsubseteq a))]$ (*a is large*)

(48) $\text{small}(a) \;:\Leftrightarrow\; \neg\text{Large}(a)$ (*a is small*)

Saying plurally of some things that they are few is even more complicated. Lewis's definition runs as follows. Suppose there are some things ζ such that its i-atomic i-parts are all disjoint from some fixed large thing z (this is to exclude i-sums that are "few in number" but have large i-atomic i-parts). Then the ζ are *few* iff there is some small x such that $\bigsqcup\zeta$ is disjoint from x, and all elements in ζ pair with some atoms of x in different individual "diatoms" (= i-sums of two LP atoms).

(49) $\text{few}(\zeta) \;:\Leftrightarrow\; \exists x\exists\alpha\,[\,\text{small}(x) \wedge x \wr \bigsqcup\zeta \wedge \forall y(y\,\dot{}\Pi\alpha \rightarrow$
$\exists z\exists u(z\,\dot{}\Pi\zeta \wedge u \cdot\!\sqsubseteq x \wedge$

$y = z \sqcup u)) \wedge \forall z(z\,\dot{}\Pi\zeta \rightarrow \exists y\exists u(y\,\dot{}\Pi\alpha \wedge u \cdot\!\sqsubseteq x \wedge y = z \sqcup u)) \wedge$

$\forall u\forall y\forall z(u \cdot\!\sqsubseteq x \wedge y\,\dot{}\Pi\alpha \wedge z\,\dot{}\Pi\alpha \wedge u \sqsubseteq y \wedge u \sqsubseteq z \rightarrow y = z)]$
(for ζ s.t. $\exists z(\text{Large}(z) \wedge \forall x'(x'\,\dot{}\Pi\zeta \rightarrow x' \wr z)))$

(50) $\text{Many}(\zeta) \;:\Leftrightarrow\; \neg\text{few}(\zeta)$ (*the ζ are many*)

There are several notions of *infinity* the most prominent being the one expressed in the axiom of infinity of set theory. That axiom is a very strong requirement though in that the set postulated to exist there is not only infinite but has an infinite rank. That is of course not available here. The other well-known concept is Dedekind infinity, which has a "flat" definition but uses functions. Lewis eventually settles on Dedekind infinity once he has succeeded in reconstructing functions mereologically; but for the time being he uses the part-whole relation to define an infinite thing as one which is the fusion of an endless ordering:

(51) $\mathrm{Inf}(a) \;:\Leftrightarrow\; \exists\alpha(a = \bigsqcup\alpha \wedge \forall x(x\,\dot{}\Pi\alpha \to \exists y(y\,\dot{}\Pi\alpha \wedge x \sqsubset y)))$ *(a is infinite)*

(52) $\mathrm{Fin}(a) \;:\Leftrightarrow\; \neg\mathrm{Inf}(a)$ *(a is finite)*

We can now formulate Lewis's hypotheses about the size of Reality:

(53) $\forall x\,[\,\mathrm{small}(x) \to \mathrm{few}(\sigma y.\, y \sqsubseteq x)\,]$ (P)

(54) $\forall\alpha\,[\,{}^{*}\mathrm{small}(\alpha) \wedge \mathrm{few}(\alpha) \to \mathrm{small}(\bigsqcup\alpha)\,]$ (U)

(55) $\exists\alpha\,[\,{}^{*}\mathrm{At}(\alpha) \wedge \mathrm{Inf}(\bigsqcup\alpha) \wedge \mathrm{small}(\bigsqcup\alpha)\,]$ (I)

(P) says that the parts of a small thing are few; according to (U), if some things are small and few, their fusion is small; and (I) postulates the existence of an infinite fusion of atoms that is still small.

So far we have not yet entered set theory proper. The decisive step towards set theory is the introduction of the temporary new primitive $\mathrm{sgt}(x, y)$: *x is a singleton of y.* This relation is going to be a function, but functions as objects are not yet available. Eventually, however, Lewis is able to show not only that ordered pairs, relations and functions can be simulated in pluralic mereology through some suitable coding, but also that a singleton function must exist under a number of plausible "megethological" assumptions.

The singleton relation is required to be both functional and one-one:

(56) $\forall x\forall y\forall z(\mathrm{sgt}(y, x) \wedge \mathrm{sgt}(z, x) \to y = z)$ (Axiom of Functionality)

(57) $\forall x\forall y\forall z(\mathrm{sgt}(x, y) \wedge \mathrm{sgt}(x, z) \to y = z)$ (Axiom of Distinctness)

Next we define the singleton predicate, the singleton and "unicle" operations,[8] and the notion of an *individual* as an object which has no singletons as parts. For reasons argued at length in his book, Lewis chooses the supremum of all individuals as the (unorthodox) null set.

(58) $\mathrm{Sgt}(x) \;:\Leftrightarrow\; \exists y\, \mathrm{sgt}(x, y)$ *(x is a singleton)*

[8]The name 'unicle' goes back to Harry Bunt's *Theory of Ensembles* (Bunt (1981)) where, as Lewis acknowledges, the idea of parts of classes was developed already in the early 1980s.

(59) $x^\circ \;:=\; \iota y.\, \mathrm{sgt}(y, x)$ (*the singleton of* x)

(60) $x_\circ \;:=\; \iota y.\, \mathrm{sgt}(x, y) \;=\; \iota y.\, y^\circ = x$ (*the unicle of* x)

(61) $\mathcal{I}x \;:\Leftrightarrow\; \forall y(y \sqsubseteq x \to \neg \mathrm{Sgt}(y))$ (x *is an individual*)

(62) $\Lambda \;:=\; \bigsqcup \sigma x.\, \mathcal{I}x$ (*the null set*)

The next axioms say first of all that there are individuals, so that the null set exists. Then they characterize the domain and the range of the singleton function. The domain consists of all individuals and all small fusions of singletons; the range consists of atoms that don't overlap individuals.

(63) $\exists x\, \mathcal{I}x$ (Individual Existence)

(64) $\forall x(x \sqsubseteq \Lambda \to \exists y.\, \mathrm{sgt}(y, x))$ (AxDom1)

(65) $\forall x(\mathrm{Sgt}(x) \to \exists y.\, \mathrm{sgt}(y, x))$ (AxDom2)

(66) $\forall \alpha\, [\, {}^*\mathrm{Sgt}(\alpha) \wedge \mathrm{small}(\bigsqcup \alpha) \to \exists y.\, \mathrm{sgt}(y, \bigsqcup \alpha)\,]$ (AxDom3)

(67) $\forall x\, [\, x \not\sqsubseteq \Lambda \wedge \neg \mathrm{Sgt}(x) \wedge \neg \exists \alpha({}^*\mathrm{Sgt}(\alpha) \wedge x = \bigsqcup \alpha \wedge \mathrm{small}(x)) \to$
$\neg \exists y.\, \mathrm{sgt}(y, x)\,]$
(AxDom4)

(68) $\forall x\, (\mathrm{Sgt}(x) \;\to\; \mathrm{At}(x))$ (AxRan1)

(69) $\forall x \forall y(x \sqsubseteq \Lambda \wedge \mathrm{Sgt}(y) \to x \wr y)$ (AxRan2)

Finally, there is the following *induction axiom* which says that all objects (that is i-atoms) are generated from the individuals by iterated formation of singletons and fusions. Due to the "second-order character" of LP the axiom can be formulated as a single sentence rather than a schema.

(70) $\forall \alpha\, [\forall u(\mathcal{I}u \to u\, {}^{\cdot}\Pi \alpha) \wedge \forall u(u\, {}^{\cdot}\Pi \alpha \wedge \mathbf{E}\, u^\circ \to u^\circ\, {}^{\cdot}\Pi \alpha) \wedge \forall \beta(\beta \Pi \alpha \to$
$\bigsqcup \beta\, {}^{\cdot}\Pi \alpha)$
$\to \forall x\, x\, {}^{\cdot}\Pi \alpha\,]$ (Axiom of Induction)

As an application of Induction we can show that no object is identical to its singleton (take $\alpha := \sigma x.x \neq x^\circ$). Then the finite iterations of the singleton operation over the null set yield infinitely many atoms, and their fusion is an infinite object in the sense of (51). Hence Reality is infinite, too.

The next battery of definitions introduces (proper) classes, urelements, mixed fusions, sets, elementhood, set-theoretic membership, and the subclass relation.

(71) $\mathcal{C}x \;:\Leftrightarrow\; \exists\alpha({}^{*}\mathrm{Sgt}(\alpha) \wedge x = \bigsqcup\alpha)$ *(x is a class)*

(72) $\mathcal{U}x \;:\Leftrightarrow\; \mathcal{I}x \wedge x \neq \Lambda$ *(x is an urelement)*

(73) $\mathcal{MF}x :\Leftrightarrow \exists y\exists z(\mathcal{C}y \wedge \mathcal{I}z \wedge x = y \sqcup z)$ *(x is a mixed fusion)*

(74) $\mathcal{S}x \;:\Leftrightarrow\; x = \Lambda \vee (\mathcal{C}x \wedge \exists y\, \mathrm{sgt}(y, x))$ *(x is a set)*

(75) $\mathcal{S}'x \;:\Leftrightarrow\; \mathcal{S}x \wedge x \neq \Lambda$ *(x is a non-null set)*

(76) $\mathcal{E}x \;:\Leftrightarrow\; \mathcal{I}x \vee \mathcal{S}x$ *(x is an element)*

(77) $\mathcal{PC}x \;:\Leftrightarrow\; \mathcal{C}x \wedge \neg\mathcal{S}x$ *(x is a proper class)*

(78) $x \in y \;:\Leftrightarrow\; \mathcal{C}y \wedge x^{\circ} \sqsubseteq y$ *(x is a member of y)*

(79) $x \subseteq y \;:\Leftrightarrow\; \mathcal{C}x \wedge \forall z(z \in x \rightarrow z \in y)$ *(x is a subclass of y)*

The following sample theorems can now be derived, among them Lewis's original *Four Theses* about the shape of Reality and the nature of classes and their parts.[9]

(80) $\mathcal{I}\Lambda \wedge \mathcal{S}\Lambda \wedge \neg\mathcal{C}\Lambda$

(81) $\forall x(\mathcal{C}x \rightarrow x = \bigsqcup \sigma y\, (\mathrm{Sgt}(y) \;\wedge\; y_{\circ} \in x)\,)$

(82) $\forall x(\mathcal{C}x \rightarrow \exists y.\, y \in x)$

(83) $\forall x(\mathcal{C}x \rightarrow (\mathcal{S}x \leftrightarrow \mathrm{small}(x)))$

(84) $\forall x(\mathcal{E}x \leftrightarrow \exists y.\, \mathrm{sgt}(y, x))$

(85) $\forall x\forall y(\mathcal{C}x \wedge \mathcal{C}y \rightarrow (y \sqsubseteq x \leftrightarrow y \subseteq x))$ (First Thesis)

(86) $\forall x(\mathcal{I}x \vee \mathcal{C}x \vee \mathcal{MF}x)$ (Division)

(87) $\forall x\forall y(\mathcal{C}x \wedge \mathcal{I}y \rightarrow x \not\sqsubseteq y)$ (Priority)

(88) $\forall\alpha({}^{*}\mathcal{I}\alpha \rightarrow \mathcal{I}(\bigsqcup\alpha))$ (Fusion)

(89) $\forall x\forall y(\mathcal{C}x \rightarrow (y \sqsubseteq x \leftrightarrow y \subseteq x))$ (Main Thesis)

[9] Actually, the proof for the Main Thesis that Lewis gives in his book doesn't go through without Weak Supplementation (45), a principle that he doesn't list as one of his mereological axioms. I have added it since it should obviously be there.

(80) says that the null set is an individual and a set, but not a class. According to (81), a class is the fusion of those singletons whose unicles are members of it; since by definition there is at least one such singleton, classes are "non-empty", whence (82). Theorem (83) says that the small classes are the sets, and (84) states that the objects with a singleton are the elements. Lewis's First Thesis (85) asserts that the property of being a part of a class x coincides with the subclass property for an object y if y is itself a class. The Main Thesis (89) eventually disposes of this condition on y. According to (86) everything is either an individual, a class or a mixed fusion formed from an individual and a class. The Priority and the Fusion Theorems (87) and (88) record the separation of classes from individuals: no class is part of any individual, and any fusion of individuals is an individual.

In the same vein the usual *axioms of set theory* can be formally proved from the things we have already (notably Hypotheses (P), (U), and (I) above), plus a few more framework principles, which I have omitted here. All these principles extract the megethological essence from their more familiar set-theoretic versions. Moreover, since the objects of quantification include (possibly proper) classes, we can prove a single sentence rather than a schema for axioms like *Aussonderung* or Replacement. Thus we have at least something of the strength of NBG set theory, but in fact, we get more. We can prove what amounts to an impredicative class abstraction principle as used in Morse-Kelley set theory:

(90) $\forall y\,(\,y \in \{\,x \mid \phi[x]\,\} \leftrightarrow \mathcal{E}y \wedge \phi[y]\,)$ (*principle of abstraction*)

No restriction is placed here on the quantified variables occurring in the formula ϕ. In order for (90) to make sense in the present context, however, we have to say what the class abstraction term $\{\,x \mid \phi[x]\,\}$ means.[10] The usual approach to abstraction terms is by elimination: they are not part of the language but mere abbreviations. In our free logic we can define certain fusion terms that play the role of class terms; no claim is made regarding their existence.

(91) $\{\,x \mid \phi[x]\,\} \;:=\; \bigsqcup \sigma x(\mathrm{Sgt}(x) \wedge \phi[x_\circ])$

If there are no singletons whose unicle satisfy the formula ϕ then the σ-term is undefined, and so is its fusion. But even if the term denotes, the corresponding class might still be too large to form a set. Thus

(92) $\mathrm{V}^\bullet \;=\; \{\,x \mid x = x\,\}$ (*the impure universal class*)

[10]Because of his informal style of presentation, Lewis is silent on this.

is the proper class of all pure or impure sets and individuals, i.e., the class of all elements. It is also the fusion of all the non-null sets: $V^\bullet = \bigsqcup \sigma x.\mathcal{S}'x$. Note that the fusion of all sets including the null set yields the whole of Reality again, since in Lewis's system, Λ happens to be the fusion of all individuals, and $V^\bullet \sqcup \Lambda = \mathcal{R}$.

The only thing that is left to complete the mereological program is to dispense with the primitive singleton function. But that can also be done by defining within LP various methods of coding, due to John Burgess, A. P. Hazen, and David Lewis, which can be found in the appendix of Lewis's book or his (1993). These methods, too, make essential use of the device of plural quantification. It can thus be seen that the mereology of pluralities, plus a second mereology of classes, are sufficient to reproduce the standard theories of iterative sets.

Acknowledgment. I wish to thank a number of people who discussed the above material with me and have helped to clarify many issues involved. First of all, there are the members of the "Ontological Reduction Group" in Munich: Volker Halbach, Stefan Iwan, Karl-Georg Niebergall, and Holger Sturm. Also, special thanks go to Ed Zalta of Stanford University for carefully reading and commenting on earlier drafts of the chapter and for inviting me to the Seattle symposium mentioned in the Preface for which he served as the APA coordinator. At this symposium I had the opportunity to discuss the content of this chapter with him and the other symposiasts, John Burgess, Michael Resnik and Peter Simons. Their comments have been very helpful to me.

I would also like to mention that I owe much to the discussion I had with the late George Boolos about our views on plural quantification when he visited Munich in January 1995; in retrospect, this was an especially precious intellectual opportunity for me.

Appendix: A Chapter in Lattice Theory

I have collected in a concise manner here some elementary definitions and facts about lattice theory that give the necessary background for the semantic and philosophical applications discussed in this book. Most of what follows, apart from the final subsection on plural lattices, is standard textbook material or mathematical folklore. I compiled this primer by drawing mainly on the following literature. First of all, there is Marcel Erné's delightful introduction (1982) to the theory of order; it is this text that I made heavy use of in the first part of the Appendix, employing in particular his convenient arrow notation. When the formulation of a result is directly taken over it will be so indicated.[11] The proofs are all straightforward, but I have filled in details at occasion. Another useful source on the introductory level is Davey and Priestley (1990). Elementary definitions can also be found in Gierz *et al.* (1980) where a variant of the arrow notation is used. Standard textbooks are Grätzer (1978); Balbes and Dwinger (1974) and, for Boolean algebras, Halmos (1963); Sikorski (1969).

I have tried to make the presentation as self-contained as possible within the limits of space available here. For a more detailed account on lattice-theoretic applications in semantics the reader should also consult Landman's excellent (1991).

15.1 Ordering Relations; Preordered Sets; Posets

Relations. Let R be a *relation in* a set L, i.e., $R \subseteq L \times L$. We write xRy instead of $\langle x, y\rangle \in R$. $R^{-1} = \{\langle y, x\rangle \mid \langle x, y\rangle \in R\}$ is the *inverse relation of* R. If R, S are two relations then their *composition* is defined by $R \circ S =$

[11] By an 'E' plus page number of that work.

$\{\langle x, y\rangle \mid \exists z(xRz \& zSy\}$. Δ_L is the "diagonal" in L, i.e., $\Delta_L = \{\langle x, x\rangle \mid x \in L\}$. Then the conditions for the most common ordering relations can be expressed succinctly in the way given below. These equivalences could in fact replace the usual definitions. To give one example, consider the condition for directedness in (8). First of all, it is trivial that $R \circ R^1 \subseteq L \times L$; hence the substantive part is the inclusion in the other direction. It says, by applying the definitions of composition and inverse relation,

$$\forall x \forall y\,(\,x, y \in L \rightarrow \exists z(xRz \,\&\, yRz)\,)$$

But this is the usual property of a directed set or relation. In a similar vein, the first-order conditions of all the other ordering relations can be made explicit.

(1) $\mathit{Refl}(R) \iff \Delta_L \subseteq R$ (*R* is *reflexive*)

(2) $\mathit{Irrefl}(R) \iff \Delta_L \cap R = \emptyset$ (*R* is *irreflexive*)

(3) $\mathit{Sym}(R) \iff R = R^{-1}$ (*R* is *symmetric*)

(4) $\mathit{Asym}(R) \iff R \cap R^{-1} = \emptyset$ (*R* is *asymmetric*)

(5) $\mathit{Antisym}(R) \iff R \cap R^{-1} \subseteq \Delta_L$ (*R* is *antisymmetric*)

(6) $\mathit{Connec}(R) \iff R \cup R^{-1} = L \times L$ (*R* is *connected*)

(7) $\mathit{Trans}(R) \iff R \circ R \subseteq R$ (*R* is *transitive*)

(8) $\mathit{Direc}(R) \iff R \circ R^{-1} = L \times L$ (*R* is *directed*)

(9) $\mathit{Filt}(R) \iff R^{-1} \circ R = L \times L$ (*R* is *filtered*)

(10) $\mathit{PreO}(R) \iff \mathit{Refl}(R) \,\&\, \mathit{Trans}(R)$ (*R* is a *preorder*)

(11) $\mathit{PO}(R) \iff \mathit{PreO}(R) \,\&\, \mathit{Antisym}(R)$ (*R* is a *partial order*)

(12) $\mathit{LinO}(R) \iff \mathit{PO}(R) \,\&\, \mathit{Connec}(R)$ (*R* is a *linear order*)

(13) $\mathit{Eq}(R) \iff \mathit{PreO}(R) \,\&\, \mathit{Sym}(R)$ (*R* is an *equivalence relation*)

Remark. 1. If xRy we will loosely use the phrases "x is *below* y", "y is *above* x", or "y *dominates* x (with respect to R)".

2. If R is an equivalence relation the equivalence classes with respect to R are standardly written as $[a]_R := \{x \in L \mid xRa\}$ for $a \in L$. The equivalence classes with respect to R form a *partition* of L, i.e., a mutually disjoint, exhaustive system of subsets of L.

Definition 3 Let L be a set, Y a subset of L, $a \in L$ and R a preorder on L. Then

1. $\uparrow Y := \{x \in L \mid yRx \text{ for some } y \in Y\}$
2. $\downarrow Y := \{x \in L \mid xRy \text{ for some } y \in Y\}$
3. $\uparrow a := \uparrow \{a\}$
4. $\downarrow a := \downarrow \{a\}$
5. $Y^{\uparrow} := \{x \in L \mid yRx \text{ for all } y \in Y\}$
6. $Y^{\downarrow} := \{x \in L \mid yRx \text{ for all } y \in Y\}$
7. $a^{\uparrow} := \{a\}^{\uparrow}$
8. $b^{\downarrow} := \{b\}^{\downarrow}$

$\uparrow Y$ is called the *upper set of* Y, $\downarrow Y$ the *lower set of* Y. $Y^{\uparrow}$ is called the *set of upper bounds of* Y, $Y^{\downarrow}$ the *set of lower bounds of* Y. Y is called *upward (downward) bounded* iff $Y^{\uparrow} \neq \emptyset$ ($Y^{\downarrow} \neq \emptyset$). Finally, Y is called *upward (downward) R-closed* or an *upper (lower) set* iff $Y = \uparrow Y$ ($Y = \downarrow Y$).

Remark. Obviously, we have $\uparrow a = a^{\uparrow}$ and $\downarrow a = a^{\downarrow}$. $a^{\uparrow}$ can be called the *upper (R-)cone of* a and $a^{\downarrow}$ the *lower (R-)cone of* a. The set of upper bounds of Y can also be expressed by $Y^{\uparrow} = \bigcap\{\uparrow y \mid y \in Y\}$; similarly, $Y^{\downarrow} = \bigcap\{\downarrow y \mid y \in Y\}$. The upper set of Y is the union of the upper sets of the elements of Y: $\uparrow Y = \bigcup\{\uparrow y \mid y \in Y\}$. It is also called the *upper R-closure of* Y. Similarly, for the lower set of Y, which is also called the *lower R-closure of* Y we have $\downarrow Y = \bigcup\{\downarrow y \mid y \in Y\}$.

Definition 4 Let L be a set, Y a subset of L, and R a preorder on L. Then $b \in Y$ is *a greatest (smallest) element of* Y iff $b \in Y \cap Y^{\uparrow}$ ($b \in Y \cap Y^{\downarrow}$). $b \in Y$ is called *maximal (minimal) in* Y iff for all $x \in Y$, if bRx then xRb (resp. if xRb then bRx). A smallest element in $Y^{\uparrow}$ (a greatest element in $Y^{\downarrow}$) is called a *supremum (infimum)* of Y.

Remark. If the relation R in Definition 4 is a partial order then the set Y has at most one greatest (smallest) element. Accordingly, suprema and infima, if they exist, are unique. In this case we use the notation $\sup_R Y$ ($\inf_R Y$) or $\sup Y$ ($\inf Y$) for short if the relation involved is clear. Sometimes it is important to be explicit about the carrier set L in which the supremum (infimum) is formed; we then write $\sup_L Y$ ($\inf_L Y$). Furthermore, because of antisymmetry, the condition for maximality (minimality) in Y can be written $\forall x(x \in Y \,\&\, bRx \Rightarrow b = x)$ (resp.

$\forall x(x \in Y \,\&\, xRb \Rightarrow b = x)$). If the carrier set L has a greatest element it is called the *top* or *unit (element) in* L and it is written $\top$ or $\mathbf{1}$; similarly, the smallest element in L, if it exists, is called the *bottom* or *zero (element) in* L and it is written $\bot$ or $\mathbf{0}$.

Definition 5 A non-empty set L is called a *preordered set* iff L carries a reflexive and transitive relation $\sqsubseteq$. A *partially ordered set* or *poset* is a non-empty set L equipped with a partial order, $\sqsubseteq$. A poset L is called *totally ordered* or a *chain* iff $\sqsubseteq$ is a linear order.

Remark. In a preordered set $\langle L, R\rangle$ there is a natural equivalence relation $\sim_R$ defined by

$$a \sim_R b \;:\Longleftrightarrow\; aRb \text{ and } bRa$$

L can be turned into a poset by forming the quotient $L/\sim_R$ with the induced ordering relation

$$[a]\tilde{R}[b] \;:\Longleftrightarrow\; aRb$$

Definition 6 Let L be a poset with $\mathbf{1}$ and $\mathbf{0}$. Then $a \in L$ is called an *atom* iff a is minimal in $L \setminus \{\mathbf{0}\}$. a is called an *antiatom* iff a is maximal in $L \setminus \{\mathbf{1}\}$.

Lemma 1 (E:86ff.) Let $\langle L, \sqsubseteq\rangle$ be a preordered set, $Y, Z \subseteq L$, and $a \in Y$. Then the following relations hold.

1. $Y \subseteq \downarrow Y$, $Y \subseteq \uparrow Y$
2. $\downarrow Y = \downarrow\downarrow Y$, $\uparrow Y = \uparrow\uparrow Y$
3. $Y \subseteq Z \implies \uparrow Y \subseteq \uparrow Z$ and $\downarrow Y \subseteq \downarrow Z$
4. $Y \subseteq Z \implies Z^{\uparrow} \subseteq Y^{\uparrow}$ and $Z^{\downarrow} \subseteq Y^{\downarrow}$
5. $Y \subseteq Y^{\uparrow\downarrow}$, $Y \subseteq Y^{\downarrow\uparrow}$
6. $Y^{\uparrow} = Y^{\uparrow\downarrow\uparrow}$, $Y^{\downarrow} = Y^{\downarrow\uparrow\downarrow}$
7. $a \in a^{\uparrow}$, $a \in a^{\downarrow}$
8. $(\downarrow Y)^{\uparrow} = Y^{\uparrow}$, $(\uparrow Y)^{\downarrow} = Y^{\downarrow}$
9. $a^{\uparrow} = a^{\downarrow\uparrow}$, $a^{\downarrow} = a^{\uparrow\downarrow}$

Proof. For *(1)*, observe that for every $y \in Y$, yRy by reflexivity, so y is below an element of Y and $Y \subseteq\downarrow Y$; similarly for $Y \subseteq\uparrow Y$. To prove *(2)*, we have $\downarrow Y \subseteq\downarrow\downarrow Y$ by (1). Conversely, let xRy for an element $y \in\downarrow Y$ implies xRy and yRz for an element $z \in Y$, and xRz by transitivity; hence $\downarrow\downarrow Y \subseteq\downarrow Y$. Similarly for the second part of assertion (2). *Ad (3):* If an x has a $y \in Y$ above it this y is also in Z by assumption, so $x \in\uparrow Z$; the second inclusion follows similarly. *Ad (4):* Everything lying above the whole of Z lies a fortiori above the whole of Y since $Y \subseteq Z$, whence $Z^{\uparrow} \subseteq Y^{\uparrow}$ and, dually, $Z^{\downarrow} \subseteq Y^{\downarrow}$. *Ad (5):* Every element of Y is a lower bound of $Y^{\uparrow}$ by definition of the high uparrow, thus $Y \subseteq Y^{\uparrow\downarrow}$ and, by a similar argument, $Y \subseteq Y^{\downarrow\uparrow}$. Assertion *(6)* follows from (5) and (4), and *(7)* holds by reflexivity. The left-to-right inclusion in *(8)* is true in virtue of (1) and (4), and transitivity is used for the other direction. Since $\uparrow a = a^{\uparrow}$ and $\downarrow a = a^{\downarrow}$, assertion *(9)* is a special case of (8). □

Proposition 1 (E:90) *Let $\langle L, \sqsubseteq\rangle$ be a preordered set, $Y \subseteq L$, and $a \in Y$. Then:*

1. *a is a maximal element of Y iff $a \in Y$ and $a^{\uparrow} \cap Y \subseteq a^{\downarrow}$;*
2. *a is a minimal element of Y iff $a \in Y$ and $a^{\downarrow} \cap Y \subseteq a^{\uparrow}$;*
3. *a is a supremum of Y iff one of the following three equivalent conditions is satisfied:*
   - *(a)* $a \in Y^{\uparrow} \cap Y^{\uparrow\downarrow}$
   - *(b)* $a^{\uparrow} = Y^{\uparrow}$
   - *(c)* $a^{\downarrow} = Y^{\uparrow\downarrow}$
4. *a is an infimum of Y iff one of the following three equivalent conditions is satisfied:*
   - *(a)* $a \in Y^{\downarrow} \cap Y^{\downarrow\uparrow}$
   - *(b)* $a^{\downarrow} = Y^{\downarrow}$
   - *(c)* $a^{\uparrow} = Y^{\downarrow\uparrow}$

Proof. Assertions (1) and (2) are the definitions rewritten in terms of the high arrow notation. For (3) we observe that a is a supremum of Y if and only if a is an upper bound of Y, i.e., $a \in Y^{\uparrow}$, and it is a least upper bound of Y, that is, a lower bound of $Y^{\uparrow}$, hence $a \in Y^{\uparrow\downarrow}$. So being a supremum of Y is equivalent to condition (3a). Now assume (3a); $a \in Y^{\uparrow}$ means yRa for all $y \in Y$, hence $Y \subseteq a^{\downarrow}$ and $a^{\uparrow} = a^{\downarrow\uparrow} \subseteq Y^{\uparrow}$ by Lemma 1. Conversely, $a \in Y^{\uparrow\downarrow}$ entails $\{a\} \subseteq Y^{\uparrow\downarrow}$ and $Y^{\uparrow} = Y^{\uparrow\downarrow\uparrow} \subseteq a^{\uparrow}$, again with

Lemma 1. This lemma also yields condition (3c) from (3b) and vice versa (apply assertions (9) and (6)). Finally, assume condition (3b) and hence, condition (3c). Since a is an element of both its upper and lower cone a is in the intersection of $Y^{\uparrow}$ and $Y^{\uparrow\downarrow}$, and this is condition (3a). A dual argument proves assertion (4) of the proposition. □

Remark. It follows that every greatest (smallest) element in Y is both maximal (minimal) in Y and a supremum (infimum) of Y. If Y has a supremum (infimum), and it is an element of Y, then it is maximal (minimal) in Y; but Y might contain maximal (minimal) elements without having a supremum (infimum).

Proposition 2 (E:91) *Let* $\langle L, \sqsubseteq \rangle$ *be a poset and* Y, Z *subsets of* L *with existing suprema. Then:*

1. $(\sup Y)^{\uparrow} = Y^{\uparrow}, \qquad (\sup Y)^{\downarrow} = Y^{\uparrow\downarrow}$
2. $\sup(\downarrow Y)$ *exists, and* $\sup Y = \sup(\downarrow Y)$
3. $\sup Y \sqsubseteq \sup Z$ *iff* $Z^{\uparrow} \subseteq Y^{\uparrow}$ *iff* $Y^{\uparrow\downarrow} \subseteq Z^{\uparrow\downarrow}$
4. $Y \subseteq Z \implies \sup Y \sqsubseteq \sup Z$
5. $a \sqsubseteq \sup Y$ *iff* $Y^{\uparrow} \subseteq a^{\uparrow}$ *iff* $a \in Y^{\uparrow\downarrow}$

Dual conditions hold if Y, Z *have an infimum.*

6. $(\inf Y)^{\downarrow} = Y^{\downarrow}, \qquad (\inf Y)^{\uparrow} = Y^{\downarrow\uparrow}$
7. $\inf(\uparrow Y)$ *exists, and* $\inf Y = \inf(\uparrow Y)$
8. $\inf Z \sqsubseteq \inf Y$ *iff* $Z^{\downarrow} \subseteq Y^{\downarrow}$ *iff* $Y^{\downarrow\uparrow} \subseteq Z^{\downarrow\uparrow}$
9. $Y \subseteq Z \implies \inf Z \sqsubseteq \inf Y$
10. $\inf Y \sqsubseteq a$ *iff* $Y^{\downarrow} \subseteq a^{\downarrow}$ *iff* $a \in Y^{\downarrow\uparrow}$

Proof. Assertion (1) is just condition (3b,c) of Proposition 1. By (1) and L. 1(8) we have $(\sup Y)^{\uparrow} = Y^{\uparrow} = (\downarrow Y)^{\uparrow}$, hence $\sup Y$ is also the supremum of the lower set of Y; this proves (2). As for the first equivalence of (3) from left to right, upper bounds of Z are above $\sup Z$ by (1), hence above $\sup Y$ and, again by (1), upper bounds of Y. Conversely, if $Z^{\uparrow} \subseteq Y^{\uparrow}$ then by (1), the upper cone of $\sup Z$ is contained in the upper cone of $\sup Y$, but that means that $\sup Z$ is above $\sup Y$. The second equivalence in (3) follows with L. 1(3,4). Assertion (4) follows from L. 1(3) and the right-to-left direction of the first equivalence of (3). To prove (5), we observe that $a \sqsubseteq \sup Y$ if and only if the upper cone of $\sup Y$, which is $Y^{\uparrow}$ by (1), is contained in the upper cone of a; because of L. 1(4,9) this is in turn the case if and only if the lower cone of a, and hence a, is contained in $Y^{\uparrow\downarrow}$. The dual assertions (6) – (10) are proved the same way. □

Proposition 3 *Let L be a poset, and $Y \subseteq L$ with existing supremum and infimum. Then we have* $\sup Y = \inf Y^{\uparrow}$ *and* $\inf Y = \sup Y^{\downarrow}$.

Proof. We have $a = \sup Y$ iff $a \in Y^{\uparrow} \cap Y^{\uparrow\downarrow} = (Y^{\uparrow})^{\downarrow} \cap (Y^{\uparrow})^{\downarrow\uparrow}$ iff $a = \inf Y^{\uparrow}$. Similarly, $a = \inf Y$ iff $a \in Y^{\downarrow} \cap Y^{\downarrow\uparrow} = (Y^{\downarrow})^{\uparrow} \cap (Y^{\downarrow})^{\uparrow\downarrow}$ iff $a = \sup Y^{\downarrow}$. □

15.2 Semilattices; Lattices

Definition 7 Let $\langle L, \sqsubseteq \rangle$ be a poset; L is called a *join semilattice (meet semilattice)* if any two elements of L have a supremum (infimum) with respect to $\sqsubseteq$. L is called a *lattice* if L is both a join semilattice and a meet semilattice.

Remark. By induction, it can be seen that L is a join (meet) semilattice iff every non-empty finite subset of L has a supremum (infimum). Also, L is a lattice if every non-empty finite subset of L has both a supremum and an infimum.

The definition of semilattice was given in order-theoretic terms, i.e., as a special kind of poset. However, there is a fundamental correspondence between this conception and the algebraic one, according to which a semilattice is a set equipped with a binary operation that for any two elements in the set returns a value which is just their supremum. This will only be the case, of course, if the operation satisfies suitable conditions. Such an algebraic structure is called "semilattice algebra." The next theorem then shows that we can freely go back and forth between the order-theoretic and the algebraic view. The same kind of correspondence holds also for lattices (see Theorem 5) and, in particular, for Boolean lattices discussed below.

Definition 8 A structure $\langle X, \circ \rangle$ is called a *semilattice algebra* or *SL-algebra* for short if $\circ : X \times X \to X$ is a binary operation on X such that the following conditions hold:

1. $\forall x \forall y \forall z\, [(x \circ y) \circ z = x \circ (y \circ z)]$ (*associativity*)
2. $\forall x \forall y\, [x \circ y = y \circ x]$ (*commutativity*)
3. $\forall x\, [x \circ x = x]$ (*idempotence*)

Theorem 4 (E:96) *Let $\langle L, \sqsubseteq \rangle$ be a join (meet) semilattice and define*

(14) $x \circ y := \sup_{\sqsubseteq}\{x, y\}$

(15) $(x \circ y := \inf_{\sqsubseteq}\{x, y\})$

Then $\langle L, \circ \rangle$ is an SL-algebra. Conversely, if $\langle L, \circ \rangle$ is an SL-algebra then L together with the relation $\sqsubseteq$ defined by

(16) $x \sqsubseteq y :\Longleftrightarrow x \circ y = y$

(17) $(x \sqsubseteq y :\Longleftrightarrow x \circ y = x)$

is a join (meet) semilattice, and we have $\sup_{\sqsubseteq}\{x, y\} = x \circ y$ (resp. $\inf_{\sqsubseteq}\{x, y\} = x \circ y$).

Proof. The operation $\circ$ as defined in Equation 14 is obviously associative, commutative, and idempotent. Thus $\langle L, \circ \rangle$ is an SL-algebra. For the other direction we have to show that the relation $\sqsubseteq$ defined by Equation 16 is a partial order. Now idempotence yields reflexivity. For transitivity, assume $x \sqsubseteq y$ and $y \sqsubseteq z$, i.e., $x \circ y = y$ and $y \circ z = z$; then, by associativity, $x \circ z = x \circ (y \circ z) = (x \circ y) \circ z = y \circ z = z$, whence $x \sqsubseteq z$. Similarly, commutativity of $\circ$ proves the antisymmetry of $\sqsubseteq$. Finally, in order to prove $\sup_{\sqsubseteq}\{x, y\} = x \circ y$ observe that $x \circ y$ is an upper bound, with respect to $\sqsubseteq$ of Equation 16, for the set $\{x, y\}$, and that it is the least upper bound. The dual assertions are proved similarly, using Equations 15 and 17. □

Definition 9 A structure $\langle X, \sqcup, \sqcap \rangle$ is called a *lattice algebra* or *L-algebra* for short if $\sqcup, \sqcap : X \times X \to X$ are binary operations on X such that the following conditions hold:

1. $\langle L, \sqcup \rangle$, $\langle L, \sqcap \rangle$ are SL-algebras;
2. $x \sqcup y = y$ iff $x \sqcap y = x$.

Theorem 5 (E:97f.) *Let L be a set and $\sqcup, \sqcap$ binary operations on L. Then the following statements are equivalent:*

1. *$\langle L, \sqcup, \sqcap \rangle$ is a lattice algebra;*
2. *there is exactly one partial order $\sqsubseteq$ on L such that $\langle L, \sqsubseteq \rangle$ is a lattice with the supremum $\sup_{\sqsubseteq}\{x, y\} = x \sqcup y$ and the infimum $\inf_{\sqsubseteq}\{x, y\} = x \sqcap y$ for all $x, y \in L$;*
3. *the operations $\sqcup, \sqcap$ satisfy the above conditions of associativity, commutativity, and the following laws of* absorption*:*

(a) $x \sqcap (x \sqcup y) = x$ *(absorption)*

(b) $x \sqcup (x \sqcap y) = x$

Proof. Assume assertion (1); then $\langle L, \sqcup \rangle$ and $\langle L, \sqcap \rangle$ are SL-algebras, and their induced partial orders coincide because of condition (2) of Definition 9. But then L is a lattice with respect to this order, where the supremum and the infimum are as given in (2), due to the second part of Theorem 4. Now assume (2); then $\sqcup$ and $\sqcap$ are associative and commutative (Theorem 4). To prove absorption, say (3a), we have to show that $x = \inf\{x, \sup\{x, y\}\}$. Now $x \sqsubseteq x$ and $x \sqsubseteq \sup\{x, y\}$ thus $x \in \{x, \sup\{x, y\}\}^{\downarrow}$. On the other hand, for $y \in \{x, \sup\{x, y\}\}^{\downarrow}$ we have $y \sqsubseteq x$, hence $x \in \{x, \sup\{x, y\}\}^{\downarrow\uparrow}$. By Proposition 1, x is the infimum in question. The other absorption law is proved similarly. Finally, in order to prove assertion (1) from assertion (3), we first have to show that the absorption laws entail idempotence of both $\sqcup$ and $\sqcap$. Indeed, $x \sqcup x = x \sqcup (x \sqcap (x \sqcup x))$ by (3a), and setting $y = x \sqcup x$, the last expression equals x by (3b). Similarly, $x \sqcap x = x$. Finally, if $x \sqcup y = y$ then $x \sqcap y = x \sqcap (x \sqcup y) = x$, and if $x \sqcap y = x$ then $x \sqcup y = (x \sqcap y) \sqcup y = y \sqcup (x \sqcap y) = y$; so the second condition of Definition 9 holds, and $\langle L, \sqcup, \sqcap \rangle$ is a lattice algebra. □

Remark. Due to this correspondence between the order-theoretic and the algebraic view no difference is commonly made between lattice algebras and lattices proper. We will follow this usage, too.

Definition 10 A join (meet) semilattice is called *complete* if every non-empty subset has a supremum (an infimum). A lattice is called *complete* if every subset has both a supremum and an infimum.

Remark. A complete join (meet) semilattice L has a unit (zero) element; in fact, $\sup L$ is a $\mathbf{1}$ if L is a join semilattice, and $\inf L$ is a $\mathbf{0}$ if L is a meet semilattice. Similarly, if L is a complete lattice L has both a zero and a unit element ($\mathbf{1} := \sup L$, $\mathbf{0} := \sup \emptyset$).

Proposition 4 (E:98) *Let L be a poset. Then the following conditions are equivalent:*

1. *L is a complete lattice;*
2. *every subset of L has a supremum;*
3. *every subset of L has an infimum;*
4. *every non-empty subset of L has both a supremum and an infimum.*

It follows that every join (meet) semilattice in which every subset has a supremum (an infimum) is already a complete lattice.

Proof. (1) $\Rightarrow$ (2) is clear. (2) $\Rightarrow$ (3): If $Y \subseteq L$ then $Y^{\downarrow} \subseteq L$, and there is an a with $a = \sup Y^{\downarrow}$. By Proposition 3, $a = \inf Y$. (3) $\Rightarrow$ (4): A dual argument produces a supremum for every subset and hence every non-empty subset of L. (4) $\Rightarrow$ (1): L is obviously a lattice. We have to show that the empty set has both a supremum and an infimum. Indeed, $\emptyset^{\downarrow} = L$ and $\sup L = \mathbf{1}$ hence $\mathbf{1} = \inf \emptyset$ by Proposition 3. Similarly, $\mathbf{0} = \inf L$ is seen to be the supremum of $\emptyset$.

Example 1 1.Every finite semilattice is complete. Every finite lattice is complete.

2. Let X be a set and 2^X its power set, ordered by set inclusion. Then $\langle 2^X, \subseteq \rangle$ is a complete lattice, with $\sup_{\subseteq} \mathcal{C} = \bigcup \mathcal{C}$ and $\inf_{\subseteq} \mathcal{C} = \bigcap \mathcal{C}$ for all $\mathcal{C} \subseteq 2^X$.

3. Another important class of complete lattices are topologies.

Definition 11 Let X be a set, 2^X its power set. A subset $\mathcal{T} \subseteq 2^X$ is called a *topology on* X if the following conditions are satisfied:

1. $X \in \mathcal{T}$ and $\emptyset \in \mathcal{T}$;
2. $A \cap B \in \mathcal{T}$ for $A, B \in \mathcal{T}$;
3. $\bigcup \mathcal{U} \in \mathcal{T}$ for $\mathcal{U} \subseteq \mathcal{T}$.

The elements of $\mathcal{T}$ are also called *($\mathcal{T}$-)open sets.*

Thus, a topology $\mathcal{T}$ is a *lattice* with respect to set inclusion, and has a unit and a zero element. Moreover, $\mathcal{T}$ is *complete*: since a topology is by definition closed under arbitrary unions, and the union operator is the supremum with respect to set inclusion, every subset U of $\mathcal{T}$ has a $\subseteq$-supremum. viz. $\sup_{\mathcal{T}} \mathcal{U} = \bigcup \mathcal{U}$; also, $\mathcal{T}$ has a zero, and completeness follows by Proposition 4. Notice, however, that while the supremum in $\mathcal{T}$ coincides with set-theoretic union the infimum in $\mathcal{T}$, that is, $\inf_{\mathcal{T}} \mathcal{U} = \sup_{\mathcal{T}} \mathcal{U}^{\downarrow}$ (see Proposition 3), *is not* in general the same as set-theoretic intersection. For instance, consider the standard topology $\mathcal{T}(\Re)$ on the real line and the set $\mathcal{U}$ of open neighborhoods of the unit interval $[0, 1]$. We have $U \in \mathcal{T}(\Re)$ for all $U \in \mathcal{U}$ but $[0, 1] \notin \mathcal{U}$. Now for $\mathcal{T} = \mathcal{T}(\Re)$, $\inf_{\mathcal{T}} \mathcal{U} = \sup_{\mathcal{T}} \mathcal{U}^{\downarrow} = \langle 0, 1 \rangle$ but $\bigcap \mathcal{U} = [0, 1]$.

Lemma 2 Let L be a complete semilattice, I a non-empty index set, and $(Y_i)_{i \in I}$ a familiy of non-empty subsets of L. Let $Y := \bigcup_{i \in I} Y_i$ and $x_i := \sup Y_i$ for all $i \in I$. Then $\sup\{x_i \mid i \in I\} = \sup Y$.

Proof. Let $x := \sup\{x_i \mid i \in I\}$ and $y := \sup Y$. We first show $x \sqsubseteq y$. For all $i \in I$ we have $Y_i \subseteq Y$, hence $x_i \sqsubseteq y$ by Proposition 2.4. Thus y is an upper bound for the set $\{ x_i \mid i \in I \}$, and so $x \sqsubseteq y$. Conversely we show that x is an upper bound for the set Y. Let $z \in Y$; then $z \in Y_i$ for some $i \in I$. So $z \sqsubseteq \sup Y_i = x_i$ and also $z \sqsubseteq \sup\{x_i \mid i \in I\} = x$. That means $z \sqsubseteq x$ for all $z \in Y$, so x is an upper bound. But then $y = \sup Y \sqsubseteq x$, and $x = y$ follows. □

Definition 12 Let L be a join (meet) semilattice and Y a non-empty subset of L. Then Y is called a *subsemilattice of* L if the supremum (infimum) in L of any two elements of Y is in Y. If L is a lattice then a non-empty $Y \subseteq L$ is a *sublattice* if both the supremum in L and the infimum in L of any two elements of Y are in Y.

Remark. If follows that subsemilattices are subsets of L that are *closed* with respect to *finite* sup (inf) formation *in* L, and sublattices are closed with respect to both finite sup and finite inf formation in L. A subset Y of a lattice L can itself be a lattice (with respect to the induced order) but sup formation and/or inf formation *in* Y might differ from from that *in* L; in this case, Y is not a sublattice of L.

Definition 13 Let $\langle L, \sqsubseteq \rangle$ be a poset. An *interval in* L is a subset $I \subseteq L$ of the form $I = \{x \in L \mid a \sqsubseteq x \sqsubseteq b\}$ for $a, b \in L$ with $a \sqsubseteq b$. I is also written $[a, b]$.

Remark. 1. Intervals in (semi)lattices are sub(semi)lattices.

2. Let L be a (semi)lattice and $(A_j)_{j \in J}$ a family of sub(semi)lattices of L. Since every single A_j is closed under the appropriate operation(s) the same is true of the intersection of the A_j, $\bigcap\{A_j \mid j \in J\}$. Now assume that all A_j contain a non-empty set $E \subseteq L$; then the intersection of the A_j is a sub(semi)lattice containing E, and it is obviously the smallest with this property.

Definition 14 Let L be a (semi)lattice and E a non-empty subset of L. Then

$$[E] := \bigcap\{Y \mid Y \text{ sub(semi)lattice of } L \text{ and } E \subseteq L\}$$

is called the *sub(semi)lattice of L generated by E*.

Definition 15 Let L be a complete join (meet) semilattice and Y a non-empty subset of L. Then Y is a *complete subsemilattice of* L if $\sup_L X \in Y$ ($\inf_L X \in Y$) for all non-empty $X \subseteq Y$. If L is a complete lattice and $Y \subseteq L$ then $Y$is a *complete sublattice of* L if $\sup_L X \in Y$ and $\inf_L X \in Y$ for all $X \subseteq Y$.

Definition 16 Let L be a complete (semi)lattice and E a non-empty subset of L. Then

$$[\![E]\!] := \bigcap\{\, Y \mid Y \text{ complete sub(semi)lattice of } L \text{ and } E \subseteq L\,\}$$

is called the *complete sub(semi)lattice of L generated by E.*

Proposition 5 *Let L be a complete semilattice, E a non-empty subset of L, and define*

(18) $M := \{\, x \in L \mid \exists F \subseteq E, F \neq \emptyset \text{ such that } x = \sup F\,\}$

Then $M = [\![E]\!]$.

Proof. If $x \in M$ there is a non-empty $F \subseteq E$ with $x = \sup F$. Let Y be a complete subsemilattice containing E; then $F \subseteq Y$, and $\sup F = x \in Y$. Conversely, assume $x \in [\![E]\!]$; then x is in every complete subsemilattice Y containing E. We show that M is one such Y. If $y \in E$ we have $\{y\} \subseteq E$ and $y = \sup\{y\} \in M$, hence $E \subseteq M$. Now consider a non-empty subset Z of M; for every $z \in Z$ there is a non-empty subset F_z of E such that $z = \sup F_z$. Set $F := \bigcup_{z \in Z} F_z$; F is a non-empty subset of E, and we have $\sup F \in M$. But $\sup F = \sup Z$ by Lemma 2, so Z has a supremum in M. This shows that M is a complete subsemilattice containing E, and therefore, $x \in M$. □

Another important type of substructure in a (semi)lattice are the ideals.

Definition 17 Let $\langle L, \sqsubseteq\rangle$ be a poset and Y a non-empty subset of L.

1. If L is a join semilattice then Y is called a *(join) ideal* iff it is a downward $\sqsubseteq$-closed subsemilattice of L.

2. If L is a meet semilattice then Y is called a *meet ideal* or a *filter* iff it is an upward $\sqsubseteq$-closed subsemilattice of L.

3. If L is a lattice then Y is called an *ideal* iff it is a join ideal with respect to the join semilattice structure of L, and Y is called a *filter* iff it is a filter with respect to the meet semilattice structure of L.

Proposition 6 (E:141f.) *Let* $\langle L, \sqsubseteq\rangle$ *be a poset and Y a non-empty subset of L.*

1. *If L is a join semilattice then Y is an ideal iff one of the following equivalent conditions holds:*

(a) *Y is a directed lower set;*

(b) $x \sqcup y \in Y \iff x \in Y$ *and* $y \in Y$ $\qquad (x, y \in L)$.

2. *If L is a lattice then Y is an ideal iff either one of the foregoing conditions is satisfied, or else one of the following two conditions:*

   (a) *Y is a sublattice of L such that* $a \in L$ *and* $y \in Y$ *imply* $a \sqcap y \in Y$*;*

   (b) i. *for all* $x, y \in Y$, $x \sqcup y \in Y$, *and*

       ii. *for all* $a \in L$ *and* $y \in Y$, $a \sqcap y \in Y$.

Proof. An ideal Y in a join semilattice L is downward $\sqsubseteq$-closed, so it is a lower set (recall that this is just another name for the same property). It is also directed (with respect to the induced order) since as a subsemilattice it contains an upper bound (in fact, a supremum) for any two of its elements. Now assume (1a) and suppose $x \sqcup y \in Y$; then $x \sqsubseteq x \sqcup y$ and $y \sqsubseteq x \sqcup y$, hence $x, y \in Y$ since Y is a lower set. Conversely, Y is directed, so if $x, y \in Y$ then there is a $z \in Y$ such that $x \sqsubseteq z$ and $y \sqsubseteq z$; but $x \sqcup y$ is the least upper bound for $\{x, y\}$ thus $x \sqcup y \sqsubseteq z$ and $x \sqcup y \in Y$ because Y is downward closed. Now condition (1b) implies that Y is a subsemilattice, and that it is downward closed: for $x \in Y$ and $y \in L$ such that $y \sqsubseteq x$ we have $y \sqcup x = x \in Y$ hence $y \in Y$. Thus Y is an ideal.

Turning to (2), it is clear that each of the conditions in (1) also characterizes an ideal in a lattice. If Y is an ideal in the lattice L then it is a sublattice since in addition to containing the supremum of a pair of elements $x, y \in Y$ it contains their infimum: $x \sqcap y \sqsubseteq x$ and downward closure entails $x \sqcap y \in Y$. By the same token we get the second part of condition (2a). Conditions (2a) and (2b) are obviously equivalent. But either one of them also suffices to show, by arguments already used, that Y is an ideal. □

Remark. Just like substructures in general the concept of an ideal (filter) is also closed under arbitrary intersections. Therefore in every (semi)lattice L there is a smallest ideal (filter) containing a non-empty set $E \subseteq L$. It is called the *ideal (filter) generated by E* and is denoted by $(E]$ (resp. $[E)$). If E is a singleton $\{a\}$ then the ideal (filter) generated by it is called a *principal ideal (filter)*, written $(a]$ (resp. $[a)$) instead of $(\{a\}]$ (resp. $[\{a\})$). We have $(a] = a^{\downarrow}$, i.e., principal ideals are lower cones, and vice versa. Similarly, $[a) = a^{\uparrow}$, that is, principal filters are upper cones, and vice versa.

Since an ideal Y is, by definition, a lower set we have $Y = \downarrow Y = \bigcup\{y^{\downarrow} \mid y \in Y\}$; thus an ideal equals the union of the lower cones of its elements.

The next structural property of the class of (semi)lattices that is worth mentioning is that we can generate new (semi)lattices by forming cartesian products and powers of (semi)lattices.

Definition 18 Let $\langle L_i, \sqsubseteq_i \rangle$ be (semi)lattices for $i = 1, 2$ and let $L = L_1 \times L_2$ be their cartesian product. Then the *product ordering relation* $\sqsubseteq$ defined by

(19) $x \sqsubseteq y \iff x_1 \sqsubseteq_1 y_1$ and $x_2 \sqsubseteq_2 y_2$

(for $x = \langle x_1, x_2 \rangle, y = \langle y_1, y_2 \rangle \in L$) is a partial order on L, and L together with $\sqsubseteq$ is a (semi)lattice, called the *direct product (semi)lattice of L_1 and L_2.*

Remark. 1. As can be seen from the definition the product order is defined "component-wise". Similarly, the supremum (and/or the infimum) in the product lattice come out component-wise; thus, for $x, y \in L_1 \times L_2$ we have

(20) $x \sqcup y \;=\; \langle x_1 \sqcup_1 y_1 \,,\, x_2 \sqcup_2 y_2 \rangle$

and an analogous equation holds for the infimum.

2. The product ordering relation preserves the partial order property but in general not additional properties; for instance, the product order of two linear orders typically ceases to be linear (example: the cartesian product of two copies of the real line equipped with the product order).

Definition 19 Let L be a (semi)lattice, I an index set and L^I the "I-fold" cartesian product of L. For $a, b \in L^I$ with $a = (a_i)_{i \in I}, b = (b_i)_{i \in I}$, define

(21) $a \trianglelefteq b \;:\iff\; a_i \sqsubseteq b_i$ for all $i \in I$

Then $\trianglelefteq$ is a partial order on L^I, and $\langle L^I, \trianglelefteq \rangle$ is a (semi)lattice, called the *direct power (semi)lattice of L with respect to I.*

Remark. If the index set I consists of two elements the power lattice L^I equals the direct product lattice of two copies of L.

Example 2 Let $\Omega_2 = \{\text{T},\text{F}\}$ be the set of classical truth values, ordered and made into a lattice by $\text{F} \sqsubseteq \text{T}$. Then the set $(\Omega_2)^E$ of extensions of 1-place predicates, considered as functions from a domain E of individuals into Ω_2, receives a lattice structure by the direct power order $\trianglelefteq$. It corresponds one-to-one to the set-theoretic lattice structure on the power set of E. The direct power construction lies at the heart of the "Boolean semantics" approach in the sense of Keenan (1981) where the Boolean lattice structure of the truth values is "lifted" to higher categories.

We next introduce *structure-preserving* mappings between pairs of posets or (semi)lattices.

Definition 20 Let $\langle L, \sqsubseteq\rangle, \langle L', \sqsubseteq'\rangle$ be posets and $h : L \to L'$ a mapping from L into L'.

1. h is called *isotone* or an *order homomorphism* iff for all $x, y \in L$,

   (22) $x \sqsubseteq y \implies h(x) \sqsubseteq' h(y)$

2. h is called a *join homomorphism* iff h preserves finite (non-empty) suprema, i.e., for all $x, y \in L$, if $x \sqcup y := \sup\{x, y\}$ and $h(x) \sqcup' h(y) := \sup\{h(x), h(y)\}$ exist, then

   (23) $h(x \sqcup y) \;=\; h(x) \sqcup' h(y)$

   Similarly, h is a *meet homomorphism* iff h preserves finite (non-empty) infima. If L, L' are join (meet) semilattices then the join (meet) homomorphism h is called a *semilattice homomorphism.* If L, L' are lattices then h is called a *lattice homomorphism* if h is both a join and a meet homomorphism.

3. h is called a *sup-homomorphism* iff h preserves arbitrary suprema, i.e., for all subsets $Y \subseteq L$, if $\sup_{\sqsubseteq} Y$ and $\sup_{\sqsubseteq'} h[Y]$ exist, then

   (24) $h(\sup_{\sqsubseteq} Y) \;=\; \sup_{\sqsubseteq'} h[Y]$

   Dually, h is a *inf-homomorphism* iff h preserves arbitrary infima. h is a *complete homomorphism* iff h is both a sup-homomorphism and an inf-homomorphism. If L, L' are join (meet) semilattices then the sup- (inf-) homomorphism h is called a *complete semilattice homomorphism.* If L, L' are complete lattices then a complete homomorphism is called a *complete lattice homomorphism.*

Remark. Let L be a lattice, $z \in L$, and $f_z, g_z : L \to L$ with $f_z(x) = x \sqcup z$ and $g_z(x) = x \sqcap z$ for all $x \in L$. Then f_z, g_z are isotone mappings.

A homomorphism gives rise to a *commuting diagram.* For instance, let $h : L \to L'$ be a join homomorphism and $\langle h, h\rangle$ that function on $L \times L$ with $\langle x, y\rangle \mapsto \langle h(x), h(y)\rangle$. Then h is said to *commute* with the join operation in the following way: no matter whether we first form the join of two elements x, y in L and then map the result $x \sqcup y$ into L', or if we first map the elements separately into L' and take their join in L' yielding $h(x) \sqcup' h(y)$, the outcome will be the same; see Figure 15.1.

Proposition 7 *A (semi)lattice homomorphism is isotone, but not vice versa.*

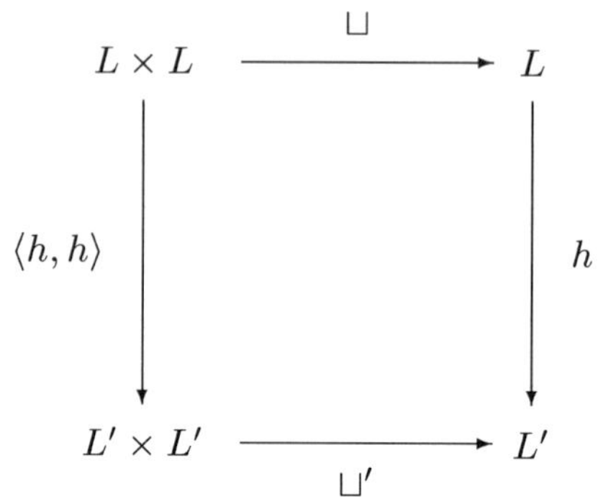

Figure 15.1: Commuting diagram for a join homomorphism.

Proof. Let L, L' be (join semi)lattices as above, and let $h : L \to L'$ be a homomorphism. If $x, y \in L$ and $x \sqsubseteq y$ then $y = x \sqcup y$, hence $h(y) = h(x \sqcup y) = h(x) \sqcup' h(y)$. But that means $h(x) \sqsubseteq' h(y)$. On the other hand, h can be an isotone mapping between two (semi)lattices without being an homomorphism. For example, let L be the four-element lattice consisting of $\mathbf{0}$, $\mathbf{1}$, and two incomparable elements x, y in between, and let L' be the four-element chain, $\mathbf{0}' \sqsubseteq' x' \sqsubseteq' y' \sqsubseteq' \mathbf{1}'$. Let h map the four elements of L onto their primed versions in L'; then h is isotone but $\mathbf{1} = h(x \sqcup y) \neq h(x) \sqcup' h(y) = x' \sqcup' y' = y'$. □

Definition 21 Let $\langle L, \sqsubseteq \rangle, \langle L', \sqsubseteq' \rangle$ be posets, and $h : L \to L'$.

1. h is called an *(order) embedding* iff for all $x, y \in L$,

   (25) $x \sqsubseteq y \iff h(x) \sqsubseteq' h(y)$

2. h is called an *(order) isomorphism* iff h is a surjective embedding.

Remark. 1. An embedding is injectice; in fact, $h(x) = h(y)$ means $h(x) \sqsubseteq' h(y)$ and $h(y) \sqsubseteq' h(x)$, whence $x \sqsubseteq y$ and $y \sqsubseteq x$ by the embedding property, thus $x = y$.

2. A mapping might be bijective and isotone but still not an order isomorphism: the mapping h in the proof of Proposition 7 is an example; what is missing is that the inverse mapping h^{-1} is not order-preserving.

Definition 22 A *semilattice embedding (lattice embedding; complete lattice embedding)* is an order embedding which is a semilattice homomorphism (lattice homomorphism; complete lattice homomorphism).

Proposition 8 *Let L, L' be a join semilattices and $h : L \to L'$ an injective join homomorphism. Then h is a semilattice embedding.*

Proof. h is a homomorphism, hence isotone. Now assume $h(x) \sqsubseteq' h(y)$ or, equivalently, $h(x) \sqcup' h(y) = h(y)$. Since h is a homomorphism this yields the equation $h(x \sqcup y) = h(y)$, and, by injectivity, $x \sqcup y = y$; hence $x \sqsubseteq y$. □

Proposition 9 *Let L, L' be join semilattices and $h : L \to L'$ a join homomorphism.*

1. *If X is a subsemilattice of L then $h[X]$ is a subsemilattice of L';*

2. *if Y is a subsemilattice of L' then $h^{-1}[Y]$ is a subsemilattice of L.*

Analogous statements hold for (complete) lattices.

Proof. For 1., it suffices note that the supremum of any two elements $h(x), h(y) \in h[X]$ in L', viz. $h(x) \sqcup' h(y)$, is already in $h[X]$; this is because the homomorphism h commutes with the join operation, so we have for $x, y \in X$: $h(x) \sqcup' h(y) = h(x \sqcup y) \in h[X]$. The other claims are proved in a similar way. □

Proposition 10 (E:116) *Let L, L' be posets and $h : L \to L'$ an order isomorphism. Then h is a complete homomorphism.*

Proof. For $Y \subseteq L$ we prove Equation 24 (the dual equation for the infima follows similarly). Set $x = \sup Y$; then $h(x)$ is an upper bound of $h[Y]$: any $z \in h[Y]$ can be written $z = h(y)$ for a $y \in Y$, but $y \sqsubseteq x$, hence $z = h(y) \sqsubseteq' h(x)$ because h is isotone. We now show that $h(x)$ is also the least upper bound of $h[Y]$. Indeed, if $z \in (h[Y])^{\uparrow}$ then for all $y \in Y$, $h(y) \sqsubseteq z$ and $h(y) \sqsubseteq' h(u)$ with $u := h^{-1}(z)$, hence $y \sqsubseteq u$ because h is an embedding. Thus $u \in Y^{\uparrow}$. But $x \in Y^{\uparrow\downarrow}$, so $x \sqsubseteq u$ and $h(x) \sqsubseteq' h(u) = z$. This shows that $h(x) \in (h[Y])^{\uparrow\downarrow}$, and Equation 24 follows.

Proposition 11 (E:116) *Let L, L' be join semilattices and $h : L \to L'$ a bijection. Then the following statements are equivalent:*

1. *h is an order isomorphism;*

2. *h is a join homomorphism;*

3. *h is a sup-homomorphism;*

4. *h is a complete homomorphism.*

Proof. The implication (1) $\Rightarrow$ (4) is Proposition 10, and the implications from (4) back to (2) via (3) are trivial. Now assume (2); h is surjective, and Proposition 8 shows that h is an embedding. Thus h is an order isomorphism. □

Corollary 1 *Let L, L' be lattices and $h : L \to L'$. Then h is an order isomorphism iff h is a bijective lattice homomorphism.*

Theorem 6 (Tarski) *Let L be a complete lattice and $h : L \to L$ an isotone transformation on L. Let $Fix(h)$ be the set of all* fixed points of h, *i.e.,*

$$Fix(h) \;:=\; \{\, x \in L \mid h(x) = x \,\}$$

Then the set $Fix(h)$ is non-empty, and it is complete with respect to the induced order.

Tarski's fixed-point theorem has many applications, for instance in modern theories of truth. For a proof see, e.g., Erné (1982):123, or Gierz *et al.* (1980):9.

15.3 Boolean Lattices

A Boolean algebra is a special kind of lattice. I am now going to provide the remaining concepts that are needed for its definition.

The first step is to define distributive lattices. Before doing so we point out (without proof) that in any lattice the following *inequalities* hold. The first three are called the *distributive inequalities*, the last the *modular inequality*.

Proposition 12 *Let $\langle L, \sqcup, \sqcap \rangle$ be a lattice. Then for all $x, y, z \in L$,*

1. $(x \sqcap y) \sqcup (x \sqcap z) \;\sqsubseteq\; x \sqcap (y \sqcup z)$

2. $x \sqcup (y \sqcap z) \;\sqsubseteq\; (x \sqcup y) \sqcap (x \sqcup z)$

3. $(x \sqcap y) \sqcup (y \sqcap z) \sqcup (z \sqcap x) \;\sqsubseteq\; (x \sqcup y) \sqcap (y \sqcup z) \sqcap (z \sqcup x)$

4. $(x \sqcap y) \sqcup (x \sqcap z) \;\sqsubseteq\; x \sqcap (y \sqcup (x \sqcap z))$

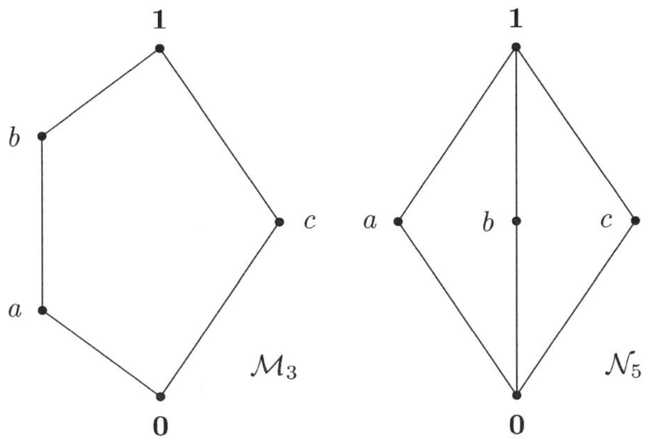

Figure 15.2: A pentagon and a diamond.

Definition 23 A lattice $\langle L, \sqcup, \sqcap \rangle$ is called *distributive* iff one of the following equivalent conditions holds:

1. $(x \sqcap y) \sqcup (x \sqcap z) \;=\; x \sqcap (y \sqcup z) \qquad (x, y, z \in L)$
2. $(x \sqcup y) \sqcap (x \sqcup z) \;=\; x \sqcup (y \sqcap z) \qquad (x, y, z \in L)$

Definition 24 A subset Y of a lattice is called a *pentagon* if Y is a sublattice isomorphic to the left diagram of Figure 15.2 which is the lattice $\mathcal{N}_5 = \{\mathbf{0}, a, b, c, \mathbf{1}\}$ with $a \sqsubseteq b$ and a, c and b, c incomparable. Y is called a *diamond* if Y is a sublattice isomorphic to the right diagram of Figure 15.2 which is the lattice $\mathcal{M}_3 = \{\mathbf{0}, a, b, c, \mathbf{1}\}$ with a, b, a, c, and b, c incomparable.

Theorem 7 *A lattice is distributive iff it neither contains a pentagon nor a diamond as sublattice.*

For a proof of this characterization theorem, see (Grätzer, 1978, p. 59f.).

Definition 25 Let $\langle L, \sqcup, \sqcap, \rangle$ be a lattice mit zero and unit element, and $x, y \in L$. x is called a *complement of* y iff $x \sqcap y = \mathbf{0}$ and $x \sqcup y = \mathbf{1}$. Let $[a, b]$ an interval in L and $x \in [a, b]$; then z is a *relative complement of* x *in* $[a, b]$ iff $x \sqcap z = a$ and $x \sqcup z = b$.

Lemma 3 Let L be a distributive lattice with $\mathbf{0}$ and $\mathbf{1}$. Then the complement y of an element x in L is uniquely determined. Notation: $y = x'$.

Lemma 4 Let L be a distributive lattice with $\mathbf{0}$ and $\mathbf{1}$. If x has a complement then it also has a relative complement in every interval containing it.

Proof. Let $x' = y$, $[a, b]$ an interval with $x \in [a, b]$, and define

(26) $z := (y \sqcup a) \sqcap b$

Then $x \sqcap z = x \sqcap (y \sqcup a) \sqcap b = [(x \sqcap y) \sqcup (x \sqcap a)] \sqcap b = (\mathbf{0} \sqcup a) \sqcap b = a$. Similarly, $x \sqcup z = b$. Thus z is the relative complement of x in $[a, b]$. □

Lemma 5 (de Morgan) Let $\langle L, \sqcup, \sqcap, \rangle$ be a distributive lattice and x, y two elements in L with complements x', y'. Then $x \sqcup y$ and $x \sqcap y$ have complements, and

1. $(x \sqcap y)' = x' \sqcup y'$
2. $(x \sqcup y)' = x' \sqcap y'$

Definition 26 A lattice with $\mathbf{0}$ and $\mathbf{1}$ is called *complemented* iff every element in L has a complement.

Definition 27 A lattice L with $\mathbf{0}$ and $\mathbf{1}$ such that $\mathbf{1} \neq \mathbf{0}$ is called *Boolean* iff it is distributive and complemented. Let $\sqcup, \sqcap, '$ the induced operations on L; then $\langle L, \sqcup, \sqcap, ', \mathbf{0}, \mathbf{1} \rangle$ is called a *Boolean algebra* (*BA* for short).

Remark. We shall freely switch back and forth between the order-theoretic viewpoint (Boolean lattice) and the algebraic viewpoint (Boolean algebra).

Example 3
1. Let X be a set and $\mathcal{F} \subseteq 2^X$ be a non-empty set of subsets of X. $\mathcal{F}$ is called a *fields of sets* if it is closed under finite union, intersection, and complementation. $\mathcal{F}$ contains X and $\emptyset$ and is a Boolean algebra with respect to those set-theoretic operations.
2. If $\mathcal{F} = 2^X$ then $\mathcal{F}$ is called the *power set algebra over X*. It is the most familiar Boolean algebra, and in fact the standard model for all *atomic* Boolean algebras (see below).

3. The set $\Omega_2 = \{\text{T},\text{F}\}$ of *classical truth values* is a Boolean algebra, with $\mathbf{1} = \text{T}$ and $\mathbf{0} = \text{F}$.

4. While a topology $\mathcal{T}$ over a set X is in general not a BA, the subsystem $\mathcal{R}$ of *regular open sets in* T is, i.e., those sets that are equal to the interiors of their closures in the topological sense. For instance, open intervals on the real line of the form $(a, b) = \{x \mid a < x < b\}$ are regular open sets since the closure of such an interval (a, b) is the closed interval $[a, b] = \{x \mid a \leq x \leq b\}$ the interior of which is, in turn, the open interval (a, b). Now when we form the union of two open intervals it might fail to be regular open again. For instance, the closure of $(0, 1) \cup (1, 2)$ is the closed interval $[0, 2]$ whose interior $(0, 2)$ contains the number 1 and thus differs from the union of intervals we started out with. But if we declare the join operation of two elements in $\mathcal{R}$ as the interior of the closure of their union, take the meet as set-theoretic intersection, and define the complement of a set as the interior of its set-theoretic complement, then $\mathcal{R}$ is a Boolean algebra.

5. Take the set F of formulas of classical propositional logic PL, and identify two formulas ϕ, ψ if their biconditional $\phi \leftrightarrow \psi$ is a theorem of PL. The corresponding equivalence classes form a BA, the *Lindenbaum algebra of PL*, with respect to the obvious Boolean operations induced by the logical connectives.

6. Consider the set of divisors of the number 30, $D_{30} = \{1, 2, 3, 5, 6, 10, 15, 30\}$. This set is partially ordered by the relation $x \sqsubseteq y$ iff x divides y. In fact, D_{30} is a Boolean algebra, where the join is the *least common multiple* (L.C.M.), the meet is the *greatest common divisor* (G.C.D.), the complement is $x' = 30/x$, and $\mathbf{0} = 1$, $\mathbf{1} = 30$.

Remark. Let L be a Boolean algebra. Then for all $x, y \in L$ we have $x \sqsubseteq y$ iff $x \sqcap y' = \mathbf{0}$.

Definition 28 Let L, L' be Boolean algebras, and let $\sqcup, \sqcap, '$ denote the Boolean operations in both L and L'. A mapping $h : L \to L'$ is called a *Boolean homomorphism* if it preserves join, meet, and complement, i.e., :

(27) $h(x \sqcup y) = h(x) \sqcup h(y)$

(28) $h(x \sqcap y) = h(x) \sqcap h(y)$

(29) $h(x') = (h(x))'$

A bijective Boolean homomorphism is a *(Boolean) isomorphism.*

Proposition 13 Let L, L' be Boolean algebras and let $h : L \to L'$. Prove:

1. If h is a Boolean homomorphism then

   (30) $h(\mathbf{1}) = \mathbf{1}', \quad h(\mathbf{0}) = \mathbf{0}'$

2. If h satisfies Equations 27, 28, and 30 then it also satisfies Equation 29; that is, h is a Boolean homomorphism.

Recall that in a lattice L with $\mathbf{0}$ an element a is called an *atom* iff a is minimal in $L \setminus \{\mathbf{0}\}$; that means that for all non-zero elements $x \in L$ with $x \sqsubseteq a$ we have $x = a$. Thus this definition carries over to Boolean algebras. Furthermore, a Boolean algebra L is called *atomic* iff every non-zero element $x \in L$ has an atom a below it, i.e., $a \sqsubseteq x$. The objects in an atomic Boolean algebra are completely determined by the atoms they dominate:

Proposition 14 *Let L be an atomic Boolean algebra, A the set of its atoms, and $x \in L$. Then x is the supremum of the atoms below it, i.e.,*

(31) $x = \sup A \cap x^{\downarrow}$

Proof. First observe that x is an upper bound for the set $A \cap x^{\downarrow}$. Now let y be another upper bound for this set, and assume $x \not\sqsubseteq y$. By a remark made above we have $x \sqcap y' \neq \mathbf{0}$. By atomicity there is an atom a below $x \sqcap y'$, and we have $a \in A \cap x^{\downarrow}$. Thus $a \sqsubseteq y$ or $a \sqcap y = a \neq \mathbf{0}$. But also $a \sqcap y \sqsubseteq (x \sqcap y') \sqcap y = x \sqcap (y \sqcap y') = \mathbf{0}$, contradiction. Therefore $x \sqsubseteq y$, and x is the least upper bound for $A \cap x^{\downarrow}$. □

We can now state and prove the fundamental representation theorem for *complete atomic Boolean algebras* (*CAB* for short). The following lemma will be used in its proof; it states that the Boolean meet operation is *sup-continuous* in that it preserves arbitrary suprema.

Lemma 6 Let L be a complete Boolean algebra, $Y \subseteq L$ and $a \in L$. Define $a \sqcap Y := \{a \sqcap y \mid y \in Y\}$. Then $a \sqcap \sup Y = \sup(a \sqcap Y)$.

Proof. We have $a \sqcap y \sqsubseteq a$ and $a \sqcap y \sqsubseteq \sup Y$, hence $a \sqcap y \sqsubseteq a \sqcap \sup Y$; therefore, $a \sqcap \sup Y \in (a \sqcap Y)^{\uparrow}$. Now let $z \in (a \sqcap Y)^{\uparrow}$, i.e., for all $y \in Y$, $a \sqcap y \sqsubseteq z$. By an earlier remark the join operation is isotone is its second argument place, so $a' \sqcup (a \sqcap y) \sqsubseteq a' \sqcup z$, and hence $y \sqsubseteq a' \sqcup y = \mathbf{1} \sqcap (a' \sqcup y) = (a' \sqcup a) \sqcap (a' \sqcup y) = a' \sqcup (a \sqcap y) \sqsubseteq a' \sqcup z$. This inequality holds for all $y \in Y$, thus we have $\sup Y \sqsubseteq a' \sqcup z$. But then, $a \sqcap \sup Y \sqsubseteq a \sqcap (a' \sqcup z) = (a \sqcap a') \sqcup (a \sqcap z) = \mathbf{0} \sqcup (a \sqcap z) \sqsubseteq z$. This shows that $a \sqcap \sup Y \in (a \sqcap Y)^{\uparrow\downarrow}$, and the assertion follows. □

Theorem 8 (Stone) *Every complete atomic Boolean algebra is isomorphic to the power set algebra over the set of its atoms.*

Proof. Let L be a CAB and A the set of its atoms. Define a function h from L into the power set algebra 2^A by

(32) $h(x) := A \cap x^{\downarrow}$

We first show that h is an embedding. Assume $x \sqsubseteq y$ and $a \in h(x)$; then $a \in A$ and $a \sqsubseteq x \sqsubseteq y$, hence $a \in h(y)$. Thus h is isotone. But h is also a meet homomorphism: for $x, y \in L$, we have $h(x \sqcap y) \subseteq h(x)$ and $h(x \sqcap y) \subseteq h(y)$ since h is isotone, hence $h(x \sqcap y) \subseteq h(x) \cap h(y)$; an $a \in h(x) \cap h(y)$ is both below x and y, hence below $x \sqcap y$, thus $a \in h(x \sqcap y)$. This fact is used for the other direction of the embedding property. Assume $h(x) \subseteq h(y)$; in order to show that $x \sqsubseteq y$ or $x \sqcap y = x$ we consider the relative complement z of $x \sqcap y$ in the interval $[\mathbf{0}, x]$. For z we have $(x \sqcap y) \sqcup z = x$; hence the proof is complete when $z = \mathbf{0}$. Now $z \sqsubseteq x$ and $h(z) \subseteq h(x) \subseteq h(y)$; therefore, $h(z) = h(z) \cap h(y) = h(z \sqcap y)$. But $z \sqcap y = (z \sqcap x) \sqcap y = (x \sqcap y) \sqcap z = \mathbf{0}$ since z is the relative complement of $x \sqcap y$ in $[\mathbf{0}, x]$. It follows that $h(z) = h(\mathbf{0}) = A \cap \mathbf{0}^{\downarrow} = \emptyset$. But then $z = \mathbf{0}$ because otherwise, by atomicity, there would be atoms below z and $h(z) \neq \emptyset$. We conclude $x \sqsubseteq y$; thus, h is an embedding and, by a remark made above, injective.

In order to show that h is surjective consider a non-empty subset $Z \subseteq A$ (for $Z = \emptyset$ we saw that $h(\mathbf{0}) = \emptyset$). Define $x := \sup Z$. We have $Z \subseteq h(x)$ since $z \in Z$ means $z \sqsubseteq x$ and $z \in h(x)$. Now let $y \in h(x)$; then $y \sqsubseteq x$ and $y = y \sqcap \sup Z = \sup(y \sqcap Z)$ by Lemma 6. But since $y \in A$ and $Z \subseteq A$, the meets $y \sqcap z$ are either all $\mathbf{0}$ (which is impossible since $y \neq \mathbf{0}$), or else are all $\mathbf{0}$ except for one such meet, say $y \sqcap z_0$ for some $z_0 \in Z$. That means that $y = z_0$, hence $y \in Z$. This shows that $Z = h(x)$, and h is surjective.

Now a surjective embedding is an order isomorphism, and by Proposition 10, h is a complete homomorphism. A fortiori, h preserves finite sups and infs, and by Proposition 13, complementation. Therefore, h is a Boolean isomorphism. □

15.4 Plural Lattices

The final part of the Appendix deals with those properties of (join) semilattices that were introduced in Chapter 6 in connection with the logic of plurality. Axiom and definition numbers refer to that chapter. The discussion here will continue to be carried out in the metalanguage, i.e., we will consider the algebraic structures characterized by the axioms of that chapter.[12]

We will assume from the outset that all semilattices are complete; recall that this means closure under suprema (or joins) of arbitrary *non-empty* subsets, so that a bottom element does not automatically come in via the supremum of the empty set. Zero elements were explicitly excluded by Axiom 19.

Now let L be a complete semilattice with join operation $\sqcup$ (corresponding to $\oplus$) and its intrinsic partial order $\sqsubseteq$ (corresponding to $\leq_i$). Call L *bottomless* if there is no zero element in L, and *atomic* if L satisfies Axiom 20. A structure satisfying Atomic Separation (Axiom 21) will be called *a-separating.* Since L is complete, Axiom 22 is also fulfilled when we interpret 'σ' as the arbitrary join operation $\bigsqcup$ in L. Since these joins are suprema L also satisfies Axioms 23 and 24. The plural star will be mapped onto that function on predicate extensions which assigns to an extension $\|P\|$ the set of all $y \in L$ such that $y = \bigsqcup(\|P\| \cap \{x \mid x \sqsubseteq y\})$. Then Axiom 25 is satisfied. Finally L is said to *have sup-prime atoms* if L satisfies the following analogue of Axiom 26, that is, with A the set of atoms in L interpreting the predicate At,

$$\text{(Ax. 26}'\text{)} \quad \forall u \in A[u \sqsubseteq \bigsqcup X \Rightarrow \exists b \in X \colon u \sqsubseteq b] \qquad (X \subseteq L)$$

We give the following definition to have a short name for the intended plural structures.

Definition 29 A *plural lattice* is a bottomless, a-separating, complete, semilattice with sup-prime atoms. Thus, a plural lattice is a CS^aP^- semilattice.

Let us take up the arrow notation for the set of all elements below a given element a.

Definition 30 $\quad a^{\downarrow} := \{x \in L \mid x \sqsubseteq a\}$

Proposition 15 *In a plural lattice L there is an atom below every element, i.e., L satisfies (Ax. 20).*

Proof. Let $x \in L$. If x is not an atom the set $x^{\downarrow} \setminus \{x\}$ is non-empty, with an element y in it, say. Then $\neg x \leq_i y$; hence, by a-separation, there is an atom u below x (which is not below y). □

Recall that in a bottomless separating semilattice there are at least two incomparable elements, which further implies that the set of atoms is non-empty. The semilattice L is called *sup-generated* if it satisfies Axiom 30,

[12] The numbering of axioms refers to Chapter 6.

i.e., if every element $a \in L$ is the supremum of all the atoms in L below a. With the above notation we can write this condition in the form

$$\text{(Ax. 30}'\text{)} \quad \forall a \in L\colon\ a = \bigsqcup a^{\downarrow} \cap A \qquad \textit{(sup-generation)}$$

Proposition 16 *An a-separating semilattice L is sup-generated, i.e., (Ax. 30′) holds. In particular, plural lattices are sup-generated.*

Proof. Let $a \in L$. Then a is trivially an upper bound for the atoms below it, so we have $\bigsqcup a^{\downarrow} \cap A \sqsubseteq a$ since the supremum is the least upper bound for a set. Conversely, assume $a \not\sqsubseteq \bigsqcup a^{\downarrow} \cap A$; then, by a-separation, there is an atom v below a which is not below $\bigsqcup a^{\downarrow} \cap A$. But this is absurd since $v \in a^{\downarrow} \cap A$. Hence we have equality. □

Definition 31 Let L be an a-separating semilattice and A the set of its atoms. Then L *has an injective supremum operator* if the following condition holds:

$$\text{(Ax. 31}'\text{)} \quad \forall X, Y \subseteq A\,[\bigsqcup X = \bigsqcup Y \Rightarrow X = Y]$$

Proposition 17 *A plural lattice has an injective supremum operator.*

Proof. Let L be a plural lattice, A the set of its atoms, $X, Y \subseteq A$, and assume $\bigsqcup X = \bigsqcup Y$. To show $X \subseteq Y$ consider $u \in X$; we have $u \sqsubseteq \bigsqcup X$ and hence $u \sqsubseteq \bigsqcup Y$. The atoms in L are sup-prime and u is an atom, so there is an element $y \in Y$ such that $u \sqsubseteq y$. But Y consists of atoms only, hence $u = y$ and $u \in Y$. The inclusion $Y \subseteq X$ is shown the same way, and the proposition follows. □

Proposition 18 *A plural lattice satisfies Axiom 29 under Definition 51.*

Proof. Let L be a plural lattice, A the set of its atoms, P a one-place predicate with extension $\|P\|$, and $a \in L$. Then a is in the extension of ${}^{D}P$ according to Definition (D.51) if

$$\forall u \in A[u \sqsubseteq a \Rightarrow u \in \|P\|]$$

But then $\{u \in A \mid u \sqsubseteq a\} = \{u \in A \mid u \sqsubseteq a \,\&\, u \in \|P\|\}$, and the suprema of the two sets are equal, too. By sup-generation, $a = \bigsqcup a^{\downarrow} \cap A$, thus $a = \bigsqcup a^{\downarrow} \cap A \cap \|P\|$. That yields the second condition on the right-hand side of Axiom 29; the first condition follows from Axiom 19. Conversely, if $a = \bigsqcup a^{\downarrow} \cap A \cap \|P\|$ by assumption and $a = \bigsqcup a^{\downarrow} \cap A$ by Axiom 30′ then $\{u \in A \mid u \sqsubseteq a\} = \{u \in A \mid u \sqsubseteq a \,\&\, u \in \|P\|\}$ follows from the injectivity of the supremum operator, and thus all atoms below a are in the extension of P, i.e., $a \in \|{}^{D}P\|$. □

We can summarize the results obtained so far in

Theorem 9 *Plural lattices satisfy Lønning's Axioms 29′, 30′ and 31′.*

Proposition 19 *A sup-generated semilattice L is also a-separating.*

Proof. Let $a, b \in L$ and $a \not\sqsubseteq b$; by sup-generation, $a = \bigsqcup a^{\downarrow} \cap A$ and $b = \bigsqcup b^{\downarrow} \cap A$. Now we have

$$a \sqsubseteq b \quad \text{iff} \quad a^{\downarrow} \cap A \subseteq b^{\downarrow} \cap A.$$

(If $a \sqsubseteq b$ and $x \in a^{\downarrow} \cap A$ then $x \sqsubseteq a$ and $x \in A$, so $x \sqsubseteq b$ and $x \in A$, hence $x \in b^{\downarrow} \cap A$. Conversely, the inclusion $a^{\downarrow} \cap A \subseteq b^{\downarrow} \cap A$ implies $\bigsqcup a^{\downarrow} \cap A \sqsubseteq \bigsqcup b^{\downarrow} \cap A$ by the monotonicity of the supremum operator, and sup-generation yields $a \sqsubseteq b$.)

Continuing with the main proof, $a \not\sqsubseteq b$ therefore means $a^{\downarrow} \cap A \not\subseteq b^{\downarrow} \cap A$. So there is an $x \in (a^{\downarrow} \cap A) \setminus (b^{\downarrow} \cap A)$, that is, $x \in A$, $x \sqsubseteq a$ and $x \not\sqsubseteq b$. So L is a-separating. □

Sup-generation and injectivity of the supremum operator also imply that the atoms are sup-prime.

Proposition 20 *A sup-generated semilattice L with an injective supremum operator has sup-prime atoms.*

Proof. We have to show the condition formulated in (Ax. 26′). Let A be the set of atoms in L, $u \in A$ and $X \subseteq L$ with $u \sqsubseteq \bigsqcup X$. For every $b \in X$ we have $b = \bigsqcup b^{\downarrow} \cap A$ by sup-generation. Define $Y_b := b^{\downarrow} \cap A$ for $b \in X$ and $Y := \bigcup_{b \in X} Y_b$. For all $b \in X$, Y_b is a subset of A, hence $Y \subseteq A$. We will show the following

Lemma 7 $\bigsqcup Y = \bigsqcup X$

To show that $\bigsqcup Y \sqsubseteq \bigsqcup X$ observe that every $b \in X$ is below $\bigsqcup X$, hence for a $v \in b^{\downarrow} \cap A$ we have $v \sqsubseteq b \sqsubseteq \bigsqcup X$ for all $b \in X$. So every $v \in \bigcup_{b \in X} b^{\downarrow} \cap A$ is below $\bigsqcup X$. Recall the definition of Y to note that therefore $\bigsqcup X$ is an upper bound for Y, hence $\bigsqcup Y \sqsubseteq \bigsqcup X$. Conversely, we have $b = \bigsqcup b^{\downarrow} \cap A$ for all $b \in X$ by sup-generation, so $b = \bigsqcup Y_b \sqsubseteq \bigsqcup \bigcup_{b \in X} Y_b = \bigsqcup Y$ for $b \in X$, whence $\bigsqcup X \sqsubseteq \bigsqcup Y$. Equality follows by antisymmetry of $\sqsubseteq$.

We can now finish the main proof. Define $Y' := Y \cup \{u\}$ with the above u. Then, since $u \sqsubseteq \bigsqcup X$, $\bigsqcup Y \sqsubseteq \bigsqcup Y' \sqsubseteq \bigsqcup X \sqsubseteq \bigsqcup Y$, so $\bigsqcup Y = \bigsqcup Y'$. Injectivity of the supremum operator yields $Y = Y'$, so we can conclude $u \in Y = \bigcup_{b \in X} Y_b$. But that means that there is a $b \in X$ such that $u \in b^{\downarrow} \cap A$, and therefore $u \sqsubseteq b$. So L has sup-prime atoms. □

We will now show that plural lattices, having sup-prime atoms, also have a property which links them to the class of free lattices.[13] This is

[13] See, for instance, (Grätzer, 1978, p. 32) and Landman (1991), Section 6.2.

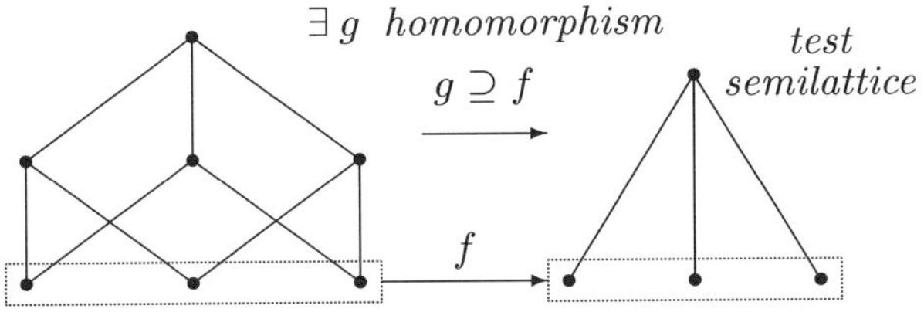

Figure 15.3: "Most general semilattice" with test semilattice

the *extension property* which was mentioned in Chapter 6 and is illustrated here in Figure 15.3. We will actually prove that the extension property is equivalent to the P-property, i.e., to sup-prime atomicity.

Recall that a semilattice homomorphism is called *complete* if it preserves the suprema of non-empty subsets.

Definition 32 An a-separating, complete semilattice L is said to have *the extension property with respect to its set of atoms* A if every mapping f from A into an arbitrary complete semilattice T can be uniquely extended to a complete semilattice homomorphism g from L into T.

Theorem 10 *Let L be a complete a-separating semilattice. Then L has the extension property with respect to its set of atoms iff L has sup-prime atoms. In particular, every plural lattice has the extension property.*

Proof. Let us first suppose that L has sup-prime atoms; let A be the set of its atoms, and f a mapping from A into an arbitrary complete "test semilattice" T. Since by Proposition 16 L is sup-generated we have $a = \bigsqcup a^{\downarrow} \cap A$ for all $a \in L$. Define a function g from L into T by

$$g(a) = \bigsqcup\{f(u) \mid u \in a^{\downarrow} \cap A\} = \bigsqcup f[a^{\downarrow} \cap A] \quad (a \in L)$$

For atoms u we have $g(u) = \bigsqcup\{f(u)\} = f(u)$, so g extends f. We are going to show that g is a complete semilattice homomorphism, that is, for all non-empty $Y \subseteq L$:

$$g(\bigsqcup Y) = \bigsqcup g[Y]$$

where $g[Y]$ denotes the set of g-values of elements of Y. We first show $\bigsqcup g[Y] \sqsubseteq g(\bigsqcup Y)$. For $y \in Y$ and every $u \in y^{\downarrow} \cap A$ we have $u \sqsubseteq y \sqsubseteq \bigsqcup Y$, hence $u \in (\bigsqcup Y)^{\downarrow} \cap A$ and $f(u) \in f[(\bigsqcup Y)^{\downarrow} \cap A]$. Therefore, $f[y^{\downarrow} \cap A] \subseteq f[(\bigsqcup Y)^{\downarrow} \cap A]$. But that means that for all $y \in Y$, $g(y) = \bigsqcup f[y^{\downarrow} \cap A] \sqsubseteq \bigsqcup f[(\bigsqcup Y)^{\downarrow} \cap A] = g(\bigsqcup Y)$. Thus, $g(\bigsqcup Y)$ is an upper bound for $g[Y]$, whence $\bigsqcup g[Y] \sqsubseteq g(\bigsqcup Y)$.

For the other direction consider a $u \in (\bigsqcup Y)^{\downarrow} \cap A$; then $u \in A$ and $u \sqsubseteq \bigsqcup Y$. By our assumption L has sup-prime atoms, so there is a $z \in Y$ such that $u \sqsubseteq z$. So $u \in z^{\downarrow} \cap A$, and $f(u) \in f[z^{\downarrow} \cap A] \subseteq \bigcup_{y \in Y} f[y^{\downarrow} \cap A]$. It follows that $f[(\bigsqcup Y)^{\downarrow} \cap A] \subseteq \bigcup_{y \in Y} f[y^{\downarrow} \cap A]$. Thus we obtain

$$\begin{aligned} g(\bigsqcup Y) &= \bigsqcup f[(\bigsqcup Y)^{\downarrow} \cap A] \\ &\sqsubseteq \bigsqcup \bigcup_{y \in Y} f[y^{\downarrow} \cap A] \\ &= \bigsqcup \{ \bigsqcup f[y^{\downarrow} \cap A] \mid y \in Y \} \\ &= \bigsqcup g[Y] \end{aligned}$$

The uniqueness of g follows directly from the fact that L is sup-generated.

Let us now suppose that L has the extension property; we have to show that L also has sup-prime atoms. Assume for *reductio* that there is a subset $X \in L$ and an atom $v \in A$ such that $v \sqsubseteq \bigsqcup X$ and for all $x \in X$, $v \not\sqsubseteq x$. Choose a three-element test semilattice $T = \{ b, c, d \}$ with $d = b \sqcup c$ and define a mapping f from A to T by setting $f(v) = c$ and $f(u) = b$ for all $u \in A$ with $u \neq v$. By the extension property f can be extended to a complete semilattice homomorphism g from L to T. Define $a := \bigsqcup X$; the we have

$$g(a) = g(\bigsqcup X) = g(\bigsqcup \{ x \mid x \in X \}) = \bigsqcup \{ g(x) \mid x \in X \}$$

Now since L is a-separating it is also sup-generated (by Proposition 16), and we have $x = \bigsqcup x^{\downarrow} \cap A$ for all $x \in X$. For any such $x \in X$ no atom below x equals v, so f sends the whole of $x^{\downarrow} \cap A$ onto $b \in T$. Then we get, because g extends f:

$$g(x) = g(\bigsqcup x^{\downarrow} \cap A) = \bigsqcup g[x^{\downarrow} \cap A] = \bigsqcup f[x^{\downarrow} \cap A] = \bigsqcup \{b\} = b$$

This holds for all $x \in X$, and hence

$$g(a) = \bigsqcup \{ g(x) \mid x \in X \} = \bigsqcup \{b\} = b$$

But on the other hand the atom v is below a, i.e., $v \in a^{\downarrow} \cap A$, and we can calculate

$$\begin{aligned} g(a) &= g(\bigsqcup a^{\downarrow} \cap A) = \bigsqcup g[a^{\downarrow} \cap A] \\ &= \bigsqcup f[a^{\downarrow} \cap A] \\ &= \bigsqcup (\, f[(a^{\downarrow} \cap A) \setminus \{v\}] \cup f(v)\,) \\ &= \bigsqcup f[(a^{\downarrow} \cap A) \setminus \{v\}] \sqcup f(v) \\ &= \bigsqcup \{b\} \sqcup c = b \sqcup c = d \end{aligned}$$

But that is a contradiction since $b \neq d$; so L has sup-prime atoms. □

We now give a formulation of the representation theorem for plural lattices that was announced in Chapter 6. According to it plural lattices are, up to isomorphism, power sets minus the empty set. Since power sets are the standard models for complete atomic Boolean algebras (CAB) it follows that at least as far as the present setup goes, the original line of taking CABs minus zero turns out to be justified on quite intuitive grounds. The proof of the theorem is mathematical folklore, but is given here for completeness of exposition.

Theorem 11 (Representation Theorem for Plural Lattices) *Let L be a plural lattice, and A the set of its atoms. Let L' be the powerset 2^A of A minus the empty set, considered as semilattice with set-theoretic union as join operation. Then there exists a complete semilattice isomorphism from L to L'.*

Proof. Define a function h from L into $2^A \setminus \{\emptyset\}$ by

$$h(a) \;=\; a^{\downarrow} \cap A \qquad (a \in L)$$

To show that h is surjective, take a $B \in 2^A \setminus \{\emptyset\}$, that is, $B \subseteq A$ and $B \neq \emptyset$. Then there is a supremum $\bigsqcup B$ for B in L since L is complete. By sup-generation, we have $\bigsqcup B = \bigsqcup (\bigsqcup B)^{\downarrow} \cap A$. But the supremum operator is injective, so

$$B \;=\; (\bigsqcup B)^{\downarrow} \cap A \;=\; h(\bigsqcup B)$$

.

Thus h is surjective. To show that h is also injective, assume $h(a) = h(b)$ for $a, b \in L$. Then $a^{\downarrow} \cap A = b^{\downarrow} \cap A$ by the definition of h, hence $\bigsqcup a^{\downarrow} \cap A = \bigsqcup b^{\downarrow} \cap A$, and by sup-generation, $a = b$.

Finally we show that h is a complete homomorphism, that is,

$$h(\bigsqcup X) \;=\; \bigcup \{h(a) \mid a \in L\}$$

for all non-empty subsets X of L. For the inclusion from left to right take a $u \in h(\bigsqcup X) = (\bigsqcup X)^{\downarrow} \cap A$. Then $u \in A$ and $u \sqsubseteq \bigsqcup X$. Since L has sup-prime atoms there is an $a \in X$ such that $u \sqsubseteq a$, so $u \in a^{\downarrow} \cap A = h(a)$, and

$u \in \bigcup\{h(a) \mid a \in L\}$. Conversely, such a u is an element of $h(a) = a^{\downarrow} \cap A$ for some $a \in X$, and $u \sqsubseteq a \sqsubseteq \bigsqcup X$. Therefore $u \in (\bigsqcup X)^{\downarrow} \cap A = h(\bigsqcup X)$, and the equality is proved. So h is a bijective complete homomorphism and hence a complete semilattice isomorphism. □

Bibliography

Aczel, Peter: 1989. Algebraic semantics for intensional logics, I. In: Gennaro Chierchia *et al.* (eds.), *Properties, Types and Meaning. Volume I: Foundational Issues*, Dordrecht: Kluwer Academic Publishers, 17–45.

Allan, Keith: 1977. Classifiers. *Language* 53:285–311.

Allan, Keith: 1980. Nouns and countability. *Language* 56:541–567.

Allgayer, Jürgen and Carola Reddig-Siekmann: 1990. What KL-ONE lookalikes need to cope with natural language. In: Karl H. Bläsius *et al.* (eds.), *Sorts and Types in Artificial Intelligence*, Berlin: Springer, Springer Lecture Notes in Artificial Intelligence No. 418, 240–285.

Altham, James E.J.: 1971. *The Logic of Plurality*. London: Methuen.

Anscombe, Gertrude E. M.: 1957. *Intention*. Oxford: Blackwell.

Aone, Chinatsu: 1991. *Treatment of Plurals and Collective-Distributive Ambiguity in Natural Language Understanding*. Ph.D. thesis, University of Texas at Austin. MCT Technical Report Number ACT-NL-155-91, Austin, Texas.

Armstrong, David M.: 1978. *Universals and Scientific Realism. 2 vols.*. Cambridge: Cambridge University Press.

Armstrong, David M.: 1989. *Universals. An Opinionated Introduction*. Boulder, CO: Westview Press.

Armstrong, David M.: 1997. *A World of States of Affairs*. Cambridge: Cambridge University Press.

Austin, John L.: 1950. Truth. *Proceedings of the Aristotelian Society. Supp. vol. xxiv* Also in J. O. Urmson and others (eds.), 1961, *J. L. Austin: Philosophical Papers* (Oxford UP: Oxford), 117–133.

Bach, Emmon: 1986. The algebra of events. *Linguistics and Philosophy* 9:5–16.

Bach, Emmon, Eloise Jelinek, Angelika Kratzer, and Barbara H. Partee (eds.): 1995a. *Quantification in Natural Languages*, vol. 1 of *Studies in Linguistics and Philosophy*. Dordrecht: Kluwer.

Bach, Emmon, Eloise Jelinek, Angelika Kratzer, and Barbara H. Partee (eds.): 1995b. *Quantification in Natural Languages*, vol. 2 of *Studies in Linguistics and Philosophy*. Dordrecht: Kluwer.

Bacon, John: 1997. Tropes. *Stanford Encyclopedia of Philosophy* Fall 1997:(http://plato.stanford.edu/archives/fall1997/entries/tropes/tropes.html). Cited October 8, 1997.

Balbes, Raymond and Philip Dwinger: 1974. *Distributive Lattices*. Columbia, Missouri: University of Missouri Press.

Ballmer, Thomas T. and Manfred Pinkal (eds.): 1983. *Approaching Vagueness*. Amsterdam: North Holland.

Bartsch, Renate: 1973. The semantics and syntax of number and numbers. In: John P. Kimball (ed.), *Syntax and Semantics. Vol. 2.*, New York: Academic Press, 51–93.

Bartsch, Renate, Johan van Benthem, and Peter van Emde Boas: 1989. *Semantics and Contextual Expression*. Dordrecht: Foris.

Barwise, Jon: 1979. On branching quantifiers in English. *Journal of Philosophical Logic* 8:47–80.

Barwise, Jon: 1987. Noun phrases, generalized quantifiers and anaphora. In: P. Gärdenfors (ed.), *Generalized Quantifiers. Linguistic and Logical Approaches*, Dordrecht: Reidel, 1–29.

Barwise, Jon and Robin Cooper: 1981. Generalized quantifiers and natural language. *Linguistics and Philosophy* 4:159–219.

Barwise, Jon, Jean M. Gawron, Gordon Plotkin, and Syun Tutiya (eds.): 1991. *Situation Theory and Its Applications. Vol. 2*. CSLI Lecture Notes No. 26. Stanford: CSLI.

Barwise, Jon and John Perry: 1983. *Situations and Attitudes*. Cambridge, MA: MIT Press.

Bäuerle, Rainer, Christoph Schwarze, and Arnim von Stechow (eds.): 1983. *Meaning, Use, and Interpretation of Language*. Berlin: De Gruyter.

Bealer, George: 1975. Predication and matter. *Synthese* 31:493–508. Also in: F. J. Pelletier (ed.), 1979, *Mass Terms: Some Philosophical Problems* (Reidel: Dordrecht), 279–294.

Bealer, George: 1982. *Quality and Concept*. Oxford: Clarendon Press.

Beeson, Michael: 1985. *Foundations of Constructive Mathematics*. Berlin: Springer.

Bencivenga, Ermanno: 1976. Set theory and free logic. *Journal of Philosophical Logic* 5:1–15.

Bencivenga, Ermanno: 1980. Free semantics for definite descriptions. *Logique et Analyse* 23:393–405.

Bencivenga, Ermanno: 1986. Free logics. In: D. Gabbay *et al.* (eds.), *Handbook of Philosophical Logic, vol. III*, Dordrecht: Reidel, chap. 6, 373–426.

Bennett, Jonathan: 1981. Killing and letting die. In: S. McMurrin (ed.), *The Tanner Lectures on Human Value II*, Salt Lake City: University of Utah Press, 45–116.

Bennett, Jonathan: 1984. Counterfactuals and temporal direction. *Philosophical Review* 93:57–91.

Bennett, Jonathan: 1985. Adverb-dropping inferences and the Lemmon criterion. In: E. LePore and B. McLaughlin (eds.), *Actions and Events. Perspectives on the Philosophy of Donald Davidson*, Oxford: Basil Blackwell, chap. 14, 193–206.

Bennett, Jonathan: 1987. Event causation: The counterfactual analysis. *Philosophical Perspectives* 1.

Bennett, Jonathan: 1988. *Events and Their Names*. Oxford: Clarendon Press.

Bennett, Michael R.: 1972. Accommodating the plural in Montague's fragment of English. In: R. Rodman (ed.), *Montague Grammar*. UCLA, Los Angeles, 25–65.

Bennett, Michael R.: 1975. *Some Extensions of a Montague Fragment of English*. Ph.D. thesis, UCLA. Distributed by the Indiana University Linguistics Club.

Bennett, Michael R.: 1979. Mass nouns and mass terms in Montague grammar. In: S. Davis *et al.* (eds.), *Linguistics, Philosophy, and Montague Grammar*, Austin, Texas: University of Texas Press, 263–285.

van Benthem, Johan: 1983a. Determiners and logic. *Linguistics and Philosophy* 6:447–478.

van Benthem, Johan: 1983b. *The Logic of Time. A Model-Theoretic Investigation into the Varieties of Temporal Ontology and Temporal Discourse.* Synthese Library vol. 156. Dordrecht: Reidel.

van Benthem, Johan and Alice ter Meulen (eds.): 1984. *Generalized Quantifiers in Natural Language.* Dordrecht: Foris.

van Benthem, Johan and Alice ter Meulen (eds.): 1997. *Handbook of Logic and Language.* Amsterdam: Elsevier.

Biermann, Anna: 1981. Die grammatische Kategorie Numerus. In: H. Seiler *et al.* (eds.), *Apprehension. Das sprachliche Erfassen von Gegenständen*, Tübingen: Narr, 229–243.

Black, Max: 1971. The elusiveness of sets. *Review of Metaphysics* 24:614–636.

Bläsius, Karl H., Ulrich Hedtstück, and Claus R. Rollinger (eds.): 1990. *Sorts and Types in Artificial Intelligence.* Lecture Notes in Artificial Intelligence 418. Berlin: Springer.

Blau, Ulrich: 1978. *Die dreiwertige Logik der Sprache. Ihre Syntax, Semantik und Anwendung in der Sprachanalyse.* Berlin: de Gruyter.

Blau, Ulrich: 1981a. Abstract objects. *Theoretical Linguistics* 8:131–144.

Blau, Ulrich: 1981b. Collective objects. *Theoretical Linguistics* 8:101–130.

Blau, Ulrich: 1983. Three-valued analysis of of precise, vague, and presupposing quantifiers. In: T. Ballmer *et al.* (eds.), *Approaching Vagueness*, Amsterdam: North Holland, 79–129.

Blau, Ulrich: 1985. Die Logik der Unbestimmheiten und Paradoxien (Kurzfassung). *Erkenntnis* 22:369–459.

Boolos, George: 1971. The iterative conception of set. *Journal of Philosophy* 68:215–231.

Boolos, George: 1975. On second-order logic. *Journal of Symbolic Logic* 72:509–527.

Boolos, George: 1984a. Nonfirstorderizability again. *Linguistic Inquiry* 15:343.

Boolos, George: 1984b. To be is to be a value of a variable (or to be some values of some variables). *Journal of Philosophy* 81:430–449.

Boolos, George: 1985a. Nominalist platonism. *Philosophical Review* 94:327–344.

Boolos, George: 1985b. Reading the 'Begriffsschrift'. *Mind* 94:331–344.

Brady, Michael and Robert C. Berwick (eds.): 1984. *Computational Models of Discourse*. Cambridge, MA: M.I.T. Press.

Brand, Myles: 1977. Identity conditions for events. *Americal Philosophical Quarterly* 14:329–377.

Bunt, Harry C.: 1979. Ensembles and the formal semantic properties of mass terms. In: Francis J. Pelletier (ed.), *Mass Terms: Some Philosophical Problems*, Dordrecht: Reidel, 279–294.

Bunt, Harry C.: 1981. *The Formal Semantics of Mass Terms*. Ph.D. thesis, University of Amsterdam.

Bunt, Harry C.: 1985a. The formal representation of (quasi-)continuous concepts. In: R. Moore *et al.* (eds.), *Formal Theories of the Common Sense World*, Norwood, NJ: Ablex, 37–70.

Bunt, Harry C.: 1985b. *Mass Terms and Model-Theoretic Semantics*. Cambridge: Cambridge University Press.

Burge, Tyler: 1972. Truth and mass terms. *Journal of Philosophy* 69:263–282.

Burge, Tyler: 1975. Mass terms, count terms, and change. *Synthese* 31:459–478. Also in: F. J. Pelletier (ed.), 1979, *Mass Terms: Some Philosophical Problems* (Reidel: Dordrecht), 199–218.

Burge, Tyler: 1977. A theory of aggregates. *Noûs* 11:97–117.

Burgess, John and Gideon Rosen: 1997. *A Subject with No Object*. Oxford: Oxford University Press.

Burke, Michael: 1992. Copper statues and pieces of copper: a challenge to the standard account. *Analysis* 52:12–17.

Burke, Michael: 1994a. Dion and Theon: an essentialist solution to an ancient puzzle. *Journal of Philosophy* 91:129–139.

Burke, Michael: 1994b. Preserving the principle of one object to a place: a novel account of the relations among objects, sorts, sortals, and persistence conditions. *Philosophy and Phenomenological Research* 54:591–624.

Campbell, Keith: 1976. *Metaphysics: An Introduction*. Encino, CA: Dickenson.

Carlson, Gregory N.: 1977. A unified analysis of the English bare plural. *Linguistics and Philosophy* 1:413–457.

Carlson, Gregory N.: 1978. *Reference to Kinds in English*. Ph.D. thesis, University of Massachusetts, Amherst. Published (1980). Garland: New York.

Carlson, Gregory N.: 1982a. Generic terms and generic sentences. *Journal of Philosophical Logic* 11:145–181.

Carlson, Gregory N. and Francis Jeffry Pelletier (eds.): 1995. *The Generic Book*. Chicago: University of Chicago Press.

Carlson, Lauri: 1980. Plural quantification. MIT, Cambridge, MA.

Carlson, Lauri: 1982b. Plural quantifiers and informational independence. *Acta Philosophica Fennica* 35:163–174.

Cartwright, Helen M.: 1963. *Classes, Quantities and Non-singular Reference*. Ph.D. thesis, University of Michigan.

Cartwright, Helen M.: 1970. Quantities. *Philosophical Analysis* 79:25–42.

Cartwright, Helen M.: 1979a. Amounts and measures of amount. In: F. J. Pelletier (ed.), *Mass Terms: Some Philosophical Problems*, Dordrecht: Reidel, 179–198.

Cartwright, Helen M.: 1979b. Some remarks about mass nouns and plurality. In: F. J. Pelletier (ed.), *Mass Terms: Some Philosophical Problems*, Dordrecht: Reidel, 31–46.

Chellas, Brian: 1975. Quantity and quantification. *Synthese* 31:487–491. Also in: F. J. Pelletier (ed.), 1979, *Mass Terms: Some Philosophical Problems* (Reidel: Dordrecht), 227–231.

Chierchia, Gennaro: 1982. Nominalization and Montague grammar. *Linguistics and Philosophy* 5:303–354.

Chierchia, Gennaro, Barbara H. Partee, and Raymond Turner (eds.): 1989a. *Properties, Types and Meaning. Volume I: Foundational Issues*. Dordrecht: Kluwer Academic Publishers.

Chierchia, Gennaro, Barbara H. Partee, and Raymond Turner (eds.): 1989b. *Properties, Types and Meaning. Volume II: Semantic Issues*. Dordrecht: Kluwer Academic Publishers.

Chihara, Charles S.: 1990. *Constructibility and Mathematical Existence*. Oxford: Clarendon.

Chisholm, Roderick M.: 1970. Events and propositions. *Noûs* 4:15–24.

Chisholm, Roderick M.: 1976. *Person and Object: A Metaphysical Study*. London: Allen and Unwin.

Chisholm, Roderick M.: 1996. *A Realistic Theory of Categories. An Essay on Ontology*. Cambridge: Cambridge University Press.

Choe, Jae Woong: 1987a. *Anti-Quantifiers and A Theory of Distributivity*. Ph.D. thesis, University of Massachusetts, Amherst.

Choe, Jae Woong: 1987b. A theory of distributivity. In: J. Groenendijk *et al.* (eds.), *Proceedings of the Sixth Amsterdam Colloquium*. ITLI, Amsterdam, 21–41.

Chomsky, Noam: 1965. *Aspects of the Theory of Syntax*. Cambridge: MIT Press.

Christian, William A.: 1959. *An Interpretation of Whitehead's Metaphysics*. New Haven: Yale University Press.

Clarke, Bowman L.: 1981. A calculus of individuals based on 'connection'. *Notre Dame Journal of Formal Logic* 22:204–218.

Clarke, D. S. Jr.: 1970. Mass terms as subjects. *Philosophical Studies* 21:25–29.

Clay, Robert E.: 1974. Relation of Leśniewski's mereology to Boolean algebra. *Journal of Symbolic Logic* 39:638–348.

Cocchiarella, Nino: 1976. On the logic of natural kinds. *Philosophy of Science* 43:202–222.

Cook, Cathleen: 1975. On the usefulness of quantities. *Synthese* 31:443–457. Also in: F. J. Pelletier (ed.), 1979, *Mass Terms: Some Philosophical Problems* (Reidel: Dordrecht), 121–135.

Cooper, Robin, Kuniaki Mukai, and John Perry (eds.): 1990. *Situation Theoriy and Its Applications. Vol. 1*. CSLI Lecture Notes No. 22. Stanford: CSLI.

Cowell, Marc: 1964. *A Reference Grammar of Syrian Arabic*. Washington, D.C.: Georgetown University Press.

Cresswell, Max: 1985. Review of Landman and Veltman (eds.), 'Varieties of formal semantics'. *Linguistics* 23.

Dahl, Östen: 1975. On generics. In: E. Keenan (ed.), *Formal Semantics of Natural Language*, Cambridge: Cambridge University Press, 99–111.

Dalrymple, Mary, Makoto Kanazawa, Yookyung Kim, Sam Mchombo, and Stanley Peters: 1997. Reciprocal expressions and the concept of reciprocity. *Linguistics and Philosophy* To appear.

Davey, B.A. and Hilary A. Priestley: 1990. *Introduction to Lattices and Order*. Cambridge: Cambridge University Press.

Davidson, Donald: 1967a. Causal relations. *Journal of Philosophy* 64:691–703. Also in: D. Davidson, 1980, *Essays on Actions and Events* (Clarendon: Oxford), 149–162.

Davidson, Donald: 1967b. The logical form of action sentences. In: Nicolas Rescher (ed.), *The Logic of Decision and Action*, Pittsburgh: The University of Pittsburgh Press, 81–95. Also in: D. Davidson, 1980, *Essays on Actions and Events* (Clarendon: Oxford), 105–122.

Davidson, Donald: 1969. The individuation of events. In: N. Rescher (ed.), *Essays in Honor of Carl G. Hempel*, Dordrecht: Reidel, 216–234. Also in: D. Davidson, 1980, *Essays on Actions and Events* (Clarendon: Oxford), 163–180.

Davidson, Donald: 1970. Events as particulars. *Noûs* 4. Also in: D. Davidson, 1980, *Essays on Actions and Events* (Clarendon: Oxford), pp. 181–187.

Davidson, Donald: 1980. *Essays on Actions and Events*. Oxford: Clarendon Press.

Davies, Martin: 1989. *Two Examiners Marked Six Scripts.* Interpretations of numerically quantified sentences. *Linguistics and Philosophy* 12:293–323.

Davis, Steven and Marianne Mithun (eds.): 1979. *Linguistics, Philosophy, and Montague Grammar*. Austin, Texas: University of Texas Press.

Denkel, Arda: 1995. Theon's tale: Does a Cambridge change result in a substantial change? *Analysis* 55:166–170.

van der Does, Jaap M.: 1991. Among collections. In: J. van der Does (ed.), *Quantification and Anaphora II*, Edinburgh: Centre for Cognitive Science, University of Edinburgh, 1–35.

van der Does, Jaap M.: 1992. *Applied Quantifier Logics. Collectives. Naked Infinitives*. Ph.D. thesis, University of Amsterdam.

van der Does, Jaap M.: 1993. Sums and quantifiers. *Linguistics and Philosophy* 16:509–550.

van der Does, Jaap M.: 1994. On complex plural noun phrases. In: M. Kanazawa *et al.* (eds.), *Dynamics, Polarity and Quantification*, Stanford: CSLI Publications, 81–115.

van der Does, Jaap M. and Jan van Eijck (eds.): 1996. *Quantifiers, Logic, and Language*. CSLI Lecture Notes No. 54. Stanford: CSLI Publications.

Dougherty, Ray C.: 1970. A grammar of coordinate conjoined structures: I. *Language* 46:850–898.

Dougherty, Ray C.: 1974. The syntax and semantics of *each other* constructions. *Foundations of Language* 12:1–47.

Dowty, David: 1979. *Word Meaning and Montague Grammar. The Semantics of Verbs and Times in Generative Semantics and in Montague's PTQ*. Dordrecht: Reidel.

Dowty, David: 1985. A unified indexical analysis of *same* and *different*. Ohio State University.

Dowty, David: 1986. A note on collective predicates, distributive predicates and *all*. In: *Proceedings of the Third Eastern States Conference on Linguistics (ESCOL 86)*. Ohio State University, 97–115.

Dowty, David: 1989. On the semantic content of the notion of 'thematic role'. In: G. Chierchia *et al.* (eds.), *Properties, Types and Meaning. Volume II: Semantic Issues*, Dordrecht: Kluwer Academic Publishers, 69–129.

Dowty, David and Belinda Brodie: 1984. The semantics of "floated" quantifiers in a transformationless grammar. In: Mark Cobler *et al.* (eds.), *Proceedings of the West Coast Conference in Formal Linguistics, vol. 3*. Stanford Linguistics Association, Stanford, 75–90.

Dowty, David, Robert Wall, and Stanley Peters: 1981. *Introduction to Montague Semantics*. Synthese Library vol. 11. Dordrecht: Reidel.

Dretske, Fred I.: 1967. Can events move? *Mind* 76:479–792.

Dretske, Fred I.: 1977. Referring to events. *Midwest Studies in Philosophy* 2:90–99.

Eberle, Kurt: 1989. *Quantifikation, Plural, Ereignisse und ihre Argumente in einer mehrsortigen Sprache der Prädikatenlogik erster Stufe.* IWBS Report 67, IBM Deutschland, Institut für Wissensbasierte Systeme, Stuttgart.

Eberle, Rolf: 1967. Some complete calculi of individuals. *Notre Dame Journal of Formal Logic* 8:267–278.

Eberle, Rolf: 1969. Denotationless terms and predicates expressive of positive qualities. *Theoria* 35, part 2:104–123.

Eberle, Rolf A.: 1970. *Nominalistic Systems*. Dordrecht: Reidel.

Ehrich, Veronika and Heinz Vater (eds.): 1988. *Temporalsemantik. Beiträge zur Linguistik der Zeitreferenz*. Tübingen.

van Eijck, Jan: 1983. Discourse representation theory and plurality. In: Alice ter Meulen (ed.), *Studies in Model-Theoretic Semantics*, Dordrecht: Foris.

Erné, Marcel: 1982. *Einführung in die Ordnungstheorie*. Mannheim: Bibliographisches Institut.

Evans, Gareth: 1978. Can there be vague objects? *Analysis* 38:208.

Faye, J. *et al.* (eds.): 1997. *Perspectives on Time*. Dordrecht: Kluwer.

Feferman, Solomon: 1995. Definedness. *Erkenntnis* 43:295–320.

Fiengo, Robert and Howard Lasnik: 1973. The logical structure of reciprocal sentences in English. *Foundations of Language* 9:447–468.

Fine, Kit: 1985. *Reasoning with Arbitrary Objects*. Oxford: Basil Blackwell.

Forrest, Peter: 1996a. From ontology to topology in the theory of regions. *The Monist* 79:34–50.

Forrest, Peter: 1996b. How innocent is mereology? *Analysis* 56:127–131.

van Fraassen, Bas C.: 1966a. The completeness of free logic. *Zeitschr. für math. Logik und Grundlagen d. Math.* 12:219–234.

van Fraassen, Bas C.: 1966b. Singular terms, truthvalue gaps, and free logic. *Journal of Philosophy* 63:481–495.

Frege, Gottlob: 1884. *Die Grundlagen der Arithmetik. Eine logisch mathematische Untersuchung über den Begriff der Zahl.* Breslau: Koebner. Bilingual edition: *The Foundations of Arithmetic. A logico-mathematical enquiry into the concept of number*. German Text with English Translation by J. L. Austin. Blackwell: Oxford ^{2}1974. Quotes are from this edition.

Frege, Gottlob: 1885. Kritische Beleuchtung einiger Punkte in E. Schröders 'Vorlesungen über die Algebra der Logik'. *Archiv für systematische Philosophie* 1:433–456. English translation in Frege (1970).

Frege, Gottlob: 1970. *Translations from the Philosophical Writings of Gottlob Frege*. Oxford: Blackwell.

French, Steven and Décio Krause: 1995. Vague identity and quantum nonindividuality. *Analysis* 55:20–26.

Furet, François: 1981. *Interpreting the French Revolution*. Cambridge: Cambridge University Press.

Gabbay, Dov and Julius M. E. Moravcsik: 1973. Sameness and individuation. *Journal of Philosophy* 70. Also in: F. J. Pelletier (ed.), 1979, *Mass Terms: Some Philosophical Problems* (Reidel: Dordrecht), 233–247.

Gabbay, Dov M. and Franz Guenthner (eds.): 1986. *Handbook of Philosophical Logic. Volume III: Alternatives to Classical Logic*. Dordrecht: Reidel.

Gabbay, Dov M. and Franz Guenthner (eds.): 1989. *Handbook of Philosophical Logic. Volume IV: Topics in the Philosophy of Language*. Dordrecht: Reidel.

Gärdenfors, Peter (ed.): 1987. *Generalized Quantifiers. Linguistic and Logical Approaches*. Dordrecht: Reidel.

Gerstner-Link, Claudia: 1995. *Über Generizität. Generische Nominalphrasen in singulären und generellen Aussagen*. Studien zur Theoretischen Linguistik, vol. 13. München: Fink.

Gerstner-Link, Claudia and Manfred Krifka: 1993. Genericity. In: Joachim Jacobs *et al.* (eds.), *Syntax: ein internationales Handbuch zeitgenössischer Forschung / An Internatinal Handbook of Contemporary Research*, Berlin: de Gruyter, 966–978.

Gierz, Gerhard, Karl Heinz Hofmann, Klaus Keimel, Jimmie D. Lawson, Michael W. Mislove, and Dana S. Scott: 1980. *A Compendium of Continuous Lattices*. Berlin-Heidelberg-New York: Springer.

Gil, David: 1982a. *Distributive Numerals*. Ph.D. thesis, UCLA.

Gil, David: 1982b. Quantifier scope, linguistic variation, and natural language semantics. *Linguistics and Philosophy* 5:421–472.

Gil, David: 1988. Georgian reduplication and the domain of distributivity. *Linguistics* 26:1039–1065.

Gil, David: 1995. Universal quantifiers and distributivity. In: E. Bach *et al.* (eds.), *Quantification in Natural Languages, vol. 1*, Dordrecht: Kluwer, chap. 10, 321–362.

Gillon, Brendan: 1987. The readings of plural noun phrases in English. *Linguistics and Philosophy* 10:199–219.

Gillon, Brendan: 1990. Plural noun phrases and their readings: A reply to Lasersohn. *Linguistics and Philosophy* 13:477–485.

Goldblatt, Robert: 1984. *Topoi. The Categorial Analysis of Logic*. Amsterdam: North-Holland. Revised edition.

Goldman, Alvin A.: 1971. The individuation of actions. *Journal of Philosophy* 68:761–774.

Goodman, Nelson: 1966. *The Structure of Appearance*. Indianapolis: Bobbs-Merrill, 2nd edn.

Grandy, Richard E.: 1975. Stuff and things. *Synthese* 31:479–485. Also in: F. J. Pelletier (ed.), 1979, *Mass Terms: Some Philosophical Problems*, 219–225.

Grätzer, George: 1978. *General Lattice Theory*. Basel: Birkhäuser.

Groenendijk, Jeroen, Theo M. V. Janssen, and Martin Stokhof (eds.): 1981. *Formal Methods in the Study of Language*, Amsterdam. Mathematical Centre.

Groenendijk, Jeroen, Dick de Jongh, and Martin Stokhof (eds.): 1986a. *Foundations of Pragmatics and Lexical Semantics*. GRASS Series no. 7. Dordrecht: Foris.

Groenendijk, Jeroen, Dick de Jongh, and Martin Stokhof (eds.): 1986b. *Studies in Discourse Representation Theory and the Theory of Generalized Quantifiers*. GRASS Series no. 8. Dordrecht: Foris.

Groenendijk, Jeroen, Martin Stokhof, and Frank Veltman (eds.): 1987. *Proceedings of the Sixth Amsterdam Colloquium*, Amsterdam. ITLI.

Grossmann, Reinhardt: 1992. *The Existence of the World. An Introduction to Ontology*. London: Routledge.

Guarino, Nicola and Roberto Poli (eds.): 1995. *Special Issue on Formal Ontology in Conceptual Analysis and Knowledge Representation*. New York: Academic Press. *International Journal of Human-Computer Studies*.

Gupta, Anil: 1980. *The Logic of Common Nouns: An Investigation in Quantified Modal Logic*. New Haven: Yale University Press.

Hacker, P. M. S.: 1982. Events and objects in space and time. *Mind* XCI:1–19.

Halmos, Paul R.: 1960. *Naive Set Theory*. Princeton: Van Nostrand.

Halmos, Paul R.: 1963. *Lectures on Boolean Algebras*. New York: Van Nostrand.

Hamm, Fritz: 1992. *Ereignisstrukturen und Nominalisierungen*. Arbeitspapiere des SFB 340 "Sprachtheoretische Grundlagen für die Computerlinguistik" Nr. 28 – 1992, Universität Stuttgart.

Hamm, Fritz and Erhard Hinrichs (eds.): 1997. *Plurality and Quantification*. Dordrecht: Kluver.

Hausser, Roland: 1974a. *Quantification in an Extended Montague Grammar*. Ph.D. thesis, University of Texas at Austin.

Hausser, Roland: 1974b. Syntax and semantics of plural. In: *Proceedings of the 10th Regional Meeting of the Chicago Linguistic Society*. 234–247.

Heim, Irene: 1983. File change semantics and the familiarity theory of definiteness. In: R. Bäuerle *et al.* (eds.), *Meaning, Use, and Interpretation of Language*, Berlin: De Gruyter, 164–198.

Heim, Irene, Howard Lasnik, and Robert May: 1991. Reciprocity and plurality. *Linguistic Inquiry* 22:63–101.

Hellman, Geoffrey: 1989. *Mathematics without Numbers. Towards a Modal-Structural Interpretation*. Oxford: Clarendon.

Hellman, Geoffrey: 1994. *Real* analysis without classes. *Philosophia Mathematica* 2:228–250.

Hendry, H. E.: 1982. Complete extensions of the calculus of individuals. *Noûs* 16:453–460.

Henkin, Leon: 1950. Completeness in the theory of types. *Journal of Symbolic Logic* 15:81–91.

Henkin, Leon: 1953. Some notes on nominalism. *Journal of Symbolic Logic* 18:19–29.

Herzog, Otthein and Claus Rollinger (eds.): 1991. *Text Understanding in LILOG: Integrating Computational Linguistics and Artificial Intelligence*. Lecture Notes in Artificial Intelligence, no. 546. Berlin-Heidelberg-New York: Springer.

Heyer, Gerhard: 1987. *Generische Kennzeichnungen. Zur Logik und Ontologie generischer Bedeutung*. München and Berlin: Philosophia.

Higginbotham, James: 1980. Reciprocal interpretation. *Journal of Linguistic Research* 1:97–117.

Higginbotham, James: 1995. Mass and count quantifiers. In: Emmon Bach *et al.* (eds.), *Quantification in Natural Languages, vol. 2*, Dordrecht: Kluwer, chap. 12, 383–419.

Hinrichs, Erhard: 1985. *A Compositional Semantics for Aktionsarten and NP Reference in English*. Ph.D. thesis, Ohio State University.

Hintikka, Jaakko: 1959. Existential presuppositions and existential commitments. *Journal of Philosophy* 56:125–137.

Hintikka, Jaakko: 1973. Partially ordered quantifiers vs. partially ordered ideas. *Dialectica* 30:89–99.

Hintikka, Jaakko, Julius M. E. Moravcsik, and Patrick Suppes (eds.): 1973. *Approaches to Natural Language*. Dordrecht: Reidel.

Hobbs, Jerry R. and Robert C. Moore (eds.): 1985. *Formal Theories of the Commonsense World*. Ablex Series in Artificial Intelligence. Norwood, NJ: Ablex.

Hoeksema, Jack: 1983. Plurality and conjunction. In: A. ter Meulen (ed.), *Studies in Model-Theoretic Semantics*, Dordrecht: Foris, 63–84.

Hoeksema, Jack: 1988. The semantics of non-Boolean 'and'. *Journal of Semantics* 6.

Hoepelman, Jakob and Christian Rohrer: 1980. On the mass–count distinction and the French imparfait and passé simple. In: C. Rohrer (ed.), *Time, Tense, and Quantifiers*, Tübingen: Niemeyer, 85–112.

Horgan, Terence: 1978. The case against events. *Philosophical Review* 87:28–47.

Hörmann, Hans: 1983. The calculating listener, or: how many are *einige*, *mehrere*, and *ein paar* (*some*, *several* and *a few*)? In: R. Bäuerle *et al.* (eds.), *Meaning, Use, and Interpretation of Language*, Berlin: De Gruyter, 221–234.

van Inwagen, Peter: 1990. Four-dimensional objects. *Noûs* 24:245–255.

van Inwagen, Peter: 1993. *Metaphysics*. Boulder, CO: Westview Press.

Janssen, Theo M. V.: 1981. Compositional semantics and relative clause formation in Montague grammar. In: J. Groenendijk *et al.* (eds.), *Formal Methods in the Study of Language, Part 1*, Mathematical Centre, 237–276.

Kadmon, Nirit: 1987. *On Unique and Non-Unique Reference and Asymmetric Quantification*. Ph.D. thesis, University of Massachusetts, Amherst.

Kadmon, Nirit: 1989. Uniqueness. *Linguistics and Philosophy* 13:273–324.

Kalish, Donald, Richard Montague, and Gary Mar: 1980. *Logic. Techniques of Formal Reasoning*. New York: Harcourt Brace Jovanovich, 2nd edn.

Kamp, Hans: 1980. Some remarks about the logic of change, Part I. In: C. Rohrer (ed.), *Time, Tense, and Quantifiers*, Tübingen: Niemeyer, 135–179.

Kamp, Hans and Uwe Reyle: 1993. *From Discourse to Logic. An Introduction to Modeltheoretic Semantics of Natural Language, Formal Logic, and Discourse Representation Theory*. Dordrecht: Kluwer.

Kanazawa, Makoto and Christopher J. Pinón (eds.): 1994. *Dynamics, Polarity and Quantification*. CSLI Lecture Notes No. 48. Stanford: CSLI Publications.

Keenan, Edward (ed.): 1975. *Formal Semantics of Natural Language*. Cambridge: Cambridge University Press.

Keenan, Edward: 1981. A Boolean approach to semantics. In: J. Groenendijk *et al.* (eds.), *Formal Methods in the Study of Language*. Mathematical Centre, Amsterdam.

Keenan, Edward: 1987. *Universal Grammar. 15 Essays*. London: Croom Helm.

Keenan, Edward and Leonard Faltz: 1985. *Boolean Semantics for Natural Language*. Dordrecht: Reidel.

Keenan, Edward and Yonathan Stavi: 1986. A semantic characterization of natural language determiners. *Linguistics and Philosophy* 9:253–326.

Kiefer, Ferenc and David M. Perlmutter (eds.): 1974. *Syntax und generative Grammatik*, vol. 1–3 of *Linguistische Forschungen*. Frankfurt/Main: Athenaion.

Kim, Jaegwon: 1969. Events and their description: some considerations. In: Nicholas Rescher (ed.), *Essays in Honor of Carl G. Hempel*, Dordrecht: Reidel, 198–215.

Kim, Jaegwon: 1973. Causation, nomic subsumption, and the concept of an event. *Journal of Philosophy* 70:217–236.

Kim, Jaegwon: 1980. Events as property exemplifications. In: Myles Brand and Douglas Walton (eds.), *Action Theory*, Dordrecht: Reidel, 159–177.

Kimball, John P. (ed.): 1973. *Syntax and Semantics. Vol. 2*. New York: Academic Press.

Kölver, Ulrike: 1982. Klassifikatorkonstruktionen in Thai, Vietnamesisch und Chinesisch. In: Hansjakob Seiler and Christian Lehmann (eds.), *Apprehension. Das sprachliche Erfassen von Gegenständen*, Tübingen: Narr, 160–185.

Krantz, David H., R. Duncan Luce, Patrick Suppes, and Amos Tversky: 1971. *Foundations of Measurement. Vol. I: Additive and Polynomial Representations*. New York: Academic Press.

Kratzer, Angelika: 1978. *Semantik der Rede. Kontexttheorie – Modalwörter – Konditionalsätze*. Königstein/Taunus: Scriptor.

Kratzer, Angelika: 1980. Die Analyse des bloßen Plural bei Gregory Carlson. *Linguistische Berichte* 70:47–50.

Kratzer, Angelika: 1989. Investigation into lumps of thought. *Linguistics and Philosophy* 12:607–653.

Kratzer, Angelika: 1995. Stage-level and individual-level predicates. In: G. N. Carlson and F. J. Pelletier (eds.), *The Generic Book*, Chicago: University of Chicago Press, chap. 2, 125–175.

Krifka, Manfred: 1987. Nominal reference and temporal constitution: Towards a semantics of quantity. In: J. Groenendijk *et al.* (eds.), *Proceedings of the Sixth Amsterdam Colloquium.* ITLI, Amsterdam, 153–173.

Krifka, Manfred: 1989a. Nominal reference and temporal constitution: Towards a semantics of quantity. In: R. Bartsch *et al.* (eds.), *Semantics and Contextual Expression*, Dordrecht: Foris, 75–115.

Krifka, Manfred: 1989b. *Nominalreferenz und Zeitkonstitution. Zur Semantik von Massentermen, Pluraltermen und Aspektklassen.* München: Fink.

Krifka, Manfred: 1991a. Four thousand ships passed through the lock: Object-induced measure functions on events. *Linguistics and Philosophy* 13:487–520.

Krifka, Manfred: 1991b. How to get rid of groups, using DRT: A case for discourse-oriented semantics. *Texas Linguistic Forum* 32:71–110.

Krifka, Manfred: 1991c. Massenausdrücke. In: A. von Stechow *et al.* (eds.), *Semantik / Semantics. Ein internationales Handbuch der zeitgenössischen Forschung / An International Handbook of Contemporary Research*, Berlin-New York: de Gruyter.

Krifka, Manfred: 1996. Parametrized sum individuals for plural anaphora. *Linguistics and Philosophy* 19.

Kripke, Saul: 1980. *Naming and Necessity.* Cambridge, MA: Harvard University Press.

von Kutschera, Franz: 1995. *Platons "Parmenides".* Berlin: de Gruyter.

Ladusaw, William: 1982. Semantic constraints on the English partitive construction. In: D. Flickinger *et al.* (eds.), *Proceedings of the First West Coast Conference on Formal Linguistics.* Stanford University, Stanford, CA, 231–242.

Lambert, Karel: 1981. On the philosophical foundations of free logic. *Inquiry* 24:147–203.

Lambert, Karel (ed.): 1991a. *Philosophical Applications of Free Logic.* New York – Oxford: Oxford University Press.

Lambert, Karel: 1991b. A theory of definite descriptions. In: K. Lambert (ed.), *Philosophical Applications of Free Logic*, Oxford: Oxford University Press, 17–27.

Lambert, Karel and Bas C. van Fraassen: 1972. *Derivation and Counterexample. An Introduction to Philosophical Logic*. Encino, CA: Dickenson.

Landman, Fred: 1987. Groups, plural, individuals and intensionality. In: J. Groenendijk *et al.* (eds.), *Proceedings of the Sixth Amsterdam Colloquium*. ITLI, Amsterdam, 197–217.

Landman, Fred: 1989. Groups. Part I, II. *Linguistics and Philosophy* 12:559–605; 723–744.

Landman, Fred: 1991. *Structures for Semantics*. Dordrecht: Kluwer Academic Publishers.

Landman, Fred and Frank Veltman (eds.): 1984. *Varieties of Formal Semantics. Proceedings of the Fourth Amsterdam Colloquium*. GRASS Series No. 3. Dordrecht: Foris.

Langendoen, Terence D.: 1978. The logic of reciprocity. *Linguistic Inquiry* 9:177–197.

van Langendonck, W.: 1972. Numerus-problemen. *Leuvense Bijdragen* 29.

Lasersohn, Peter N.: 1988. *A Semantics for Groups and Events*. Ph.D. thesis, Ohio State University, Columbus.

Lasersohn, Peter N.: 1989. On the readings of plural noun phrases. *Linguistic Inquiry* 20:130–134.

Lasersohn, Peter N.: 1990. Group action and spatio-temporal proximity. *Linguistics and Philosophy* 13:179–206.

Laycock, Henry: 1972. Some questions of ontology. *Philosophical Review* 81:3–42.

Laycock, Henry: 1975. Theories of matter. *Synthese* 31:411–442. Also in: F. J. Pelletier (ed.), 1979, *Mass Terms: Somc Philosophical Problems* (Reidel: Dordrecht), 89–120.

Lejewski, Czesław: 1958. On Leśniewski's ontology. *Ratio* 1:150–176.

Lejewski, Czesław: 1960/61. Studies in the axiomatic foundations of boolean algebra. *Notre Dame Journal of Formal Logic* 1 (1960); 2 (1961):23–47, 91–106; 79–93.

Lemmon, E. John: 1967. Comments. In: Nicholas Rescher (ed.), *The Logic of Decision and Action*, Pittsburgh: University of Pittsburgh Press, 96–103.

Lenzen, Wolfgang: 1990. *Das System der Leibnizschen Logik*. Berlin: de Gruyter.

Leonard, Henry and Nelson Goodman: 1940. The calculus of individuals and its uses. *Journal of Symbolic Logic* 5:45–55.

Leonard, Henry S.: 1956. The logic of existence. *Philosophical Studies* 7:49–64.

LePore, Ernest (ed.): 1987. *New Directions in Semantics*. London: Academic Press.

LePore, Ernest and Brian McLaughlin (eds.): 1985. *Actions and Events: Perspectives on the Philosophy of Donald Davidson*. Oxford: Basil Blackwell.

Leśniewski, Stanisław: 1916. Podstawy ogólnej teoryi mnogości I. (Foundations of a General Theory of Collective Sets. Tech. Rep. 2, Prace Polskiego Koła Naukowego w Moskwie. Sekcya matematyczno-przyrodnicza, Moscow. English translation in: S. Leśniewski, *Collected Works*, vol. I, 129–173.

Leśniewski, Stanisław: 1927-31. O podstawack matematyki. *Prezeglad Filosoficzny* 30 (1927); 31 (1928); 32 (1929); 33 (1930); 34 (1931):164–206; 261–291; 60–101; 77–105; 142–170.

Leśniewski, Stanisław: 1929. Grundzüge eines neuen Systems der Grundlagen der Mathematik. *Fundamenta Mathematica* 14:1–81.

Lewis, David: 1975. Adverbs of quantification. In: Edward Keenan (ed.), *Formal Semantics of Natural Language*, Cambridge: Cambridge University Press, 3–15.

Lewis, David: 1983. *Philosophical Papers*, vol. I. Oxford: Oxford University Press.

Lewis, David: 1986a. Events. In: *Philosophical Papers. Volume II*, Oxford: Oxford University Press, 241–269.

Lewis, David: 1986b. *On the Plurality of Worlds*. Oxford: Basil Blackwell.

Lewis, David: 1986c. *Philosophical Papers*, vol. II. Oxford: Oxford University Press.

Lewis, David: 1988. Vague identity: Evans misunderstood. *Analysis* 48:130–135.

Lewis, David: 1991. *Parts of Classes*. Oxford: Basil Blackwell.

Lewis, David: 1993. Mathematics is megethology. *Philosophia Mathematica* 1:3–23.

Link, Godehard: 1974. Quantoren-Floating im Deutschen. In: F. Kiefer *et al.* (eds.), *Syntax und generative Semantik, vol. 2*, Frankfurt: Athenaion, 105–127.

Link, Godehard: 1979. *Montague–Grammatik. Die Logischen Grundlagen*. München: Fink.

Link, Godehard: 1983a. The logical analysis of plurals and mass terms: A lattice-theoretical approach. In: R. Bäuerle *et al.* (eds.), *Meaning, Use, and Interpretation of Language*, Berlin: De Gruyter, 302–323. Chapter 1 of this book.

Link, Godehard: 1983b. Logical semantics for natural language. *Erkenntnis* 19:261–283.

Link, Godehard: 1984. Hydras: On the logic of relative constructions with multiple heads. In: F. Landman *et al.* (eds.), *Varieties of Formal Semantics. Proceedings of the Fourth Amsterdam Colloquium*, Dordrecht: Foris, 245–257. Chapter 3 of this book.

Link, Godehard: 1987a. Algebraic semantics of event structures. In: J. Groenendijk *et al.* (eds.), *Proceedings of the Sixth Amsterdam Colloquium*. ITLI, Amsterdam, 243–262. Chapter 11 of this book.

Link, Godehard: 1987b. Generalized quantifiers and plurals. In: P. Gärdenfors (ed.), *Generalized Quantifiers. Linguistic and Logical Approaches*, Dordrecht: Reidel, 151–180. Chapter 4 of this book.

Link, Godehard: 1987c. Prespie in pragmatic wonderland, or: The projection problem for presuppositions revisited. In: J. Groenendijk *et al.* (eds.), *Foundations of Pragmatics and Lexical Semantics*, Dordrecht: Foris, 101–126.

Link, Godehard: 1991a. Plural. In: A. von Stechow and D. Wunderlich (eds.), *Semantik. Ein internationales Handbuch der zeitgenössischen Forschung*, Berlin: de Gruyter, chap. 19, 418–440. Chapter 2 of this book.

Link, Godehard: 1991b. Quantity and number. In: D. Zaefferer (ed.), *Semantic Universals and Universal Semantics*, Dordrecht: Foris, 133–149. Chapter 9 of this book.

Link, Godehard: 1995a. Algebraic semantics for natural language: some philosophy, some applications. *International Journal of Human-Computer Studies* 43:765–784. Chapter 8 of this book.

Link, Godehard: 1995b. Generic information and dependent generics. In: G. Carlson *et al.* (eds.), *The Generic Book*, Chicago: University of Chicago Press, chap. 9, 358–382.

Link, Godehard: 1997. Ten years of research on plurals — where do we stand? In: F. Hamm *et al.* (eds.), *Plurality and Quantification*, Dordrecht: Kluwer, 19–54. Chapter 7 of this book.

Link, Godehard and Hinrich Schütze: 1991. The treatment of plurality in L-LILOG. In: O. Herzog *et al.* (eds.), *Text Understanding in LILOG: Integrating Computational Linguistics and Artificial Intelligence*, Berlin: Springer, Lecture Notes in Artificial Intelligence, no. 546, 342–352.

Lønning, Jan Tore: 1982. *Kvantiteter og Kvantorer*. Ph.D. thesis, Mathematisches Institut Oslo.

Lønning, Jan Tore: 1987a. Collective readings of definite and indefinite noun phrases. In: P. Gärdenfors (ed.), *Generalized Quantifiers. Linguistic and Logical Approaches*, Dordrecht: Reidel, 203–235.

Lønning, Jan Tore: 1987b. Mass terms and quantification. *Linguistics and Philosophy* 10:1–52.

Lønning, Jan Tore: 1989. *Some Aspects of the Logic of Plural Noun Phrases*. Ph.D. thesis, University of Oslo. COSMOS-Report No. 11, Department of Mathematics, University of Oslo.

Lønning, Jan Tore: 1991. Among readings. Some comments on "Among collections". In: J. van der Does (ed.), *Quantification and Anaphora II*, Edinburgh: Centre for Cognitive Science, University of Edinburgh, DYANA Deliverable 2.2.b, 37–51.

Lønning, Jan Tore: 1997. Plurals and collectivity. In: J. van Benthem *et al.* (eds.), *Handbook of Logic and Language*, Elsevier: Elsevier, 1009–1053.

Löbner, Sebastian: 1985. Definites. *Journal of Semantics* 4:279–326.

Löbner, Sebastian: 1987. Natural language and generalized quantifier theory. In: P. Gärdenfors (ed.), *Generalized Quantifiers. Linguistic and Logical Approaches*, Dordrecht: Reidel, 181–201.

Lombard, Lawrence B.: 1974. A note on level-generation and the time of a killing. *Philosophical Studies* 26:151–152.

Lombard, Lawrence B.: 1986. *Events. A Metaphysical Study*. London: Routledge and Kegan Paul.

Lorimer, D. L. R.: 1915. Pashtu. Part I.

Lowe, E. Jonathan: 1989. *Kinds of Being: A Study of Individuation, Identity and the Logic of Sortal Terms*. Oxford: Blackwell.

Lowe, E. Jonathan: 1994. Vague identity and quantum indeterminacy. *Analysis* 54:110–114.

Lowe, E. Jonathan: 1995. Coinciding object: in defence of the 'standard account'. *Analysis* 55:171–178.

Luschei, E. C.: 1962. *The Logical Systems of Leśniewski*. Amsterdam: North Holland.

MacDonald, C. A.: 1978. On the unifier-multiplier controversy. *Canadian Philosophical Quarterly* 8:707–714.

Mackie, John L.: 1974. *The Cement of the Universe*. Oxford: Oxford University Press.

Maddy, Penelope: 1989. The roots of contemporary platonism. *Journal of Symbolic Logic* 54:1121–1144.

Maddy, Penelope: 1990. *Realism in Mathematics*. Oxford: Clarendon.

Martin, C. B.: 1980. Substance substantiated. *Australasian Journal of Philosophy* 58:3–10.

Martin, Richard M.: 1943. A homogeneous system for formal logic. *Journal of Symbolic Logic* 8:1–23.

Martin, Richard M.: 1978. *Events, Reference and Logical Form*. Washington, D.C.: Catholic University of America Press.

Martino, Enrico: 1995. The megethological foundation of set theory (abstract). In: *Proceedings of the Tenth International Congress of Logic, Methodology and Philosophy of Science*. Florence, Italy.

Massey, Gerald J.: 1976. Tom, Dick and Harry, and all the King's men. *American Philosophical Quarterly* 13:89–107.

Mayo, Bernard: 1961. Objects, events and complementarity. *Philosophical Review* 70:340–361.

McCann, Hugh J.: 1982. The trouble with level-generation. *Mind* 91:481–500.

McCann, Hugh J.: 1983. Individuating actions: the fine-grained approach. *Canadian Journal of Philosophy* 13:493–512.

McGilvray, James A.: 1983. Pure process(es)? *Philosophical Studies* 43:243–251.

McMullin, Ernan (ed.): 1963. *The Concept of Matter*. Notre Dame: University of Notre Dame Press.

Meixner, Uwe: 1997a. *Axiomatic Formal Ontology*. Dordrecht: Kluwer.

Meixner, Uwe: 1997b. *Ereignis und Substanz*. Paderborn: Schöningh.

Mertz, Donald W.: 1996. *Moderate Realism and its Logic*. New Haven: Yale University Press.

ter Meulen, Alice: 1980. *Substance, Quantities and Individuals: A Study in the Formal Semantics of Mass Terms*. Ph.D. thesis, Stanford University. Indiana University Linguistics Club, Bloomington.

ter Meulen, Alice: 1981. An intensional logic for mass terms. *Philosophical Studies* 40:105–125.

ter Meulen, Alice (ed.): 1983. *Studies in Model-Theoretic Semantics*. Dordrecht: Foris.

ter Meulen, Alice: 1984. Events, quantities and individuals. In: J. van Benthem *et al.* (eds.), *Generalized Quantifiers in Natural Language*, Amsterdam: Elsevier, 259–279.

de Mey, Sjaak: 1984. Mass terms and count terms. Institute for General Linguistics. University of Groningen.

Moltmann, Friederike: 1996. *Parts and Wholes in Semantics*. Oxford: Oxford University Press.

Monk, J. Donald: 1976. *Mathematical Logic*. Berlin - Heidelberg - New York: Springer.

Montague, Richard: 1969. On the nature of certain philosophical entities. *The Monist* 53:159–194. Also in: R. H. Thomason (ed.), 1974, *Formal Philosophy. Selected Papers of Richard Montague* (Yale University Press: New Haven and London), chapter 5, 148–187.

Montague, Richard: 1970. Universal grammar (UG). *Theoria* 36:373–398. Also in: R. H. Thomason (ed.), 1974, *Formal Philosophy. Selected Papers of Richard Montague* (Yale University Press: New Haven and London), chapter 7, 222-246.

Montague, Richard: 1973a. The proper treatment of quantification in ordinary English (PTQ). In: Jaakko Hintikka *et al.* (eds.), *Approaches to Natural Language*, Dordrecht: Reidel, 221–242. Also in: R. H. Thomason (ed.), 1974, *Formal Philosophy. Selected Papers of Richard Montague* (Yale University Press: New Haven and London), chapter 8, 247–270.

Montague, Richard: 1973b. Reply to Moravcsik. In: J. Hintikka *et al.* (eds.), *Approaches to Natural Language*, Dordrecht: Reidel, 289–294. Reprinted as 'The proper treatment of mass terms in English', in F. J. Pelletier (ed.), 1979, *Mass Terms: Some Philosophical Problems* (Reidel: Dordrecht), 173 – 178.

Montague, Richard: 1974. *Formal Philosophy. Selected Papers of Richard Montague. Edited and with an introduction by Richmond H. Thomason.* New Haven and London: Yale University Press.

Moravcsik, Julius M. E. : 1973. Mass terms in English. In: Jaakko Hintikka *et al.* (eds.), *Approaches to Natural Language*, Dordrecht: Reidel, 263–285.

Morton, Adam: 1975. Complex individuals and multigrade relations. *Noûs* 9:309–318.

Mourelatos, Alexander: 1978. Events, processes, and states. *Linguistics and Philosophy* 2:415–434.

Noonan, Harold W.: 1995. E. J. Lowe on vague identity and quantum indeterminacy. *Analysis* 55:14–19.

Ojeda, Almerindo E.: 1993. *Linguistic Individuals.* CSLI Lecture Notes No. 31. Stanford: CSLI Publications.

Orey, S.: 1959. Model theory for higher order predicate calculus. *Transactions from the American Mathematical Society* 92:72–84.

Parsons, Charles: 1983. *Mathematics in Philosophy.* Ithaca, NY: Cornell University Press.

Parsons, Terence: 1970. An analysis of mass terms and amount terms. *Foundations of Language* 6:362–388. Also in: F. J. Pelletier (ed.), 1979, *Mass Terms: Some Philosophical Problems* (Reidel: Dordrecht), 137–166.

Parsons, Terence: 1975. Afterthoughts on mass terms. *Synthese* 31:517–521. Also in: F. J. Pelletier (ed.), 1979, *Mass Terms: Some Philosophical Problems* (Reidel: Dordrecht), 167–171.

Parsons, Terence: 1985. Underlying events in the logical analysis of English. In: E. LePore *et al.* (eds.), *Actions and Events: Perspectives on the Philosophy of Donald Davidson*, Oxford: Basil Blackwell, 235–267.

Parsons, Terence: 1990. *Events in the Semantics of English: A Study in Subatomic Semantics*. Cambridge, MA: MIT Press.

Partee, Barbara: 1984. Compositionality. In: F. Landman *et al.* (eds.), *Varieties of Formal Semantics. Proceedings of the Fourth Amsterdam Colloquium*, Dordrecht: Foris, GRASS Series No. 3, 281–311.

Partee, Barbara: 1986. Noun phrase interpretation and type shifting principles. In: J. Groenendijk *et al.* (eds.), *Studies in Discourse Representation Theory and the Theory of Generalized Quantifiers*, Dordrecht: Foris, GRASS Series vol. 8, 115–143.

Partee, Barbara and Mats Rooth: 1983. Generalized conjunction and type ambiguity. In: R. Bäuerle *et al.* (eds.), *Meaning, Use, and Interpretation of Language*, Berlin: de Gruyter, 361–383.

Pelletier, Francis Jeffry: 1975. Non-singular reference: some preliminaries. *Philosophia* 5. Also in: F. J. Pelletier (ed.), 1979, *Mass Terms: Some Philosophical Problems* (Reidel: Dordrecht), 1–14.

Pelletier, Francis Jeffry: 1979a. A bibliography of recent work on mass terms. In: F. J. Pelletier (ed.), *Mass Terms: Some Philosophical Problems*, Dordrecht: Reidel, 295–298.

Pelletier, Francis Jeffry (ed.): 1979b. *Mass Terms: Some Philosophical Problems*. Dordrecht: Reidel.

Pelletier, Francis J. and Lenhart K. Schubert: 1987a. Generically speaking. with remarks on the interpretation of pronouns and tenses. Tech. Rep. 3, Department of Computer Science. University of Alberta, Edmonton.

Pelletier, Francis J. and Lenhart K. Schubert: 1987b. Problems in the representation of the logical form of generics, plurals, and mass nouns. In: E. LePore (ed.), *New Directions in Semantics*, London: Academic Press, 385–451.

Pelletier, Francis J. and Lenhart K. Schubert: 1989. Mass expressions. In: D. Gabbay *et al.* (eds.), *Handbook of Philosophical Logic. Volume IV: Topics in the Philosophy of Language*, Dordrecht: Reidel, 327–407.

Pollard, Stephen: 1986. Plural quantification and the iterative concept of set. *Philosophy Research Archives* 11:579–587.

Pollard, Stephen: 1988. Plural quantification and the axiom of choice. *Philosophical Studies* 54:393–397.

Pollard, Stephen: 1990. *Philosophical Introduction to Set Theory*. Notre Dame, IN: University of Notre Dame Press.

Postal, Paul: 1974. On certain ambiguities. *Linguistic Inquiry* 5:367–425.

Quine, Willard van Orman: 1960. *Word and Object*. Cambridge, MA: M.I.T. Press.

Quine, Willard van Orman: 1969. *Ontological Relativity and Other Essays*. New York: Columbia University Press.

Quine, Willard van Orman: 1972. *Methods of Logic*. New York: Holt, Rinehart and Winston, 3rd edn.

Quine, Willard van Orman: 1981. *Theories and Things*. Cambridge, MA: Harvard University Press.

Quine, Willard van Orman: 1985. Events and reification. In: E.LePore *et al.* (eds.), *Actions and Events: Perspectives on the Philosophy of Donald Davidson*, Oxford: Basil Blackwell, 162–171.

Quine, Willard van Orman: 1991. Immanence and validity. *Dialectica* 45:219–230.

Quinton, Anthony M.: 1979. Objects and events. *Mind* 88:197–214.

Ramsey, Frank Plumpton: 1925. Universals. *Mind* 34:401–417. Also in: F. P. Ramsey, *The Foundations of Mathematics and Other Essays*, edited by R. B. Braithwaite, Routledge and Kegan Paul, London 1931, 112–134; and in: F. P. Ramsey, *Philosophical Papers*, edited by D. H. Mellor, Cambridge University Press, Cambridge 1990, 8–30.

Rasiowa, Helena and Roman Sikorski: 1968. *The Mathematics of Metamathematics*. Warszawa: PWN — Polish Scientific Publishers. 2nd revised edition.

Reichenbach, Hans: 1947. *Elements of Symbolic Logic*. New York: Macmillan.

Rescher, Nicholas: 1959. On the logic of existence and denotation. *Philosophical Review* 69:157–180.

Rescher, Nicholas: 1962. Plurality quantification. *Journal of Symbolic Logic* 38:373–374.

Rescher, Nicholas (ed.): 1967. *The Logic of Decision and Action.* Pittsburgh: University of Pittsburgh Press.

Rescher, Nicolas (ed.): 1969. *Essays in Honor of Carl G. Hempel.* Dordrecht: Reidel.

Resnik, Michael: 1988. Second-order logic still wild. *Journal of Philosophy* 85:75–87.

Resnik, Michael: 1997. *Mathematics as a Science of Patterns.* Oxford: Oxford University Press.

Reyle, Uwe: 1986. *Zeit und Aspekt bei der Verarbeitung natürlicher Sprachen.* Ph.D. thesis, Universität Stuttgart.

Roberts, Craige: 1987a. Distributivity. In: J. Groenendijk *et al.* (eds.), *Proceedings of the Sixth Amsterdam Colloquium.* ITLI, Amsterdam, 291–309.

Roberts, Craige: 1987b. *Modal Subordination, Anaphora, and Distributivity.* Ph.D. thesis, University of Massachusetts, Amherst.

Rodman, R.: 1972. Montague grammar. UCLA.

Roeper, Peter: 1983. Semantics for mass terms with quantifiers. *Noûs* 17:251–265.

Roeper, Peter: 1985. Generalization of first-order logic to non-atomic domains. *Journal of Symbolic Logic* 50:815–838.

Roeper, Peter: 1987. Principles of abstraction for events and processes. *Journal of Philosophical Logic* 16:273–307.

Roeper, Peter: 1997. Region-based topology. *Journal of Philosophical Logic* 26:251–309.

Rohrer, Christian (ed.): 1980. *Time, Tense, and Quantifiers.* Linguistische Arbeiten 83. Tübingen: Niemeyer.

Rooth, Mats: 1987. Noun phrase interpretation in Montague grammar, file change semantics, and situation semantics. In: P. Gärdenfors (ed.), *Generalized Quantifiers. Linguistic and Logical Approaches*, Dordrecht: Reidel, 237–268.

Rorty, Richard: 1963. Matter and event. In: E. McMullin (ed.), *The Concept of Matter*, Notre Dame, IN: University of Notre Dame Press, 497–524.

Russell, Bertrand: 1903. *The Principles of Mathematics*. London: George Allen & Unwin. Second edition 1937.

Russell, Bertrand: 1914. *Our Knowledge of the External World*. London: Open Court.

Russell, Bertrand: 1927. *The Analysis of Matter*. London: K. Paul, Trench, Trubner.

Scha, Remko: 1981. Distributive, collective and cumulative quantification. In: J. Groenendijk *et al.* (eds.), *Formal Methods in the Study of Language*. Mathematical Centre, Amsterdam, 483–512.

Scha, Remko and D. Stallard: 1988. Multi-level plurals and distributivity. In: *Proceedings of the ACL meeting, June 1988*. 17–24.

Schein, Barry: 1993. *Plurals and Events*. Cambridge, MA: MIT Press.

Schmitt, Frederick: 1983. Events. *Erkenntnis* 20:281–293.

Schütze, Hinrich: 1989. Pluralbehandlung in natürlichsprachlichen Wissensverarbeitungssystemen. Diplomarbeit, Institut für Informatik, University of Stuttgart.

Schwarzschild, Roger: 1990. Against groups. In: M. Stokhof *et al.* (eds.), *Proceedings of the Seventh Amsterdam Colloquium*. ITLI, Amsterdam, 475–493.

Schwarzschild, Roger: 1991. *On the Meaning of Definite Plural Noun Phrases*. Ph.D. thesis, University of Massachusetts, Amherst.

Schwarzschild, Roger: 1992. Types of plural individuals. *Linguistics and Philosophy* 15:641–675.

Schwarzschild, Roger: 1996. *Pluralities*. Dordrecht: Kluwer.

Scott, Dana: 1967. Existence and descriptions in formal logic. In: Ralph Shoenman (ed.), *Bertrand Russell: Philosopher of the Century*, London: Allen and Unwin, 181–200. Also in K. Lambert (ed.), 1991, *Philosophical Applications of Free Logic* (Oxford UP: Oxford), 28–48.

Seibt, Johanna: 1990. *Properties as Processes. A Synoptic Study of Wilfrid Sellars' Nominalism*. Atascadero, CA: Ridgeview Publ. Co.

Seibt, Johanna: 1997. Existence in time: From substance to process. In: J. Faye *et al.* (eds.), *Perspectives on Time*, Dordrecht: Kluwer, 143–182.

Seibt, Johanna: 1998. *Processes. Foundations of a New Ontology.* (In preparation).

Seiler, Hansjakob and Christian Lehmann (eds.): 1982. *Apprehension. Das sprachliche Erfassen von Gegenständen.* Tübingen: Narr.

Seligman, Jerry and Dag Westerståhl (eds.): 1996. *Logic, Language, and Computation, vol. 1.* CSLI Lecture Notes No. 58. Stanford: CSLI Publications.

Sellars, Wilfrid: 1962. Time and the world order. In: H. Feigl *et al.* (eds.), *Minnesota Studies in the Philosophy of Science, Vol. III*, University of Minnesota Press, 527–618.

Sellars, Wilfrid: 1981. Foundations for a metaphysics of pure process. *The Monist* 64:3–90.

Serzisko, Fritz: 1980. Sprachen mit Zahlklassifikatoren: Analyse und Vergleich. Arbeiten des Kölner Universalienprojekts 37, University of Cologne.

Shapiro, Stewart: 1991. *Foundations without Foundationalism. A Case for Second-order Logic.* Oxford: Clarendon Press.

Shapiro, Stewart: 1997. *Philosophy of Mathematics: Structure and Ontology.* Oxford: Oxford University Press.

Sharvy, Richard: 1978. Maybe English has no count nouns: Notes on Chinese semantics. an essay in metaphysics and linguistics. *Studies in Language* 2:345–365.

Sharvy, Richard: 1979. The indeterminacy of mass predication. In: F. J. Pelletier (ed.), *Mass Terms: Some Philosophical Problems*, Dordrecht: Reidel, 47–54.

Sharvy, Richard: 1980. A more general theory of definite descriptions. *Philosophical Review* 89:607–624.

Sharvy, Richard: 1983. Mixtures. *Journal of Philosophy* 80:227–239.

Sikorski, Roman: 1969. *Boolean Algebras.* Berlin: Springer, 3rd edn.

Simons, Peter: 1982. Plural reference and set theory. In: B. Smith (ed.), *Parts and Moments. Studies in Logic and Formal Ontology*, München: Philosophia, 199–256.

Simons, Peter: 1983. Class, mass and mereology. *History and Philosophy of Logic* 4:157–180.

Simons, Peter: 1987. *Parts. A Study in Ontology.* Oxford: Clarendon Press.

Simons, Peter: 1991a. Free part-whole theory. In: K. Lambert (ed.), *Philosophical Applications of Free Logic*, Oxford: Oxford University Press, chap. 16, 285–305.

Simons, Peter: 1991b. On being spread out in time: Temporal parts and the problem of change. In: W. Spohn (ed.), *Existence and Explanation*, Dordrecht: Kluwer, 131–147.

Smart, J. J. C.: 1982. Sellars on process. *The Monist* 65:302–314.

Smith, Barry (ed.): 1982. *Parts and Moments. Studies in Logic and Formal Ontology.* München: Philosophia.

Smith, Brian C.: 1996. *On the Origin of Objects.* Cambridge, MA: MIT Press.

Smith-Stark, Cedric: 1974. The plurality split. In: *Chicago Linguistics Society Papers.* vol. 10, 657–671.

Spohn, Wolfgang: 1984. *Eine Theorie der Kausalität.* Habilitationsschrift, University of Munich.

Spohn, Wolfgang (ed.): 1991. *Existence and Explanation. Essays Presented in Honor of Karel Lambert.* Dordrecht: Kluwer.

von Stechow, Arnim: 1980. Modification of noun phrases. A challenge for compositional semantics. *Theoretical Linguistics* 7:57–110.

von Stechow, Arnim and Dieter Wunderlich: 1991. *Semantik / Semantics. Ein internationales Handbuch der zeitgenössischen Forschung / An International Handbook of Contemporary Research.* Berlin-New York: de Gruyter.

von Stechow, Arnim and Thomas E. Zimmermann: 1984. Term answers and contextual change. *Linguistics* 22:3–40.

Stockwell, Robert P. *et al.* (eds.): 1973. *The Major Syntactic Structures of English.* New York: Holt, Rinehart and Winston.

Stokhof, Martin and L. Torenvliet (eds.): 1990. *Proceedings of the Seventh Amsterdam Colloquium*, Amsterdam. ITLI.

Strawson, Peter F.: 1959. *Individuals: An Essay in Descriptive Metaphysics*. London: Methuen.

Suppes, Patrick: 1984. *Probabilistic Metaphysics*. Oxford: Blackwell.

Swoyer, Chris: 1994. Leibniz's calculus of real addition. *Studia Leibnitiana* XXVI/I:1–30.

Swoyer, Chris: 1995a. Leibniz on intension and extension. *Nous* 29:96–114.

Swoyer, Chris: 1995b. Leibnizian expression. *Journal of the History of Philosophy* 33:65–99.

Tarski, Alfred, Andrzej Mostowski, and Raphael M. Robinson: 1953. *Undecidable Theories*. Amsterdam: North-Holland.

Thalberg, Irving: 1978. The irreducibility of events. *Analysis* 38:1–9.

Thalberg, Irving: 1985. A world without events? In: B. Vermazen and M. Hintikka (eds.), *Essays on Davidson. Actions and Events*, Oxford: Clarendon, chap. IX, 137–155.

Thomason, Richmond H.: 1977. Multiple quantification, questions, and Bach-Peters sentences. Some preliminary notes. University of Pittsburgh.

Thomason, Richmond H.: 1984. Some issues concerning the interpretation of derived and gerundive nominals. *Linguistics and Philosophy* 8:73–78.

Thomson, Judith J.: 1971. Individuating actions. *Journal of Philosophy* 68:771–781.

Thomson, Judith J.: 1977. *Act and Other Events*. Ithaca, NY: Cornell University Press.

Vendler, Zeno: 1967a. Causal relations. *Journal of Philosophy* 64:704–713.

Vendler, Zeno: 1967b. *Facts and Events*, Ithaca, NY: Cornell University Press, 122–146.

Vendler, Zeno: 1967c. *Linguistics* in *Philosophy*. Ithaca, NY: Cornell University Press.

Verkuyl, Henk J.: 1972. *On the Compositional Nature of Aspects*. *Foundations of Language* Supplementary Series 15. Dordrecht: Reidel.

Verkuyl, Henk J.: 1981. Numerals and quantifiers in Xbar-syntax and their semantic interpretation. In: J. Groenendijk *et al.* (eds.), *Formal Methods in the Study of Language*. Mathematical Centre, Amsterdam, 567–599.

Verkuyl, Henk J.: 1988. Aspectual asymmetry and quantification. In: V. Ehrich *et al.* (eds.), *Temporalsemantik. Beiträge zur Linguistik der Zeitreferenz*. Tübingen, 220–259.

Verkuyl, Henk J.: 1991. NP stucture. OTS Prepublication Series, Rijksuniversiteit Utrecht.

Verkuyl, Henk J.: 1994. Distributivity and collectivity: a couple at odds. In: M. Kanazawa *et al.* (eds.), *Dynamics, Polarity and Quantification*, Stanford: CSLI Publications, CSLI Lecture Notes No. 48, 49–80.

Verkuyl, Henk J. and Jaap M. van der Does: 1996. The semantics of plural noun phrases. In: J. van der Does *et al.* (eds.), *Quantifiers, Logic, and Language*, Stanford: CSLI Publications, chap. 15, 337–374.

Vermazen, Bruce and Merrill Hintikka (eds.): 1985. *Essays on Davidson: Actions and Events*. Oxford: Clarendon.

Wald, Jan David: 1977. *Stuff and Words: A Semantic and Linguistic Analysis of Non-Singular Reference*. Ph.D. thesis, Brandeis University.

Ware, Robert: 1975. Some bits and pieces. *Synthese* 31:15–29.

Webber, Bonnie: 1984. So what can we talk about now? In: M. Brady *et al.* (eds.), *Computational Models of Discourse*, Cambridge, MA: M.I.T. Press, 331–371.

Westerståhl, Dag: 1988. Quantifiers in formal and natural language. In: D. Gabbay *et al.* (eds.), *Handbook of Philosophical Logic. Volume IV: Topics in the Philosophy of Language*, Dordrecht: Reidel, 1–131.

Whitehead, Alfred N.: 1919. *An Enquiry Concerning the Principles of Natural Knowledge*. Cambridge: Cambridge University Press.

Whitehead, Alfred N.: 1929. *Process and Reality. An Essay in Cosmology*. New York: Macmillan. Corrected edition, edited by D. R. Griffin and D. W. Sherburne. The Free Press, New York 1978.

Wiener, Norber: 1912-14. A contribution to the theory of relative position. *Proceedings of the Cambridge Philosophical Society* 17:441–449.

Wiggins, David: 1968. On being in the same place at the same time. *Philosophical Review* 77:90–95.

Williams, Donald C.: 1966. *Principles of Empirical Realism*. Springfield, IL: Thomas.

Wilson, N. L.: 1974. Facts, events and their identity conditions. *Philosophical Studies* 25:303–321.

Yovel, Yirmiahu (ed.): 1978. *Philosophy of History and Action*. Dordrecht: Reidel.

Zaefferer, Dietmar: 1984. *Frageausdrücke und Fragen im Deutschen. Zu ihrer Syntax, Semantik und Pragmatik*. München: Fink.

Zaefferer, Dietmar (ed.): 1991. *Semantic Universals and Universal Semantics*. GRASS Series No 12. Dordrecht: Foris.

Zaefferer, Dietmar: 1992. *And* and *or*: meet and/or join? a problem in the semantics of (non-)propositional connectives. Paper read at the International Congress of Linguistics, Montreal, August 1992.

Zalta, Edward N.: 1983. *Abstract Objects. An Introduction to Axiomatic Metaphysics*. Dordrecht: Reidel.

Zalta, Edward N.: 1997a. A (Leibnizian) theory and calculus of concepts. ms. Department of Philosophy, Stanford University.

Zalta, Edward N.: 1997b. *Principia Metaphysica*. Online version of June 13, 1997; see http://mally.stanford.edu/principia.ps.gz.

Zemach, Eddy: 1970. Four ontologies. *Journal of Philosophy* 62:213–247.

Zemach, Eddy M.: 1978. Events. In: Y. Yovel (ed.), *Philosophy of History and Action*, Dordrecht: Reidel, 85–95.

Subject Index

Name Index